# COMPUTATIONAL APPROACHES TO ARCHAEOLOGICAL SPACES

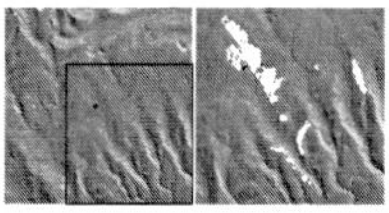

PUBLICATIONS OF THE
INSTITUTE OF ARCHAEOLOGY, UNIVERSITY COLLEGE LONDON

Series Editor: Ruth Whitehouse
Director of the Institute: Stephen Shennan
Founding Series Editor: Peter J. Ucko

The Institute of Archaeology of University College London is one of the oldest, largest, and most prestigious archaeology research facilities in the world. Its extensive publications program includes the best theory, research, pedagogy, and reference materials in archaeology and cognate disciplines, through publishing exemplary work of scholars worldwide. Through its publications, the Institute brings together key areas of theoretical and substantive knowledge, improves archaeological practice, and brings archaeological findings to the general public, researchers, and practitioners. It also publishes staff research projects, site and survey reports, and conference proceedings. The publications program, formerly developed in-house or in conjunction with UCL Press, is now produced in partnership with Left Coast Press, Inc. The Institute can be accessed online at www.ucl.ac.uk/archaeology.

*Recent Titles*
Andrew Bevan and Mark Lake (Eds.), *Computational Approaches to Archaeological Spaces*
Sue Colledge, James Conolly, Keith Dobney, Katie Manning, and Stephen Shennan (Eds.), *The Origins and Spread of Domestic Animals in Southwest Asia and Europe*
Julia Shaw, *Buddhist Landscapes of Central India*
Ralph Haeussler, *Becoming Roman?*
Ethan E. Cochrane and Andrew Gardner, *Evolutionary and Interpretive Archaeologies*
Andrew Bevan and David Wengrow (Eds.), *Cultures of Commodity Branding*
Peter Jordan (Ed.), *Landscape and Culture in Northern Eurasia*
Peter Jordan and Marek Zvelebil (Eds.), *Ceramics before Farming*
Marcos Martinón-Torres and Thilo Rehren (Eds.), *Archaeology, History, and Science*
Miriam Davis, *Dame Kathleen Kenyon*
Elizabeth Pye (Ed.), *The Power of Touch*
Russell McDougall and Iain Davidson (Eds.), *The Roth Family, Anthropology, and Colonial Administration*
Eleni Asouti and Dorian Q. Fuller, *Trees and Woodlands of South India*
Tony Waldron, *Paleoepidemiology*
Janet Picton, Stephen Quirke, and Paul C. Roberts (Eds.), *Living Images*
Timothy Clack and Marcus Brittain (Eds.), *Archaeology and the Media*
Sue Colledge and James Conolly (Eds.), *The Origins and Spread of Domestic Plants in Southwest Asia and Europe*
Gustavo Politis, *Nukak*
Sue Hamilton, Ruth Whitehouse, and Katherine I. Wright (Eds.), *Archaeology and Women*

*Critical Cultural Heritage Series, Beverley Butler (Ed.)*
Charlotte Joy, *The Politics of Heritage Management in Mali*
Layla Renshaw, *Exhuming Loss*
Katharina Schramm, *African Homecoming*
Mingming Wang, *Empire and Local Worlds*
Dean Sully (Ed.), *Decolonizing Conservation*
Ferdinand de Jong and Michael Rowlands (Eds.), *Reclaiming Heritage*

Information on older titles in this series can be obtained from the Left Coast Press, Inc. website www.LCoastPress.com.

# COMPUTATIONAL APPROACHES TO ARCHAEOLOGICAL SPACES

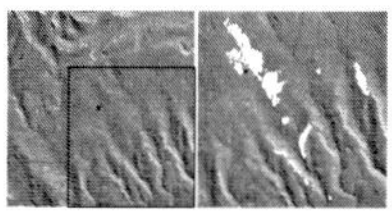

Edited by

Andrew Bevan
Mark Lake

Left Coast Press Inc.

Walnut Creek, California

LEFT COAST PRESS, INC.
1630 North Main Street, #400
Walnut Creek, CA 94596
www.LCoastPress.com



ISBN 978-1-61132-346-7 hardback
ISBN 978-1-61132-348-1 institutional eBook
ISBN 978-1-61132-727-4 consumer eBook

Library of Congress Cataloging-in-Publication Data:

Computational approaches to archaeological spaces/edited by Andrew Bevan, Mark Lake.
pages cm. — (Publications of the Institute of Archaeology, University College London)
Includes bibliographical references and index.
ISBN 978-1-61132-346-7 (hardback : alk. paper) — ISBN 978-1-61132-348-1 (institutional eBook) — ISBN 978-1-61132-727-4 (consumer eBook)
1. Spatial analysis (Statistics) in archaeology. 2. Archaeology—Computer simulation. 3. Virtual reality in archaeology. I. Bevan, Andrew, 1974- author, editor of compilation. II. Lake, Mark (Mark W.) author, editor of compilation.
CC79.S63.C66 2013
930.1—dc23
2013007682

Printed in the United States of America

∞™ The paper used in this publication meets the minimum requirements of American National Standard for Information Sciences—Permanence of Paper for Printed Library Materials, ANSI/NISO Z39.48–1992.

# Contents

# Illustrations

## Tables

## Figures

## Color Figures (following page 168)

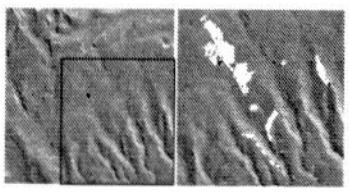

# Acknowledgments

The initial impetus for this volume was an international seminar series, *Contemporary Roles for Spatial Analysis in Archaeology*, hosted by the UCL Institute of Archaeology in the spring of 2010. We would like to thank all the speakers in the series for their lively presentations, irrespective of whether they were able to contribute to this collection. We also thank the Director of the Institute of Archaeology, Stephen Shennan, for his encouragement and for arranging financial assistance to defray the travel expenses incurred by participants from mainland Europe and North America.

Approximately half of the papers gathered here were presented at the original seminar series; the others were commissioned to provide fuller coverage of the field. We are grateful to those later invitees for agreeing to join the project and to all authors for their forbearance as we assembled the manuscripts for production. We also thank the UCL Institute of Archaeology Publications Committee and Left Coast Press, Inc., for their part in the process, along with the anonymous reviewers whose constructive comments were of value to all concerned.

*Andrew Bevan and Mark Lake*
*July 2013*

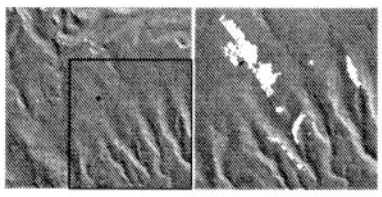

# 1 Introduction: Archaeological Inferences and Computational Spaces

*Andrew Bevan and Mark Lake*

Spatial analysis in archaeology today encompasses a wide range of experiential, fieldwork-based, deterministic, and/or stochastic approaches that vary considerably in their intended purpose and theoretical underpinnings. Although Hodder and Orton (1976) and Clarke (1977) both provided book-length treatment of the application of statistical and/or simulation approaches to spatial analysis in archaeology many years ago, the rapid uptake of computational methods such as Geographical Information Systems (GIS) and related methods in archaeology from the late 1980s and early 1990s marks a disciplinary step change, for enthusiasts and critics alike, increasing by an order of magnitude the quantity of spatial data that could be managed and analyzed, especially for those working at the scale of entire archaeological landscapes. The 1990s saw a string of methodologically focused edited volumes covering the development of archaeological GIS (Aldenderfer and Maschner, 1996; Allen, Green, and Zubrow, 1990; Lock and Stančič, 1995; Maschner, 1996). The most recent in this vein is now over 10 years old (Lock, 2000), perhaps partly reflecting the more mainstream adoption of standard GIS techniques, as also evidenced by the appearance of textbooks on archaeological GIS (Chapman, 2006; Conolly and Lake, 2006; Wheatley and Gillings, 2002) and recommendations for archaeological predictive modeling (Kamermans, van Leusen, and Verhagen, 2009; Mehrer and Wescott, 2005; Wescott and Brandon, 2000).

Since 2000 a major focus of methodological development in GIS has been integration with computer simulation, particularly agent-based modeling; the

*Computational Approaches to Archaeological Spaces* by Andrew Bevan and Mark Lake, Eds., 17–26 

papers in Gimblett (2002) provide heritage management applications of this technology, while more research-oriented examples can be found in Kohler and Gumerman (2000). More generally, however, the last decade has seen something of a dichotomy in the presentation of new computational techniques as either forming part of something labeled Digital Archaeology (for example, Evans and Daly, 2006; more widely, the Digital Humanities, Bodenhamer, Corrigan, and Harris, 2010) or as the handmaiden of more explicitly model-based (Kohler and van der Leeuw, 2007) or spatial analytical (Robertson et al., 2006) approaches to archaeological inference (see also McCoy and Ladfoged, 2009). Gary Lock (2003) convincingly argued that the development of computer applications in archaeology has always reflected the entanglement of both methodological and wider theoretical advances, as documented in detail by others for the specific examples of GIS-based visibility analysis (Lake and Woodman, 2003) and computer simulation (several papers in Costopoulos and Lake, 2010). We are thus very conscious that any collection of computational approaches to archaeology will likewise be of the moment, so it is worth elaborating what makes this a particularly exciting time to be presenting such a collection.

Two broad sets of developments in the computational and social sciences over the last few years have conspired to create something of a research watershed and topical point of departure for a volume of this kind. The first set combines (1) some practical thresholds now reached in an otherwise continuous trajectory of growth in modern computing, (2) the sharply increased amounts of digital data now available, and (3) the wider climate of more open access to both data and software source code. A popular metric of the modern age has been the near exponential upward trend in computational power over the last six decades (so-called Moore's Law). However, there are good reasons to point to the last 10 years as unusually important, particularly with respect to the demands of research within the social sciences and humanities. On the one hand, laptop and desktop computers have reached a stage of development where they can be used for the vast bulk of programming, statistical analysis, modeling, and representational tasks in these disciplines. This is the hardware side of a wider democratization process in computational archaeology. On the other hand, high-performance computing, often leveraging clusters of otherwise idle processors rather than dedicated supercomputers, is enabling (1) far more intensive forms of Monte Carlo simulation with carefully constrained random samples and explicit hypothesis-testing, as well as (2) far higher-fidelity renderings of visually sophisticated spaces than has hitherto been possible (in this volume, see Bevan et al., Chapter 2; Lake and Ortega, Chapter 9; Earl et al., Chapter 11).

Alongside this added capacity has also come a flood of digital information. Perhaps the most glamorous agents of this information revolution have been the automated or semiautomated collection routines enabled by Global

Positioning Systems (GPS), mobile phones, social networking sites, and space-borne remote sensing instruments. Such datasets are often referred to as "big data" given that, at their most extreme, they can be large enough to demand innovative storage, retrieval, and analysis practices (though rarely so daunting in archaeology), but, in addition, there have also been major efforts at digitizing archival or legacy datasets, as well as a steady trickle of manually inputted information (for example, via citizen-led projects such as Wikipedia, Google Earth Community, and, in archaeology, the UK Portable Antiquities Scheme—such as Bevan, 2012a).

Within archaeology, some datasets offer richly structured formats and metadata in order to facilitate close semantic and technical interoperability with so-called Web 2.0 technologies, whereas others are archived in simple formats for direct download from institutional or national data repositories, and a third unknown, but most likely numerically dominant, group of datasets continues to float in a netherworld of eccentric colleague-to-colleague gifting. Even so, archaeology is increasingly in step with a wider social, political, and economic climate that emphasizes open approaches to both software development and data dissemination (Bevan, 2012b; Ducke, this volume; Lake, 2012) that stress transparency, universal access, and the promise of unanticipated onward use. Increasing instances of the tailoring of software applications to archaeological purposes is one outcome of this wider climate (for an early emphasis, see Lake, Woodman, and Mithen, 1998), and many of the papers in this volume reflect such efforts at customization.

The second set of developments that arguably make a volume of this kind a timely venture relate to the increasing maturity with which researchers have, over the last 10 years or so, been discussing some key trade-offs in analytical and interpretative approach: for example, with respect to the importance of realism versus abstraction, the primacy of bottom-up versus top-down processes, and the relative heuristic value of lumping or splitting human behavior. In this regard, and as Mark Lake and Damon Ortega argue in a later chapter, there has been a tendency for the subdiscipline of computational spatial approaches in archaeology to recapitulate broader disciplinary rifts. A good example is the dichotomy between space and place that pervades not only archaeology but also academic research more generally. Contrary to the traditional view of this as a confrontation between tribalized cultures and a few culture heroes (mirrored in the often stereotyped priorities of processual versus post-processual archaeology), we would argue that many researchers now consider more quantitative and more experiential approaches—for instance, as valid alternatives that together generate a useful dialectic. Indeed, the strong overlaps between, for example, complexity science and more avowedly humanistic approaches to space are visible in common interests with respect to historical contingency and path dependence or to behavioral emergence and structuration (see Portugali, 2006). Where there is diversity of approach and opinion moreover, we would suggest that this diversity should be thought of less as a set of generational pendulum swings or paradigm shifts and increasingly as desirable to-and-fro to encourage within a research institution, over the course of an individual research career or indeed within the same edited volume.

It should be no surprise therefore that contemporary practicalities and cross-cutting methods are salient features of the chapters that follow, but they manifest themselves slightly differently depending on different researchers' analytical and interpretative agendas. Certain differences of intention persist and suggest a structuring framework for this volume that (1) begins with a focus on the continuing importance of spatial and spatiotemporal pattern recognition in the archaeological record, (2) continues by considering more wholly model-based approaches that fix ideas and build theory, and then (3) goes on to address those applications where situated human experience and perception are a core interest.

*1. Spatial Analysis*

With these three emphases—loosely speaking, spatial analysis, spatial modeling, and spatial experience—in mind, the authors of the first three chapters in this volume all have a common interest in more inductive spatial approaches to archaeological inference. Together they take us across a sweep of different kinds of pattern recognition: moving from the spatial analysis of data recovered from primary fieldwork (for example, via case studies spanning a major research excavation, a field survey, and joint assessment of records from rescue archaeology; Bevan and associates) to extraction of archaeological features from ground-based or airborne remote-sensing imagery (Kvamme), to consideration of the collected records from museum archives and ethnographic observation (Clark).

Bevan, Crema, Li, and Palmisano (Chapter 2) return to a long-cherished tool of empirical archaeology, the distribution map, and argue that a range of new methods and theoretical perspectives should encourage renewed efforts to characterize such spatial patterns and the underlying generative processes behind them, even if we will often still be left with multiple competing explanations at the end of this effort. Their discussion ranges from patterns exhibited by typological subgroups of bronze weapons in the tomb complex of China's First Emperor, to early Iron Age settlements in the uplands of the West Bank, to the pithouses of Japan's early complex hunter-gatherers. Throughout, they stress the importance of grappling with an archaeological record that is chronologically fuzzy and spatially uneven.

Kvamme (Chapter 3) offers a focus on two aspects of spatial and computational research that are common in archaeology but otherwise rare in this volume: the demands of cultural resource management (CRM) and the digital riches offered by remote-sensing imagery. The challenge he articulates for the archaeological community is to come up with automated or semiautomated pattern recognition techniques that can leverage the vast amounts of remote-sensing data that are now available to provide cost-effective CRM solutions across the large swaths of the planet currently under threat from development and destruction. The challenge is therefore one of seeking to upscale the techniques that would otherwise provide nice vignettes at the level of individual sites to ones that can be used routinely or can have a more global reach (see also Ducke, Score, and Reeves, 2011; Menze and Ur, 2012).

Clark (Chapter 4) argues for a more joined-up approach to multivariate analysis of large cross-cultural datasets. Drawing in particular on analytical advances in ecology, he suggests that we can squeeze far more out of complex archaeological datasets (1) by collapsing such data into distance matrices expressing dissimilarities among different sample sites (for instance, taken from different settlements or different perceived culture groups) via one or more ordination techniques, (2) comparing them with patterns of geographic or social proximity, and (3) conducting careful outlier analysis. His case study using an ethnographic example from the Pacific coast of North America highlights how evidence as seemingly distinct as social structure, material technology, subsistence practices, and ordinary distance can be brought together to offer useful analytical traction on one another.

The level of simplification and abstraction advocated in these three chapters is often considerable. Bevan and colleagues choose to ignore the complex 2D or 3D morphologies and alignments of their artifacts, houses, and settlements in favor of pure point locations. Clark chooses to collapse the complex relationships that exist among a host of different cultural variables into a series of distance matrices. Both chapters adopt relatively simple, Euclidean representations of geographic distances and social proximities that are in reality far more complex. Kvamme argues for the use of similarly idealized geometric types, such as the circle, as benchmarks for comparison with the features that appear in the raster grids produced by remote sensing. In all three cases, these trade-offs are deliberate and ones that are meant to be effective only at certain spatial scales and not at others (see also Lock and Molyneaux, 2006).

### *2. Spatial Modeling*

In partial contrast to these first three chapters, the subsequent four by Rivers and colleagues, Premo, Barton, and Herzog can all be characterized as more deductive and model-based in that they start with explicit theories or more informal understandings of human behavior and then seek to establish how these might play out under specific circumstances that did or could have obtained in the past. It is a measure of the diversity within this broad approach, however, that the conceptualization of space in these four chapters ranges from the more abstract topological model favored by Rivers, Knappett, and Evans to the geographically referenced, continuous field, elevation models used by Barton and Herzog. Another axis on which these chapters vary is the extent to which they focus on individuals or groups as the unit of analysis.

Rivers, Knappett, and Evans (Chapter 5) offer a graph theoretic model of the evolution of trade and exchange links in the Bronze Age Aegean. They formalize the spatial relationships assumed to be in operation between Bronze Age sites as an evolving set of weighted networks and in so doing provide important discussion of the sense in which network models can incorporate properties of the actual geographical space in which the network is ultimately embedded. Their exploration of the different insights provided by looking at the most likely configuration of links and nodes, on the one hand, versus the most efficient one, on the other, brings together ideas from statistical physics and network theory, as well as from the kinds of spatial interaction model long popular in urban geography.

Premo (Chapter 6) conceptualizes space as a continuous field sampled on a grid. This form of sample is, of course, the classic GIS raster map, but in this case the content of the map is artificial rather than a representation of any particular segment of the earth's surface. Premo's approach is congruent with the theory-building ambition of his chapter, which is to understand whether geographically fragmented woodland resources could have provided a selective environment favoring the evolution of altruistic food-sharing by early hominins. Premo attempts to answer this question using an agent-based computer simulation, a technique that he argues is particularly appropriate for exploring the interplay between individual decision making and the interests of the group as a whole.

The question of what is an appropriate level of abstraction is currently the subject of lively debate among proponents of computer-simulation modeling, and Barton directly addresses this issue in his chapter. Whereas Premo's theory-building model maintains a comparatively high level of abstraction, Barton (Chapter 7) is more concerned to link agent-based simulation to real geographical spaces and some of the processes that occur in those spaces, not least because of his conviction that "the human past is essential to understanding human society today and planning for our long-term future." He explores this possibility through a discussion of four case studies drawn from Europe and the United States of America and covering archaeological interests ranging from the Upper Pleistocene though early agricultural societies to what he describes as a "palimpsest of prehistoric artifactual residues."

If Barton's models typically require accurate representations of the earth's surface, the same is equally or even more true of the input to least-cost path analysis, discussed in detail by Herzog (Chapter 8). She presents the issues that arise when attempting to implement GIS-based models of the impact of terrain and land-cover on human—typically, but not exclusively, pedestrian—movement through landscapes. Quite apart from important technical issues that arise from the fact that these models represent imperfect and ultimately discontinuous approximations of reality, there are wider issues concerning the necessary assumptions surrounding the principles governing path choice and indeed the scale of analysis at which they are applicable, with several coarser grained, less goal-directed options available as alternative models of movement (for example, McRae et al., 2008).

*3. Spatial Experience*

In the mid-1990s some leading proponents of GIS (such as Wheatley, 1993) expressed disquiet that the increasing uptake of the technology had revived a functionalist approach to archaeological explanation, which had supposedly been discredited by the humanistic critique of processual archaeology. Although this detraction may have been more of a concern in some parts of the world than others—notably in the United Kingdom—it certainly led to a spirited internal critique of the explanatory power of GIS models and, ultimately, a rash of studies focusing on the human sensory experience of landscape, particularly in terms of visibility. This interest in turn gave way to a degree of disillusionment as many researchers realized that GIS-based analysis of visibility was often too coarse-grained to sustain the kinds of—often very particularistic—interpretation they sought and/or was excessively computationally demanding, although the work of Marcos Llobera (for example, 2003, 2007) is a notable exception.

In Chapter 9, Lake and Ortega recount the vicissitudes of GIS-based visibility analysis in more detail before suggesting that contemporary computing power may now allow us to overcome some of the deficiencies of earlier visibility studies. They hazard to predict, on the one hand, a renewal of interest in large-scale comparative studies of the visual properties of archaeological sites, and, on the other hand, the more frequent use of virtual and augmented reality to provide a means for understanding how these visual properties worked in terms of embodied experience. Lake and Ortega provide an example of the first approach and use the Monte Carlo simulation of nearly 30,000 viewsheds to investigate the extent to which the builders of prehistoric stone circles in England and Scotland deliberately chose topographical settings with particular properties.

Paliou's Chapter 10 provides a bridge between the work of Lake and Ortega, on the one hand, and that of Earl and associates (Chapter 11) on the other. It does so in two ways that reflect important axes of variability in computational approaches to visual space. First, through a review of the concept of the "visualscape" (see Llobera, 2003), Paliou explicitly discusses the relationship between the kind of 2-dimensional visibility analysis exemplified by Lake and Ortega's work and the possibilities offered outside conventional GIS for fully 3-dimensional study of visual space. Second, she describes recent work in geography and urban studies that could provide the basis for the analytical, as opposed to purely experiential, use of 3-dimensional visual models. In particular, she introduces archaeologists to the iso-visi-matrix and the existence of 3D visibility graph analysis.

The work of Earl and his colleagues exemplifies the 3-dimensional end of the spectrum discussed by Paliou. They describe the construction of a physically accurate 3-dimensional reconstruction of the Roman Basilica Portuense, which is then used as the framework for an analysis of how people would have perceived the space in terms of the distribution of areas of light and dark. Perhaps the key thing that takes Earl and associates' approach well beyond the sophisticated but ultimately inferentially naive production of an "attractive reconstruction" is their attempt to instill methodological rigor in virtual reality modeling. Thus, for example, they not only acknowledge the uncertainty inherent in reconstructing the upper part of the basilica (a quite different but equally important challenge to the temporal uncertainty discussed by Bevan and colleagues), but they also actually devise a method to address it: the use of procedural modeling to automatically generate alternative scenarios within the parameters set by surviving basilicas.

---

Benjamin Ducke's contribution (Chapter 12) brings this volume to a close by returning to some of the opportunities and challenges of contemporary computer science that we raised earlier in this introduction. In particular, he offers a persuasive argument in favor of open source software (a.k.a. Free/Libre and Open Source Software, F/LOSS). It remains remarkable that archaeology has so far wedded itself so thoroughly to commercial solutions. Although there are archaeologists who use open source software for spatial modeling (of which there are many examples in this volume), the vast majority of teaching and research has so far been conducted with a limited number of (typically closed source and costly) packages. This situation is reinforced, for better or worse, by the institutional and infrastructural umbrellas within which archaeologists work, whether these be commercial archaeology units, provincial public sector organizations (for instance, U.K. county councils), museums, or universities. Hence archaeology exhibits an interesting combination of enthusiasm and resistance to these open source initiatives, even if the slow trend is arguably in their favor.

The suite of chapters described above certainly cannot hope to capture the full range of computational spatial methods in use in archaeological research today, and we might further highlight important domains of enquiry such as mapping radiocarbon date distributions (Collard et al., 2010) or new spatial acquisition methods that are only beginning to be turned to research-driven agendas, such as large-scale geophysics, Structure-from-Motion and multitemporal satellite datasets (Ducke, Score, and Reeves, 2011; Gaffney et al., 2012; Menze and Ur, 2012). Keeping such additional growth areas in mind, this volume nonetheless offers a representative snapshot of how historical built spaces, past cultural landscapes, and archaeological distributions are currently being explored at a particularly compelling moment in the history of computational social science.

## References

Aldenderfer, M., and Maschner, H. D. G. (Eds.). 1996. *Anthropology, Space, and Geographic Information Systems*. New York: Oxford University Press.

Allen, K. M. S., Green, S. W., and Zubrow, E. B. W. (Eds.). 1990. *Interpreting Space: GIS and Archaeology*. London: Taylor and Francis.

Bevan, A. 2012a. Spatial methods for analysing large-scale artefact inventories, *Antiquity* 86(332): 492–506.

———. 2012b. Value, authority and the open society: Some implications for digital and online archaeology, in Bonacchi, C. (Ed.), *Archaeology and Digital Communication: Towards Strategies of Public Engagement*, pp. 1–14. London: Archetype.

Bodenhamer, D. J., Corrigan, J., and Harris, T. M. (Eds.). 2010. *The Spatial Humanities: GIS and the Future of Humanities Scholarship*. Bloomington: Indiana University Press.

Chapman, H. 2006. *Landscape Archaeology and GIS*. Stroud: Tempus.

Clarke, D. L. (Ed.). 1977. *Spatial Archaeology*. London: Academic Press.

Collard, M., Edinborough, K., Shennan, S., and Thomas, M. G. 2010. Radiocarbon evidence indicates that migrants introduced farming to Britain, *Journal of Archaeological Science* 37: 866–70.

Conolly, J., and Lake, M. 2006. *Geographical Information Systems in Archaeology*, Cambridge: Cambridge University Press.

Costopoulos, A., and Lake, M. W. (Eds.). 2010. *Simulating Change: Archaeology into the 21st Century*. Salt Lake City: University of Utah Press.

Ducke, B., Score, D., and Reeves, J. 2011. Multiview 3D reconstruction of the archaeological site at Weymouth from image series, *Computers and Graphics* 35: 375–82.

Evans, T. L., and Daly, P. T. (Eds.). 2006. *Digital Archaeology: Bridging Method and Theory*. London: Routledge.

Gaffney, C., Gaffney, V., Neubauer, W., Baldwin, E., Chapman, H., Garwood, P., Moulden, H., Sparrow, T., Bates, R., Löcker, K., Hinterleitner, A., Trinks, I., Nau, E., Zitz, T., Floery, S., Verhoeven, G., and Doneus, M. 2012. The Stonehenge Hidden Landscapes Project, *Archaeological Prospection* 19: 147–55. doi: 10.1002/arp.1422

Gimblett, H. R. (Ed.). 2002. *Integrating Geographic Information Systems and Agent-Based Modeling Techniques for Simulating Social and Ecological Processes*. Oxford: Oxford University Press.

Hodder, I., and Orton, C. 1976. *Spatial Analysis in Archaeology*. Cambridge: Cambridge University Press.

Kamermans, H., van Leusen, M., and Verhagen, P. (Eds.). 2009. *Archaeological Prediction and Risk Management: Alternatives to Current Practice*. Leiden: Leiden University Press.

Kohler, T. A., and Gumerman, G. J. (Eds.). 2000. *Dynamics in Human and Primate Societies: Agent-Based Modeling of Social and Spatial Processes*. Oxford: Oxford University Press.

Kohler, T. A., and van der Leeuw, S. E. (Eds.). 2007. *The Model-Based Archaeology of Socionatural Systems*. Santa Fe, NM: School for Advanced Research Press.

Lake, M. 2012. Open archaeology, *World Archaeology* 44: 471–78.

Lake, M. W., and Woodman, P. E. 2003. Visibility studies in archaeology: A review and case study, *Environment and Planning B: Planning and Design* 30: 689–707.

Lake, M. W., Woodman, P. E., and Mithen, S. J. 1998. Tailoring GIS software for archaeological applications: An example concerning viewshed analysis, *Journal of Archaeological Science* 25: 27–38.

Llobera, M. 2003. Extending GIS-based visual analysis: The concept of visualscapes, *International Journal of Geographical Information Science* 17: 25–48.

________. 2007. Reconstructing visual landscapes, *World Archaeology* 39: 51–69.

Lock, G. (Ed.). 2000. *Beyond the Map: Archaeology and Spatial Technologies*. Amsterdam: IOS Press.

________. 2003. *Using Computers in Archaeology: Towards Virtual Pasts*. London: Routledge.

Lock, G., and Molyneaux, B. (Eds.). 2006. *Confronting Scale in Archaeology: Issues of Theory and Practice*. New York: Springer.

Lock, G., and Stančič, Z. (Eds.). 1995. *Archaeology and Geographical Information Systems: A European Perspective*. London: Taylor and Francis.

Maschner, H. D. G. (Ed.). 1996. *New Methods, Old Problems: Geographic Information Systems in Modern Archaeological Research*. Carbondale: Southern Illinois University Center for Archaeological Investigations.

McCoy, M. D., and Ladefoged, T. N. 2009. New developments in the use of spatial technology in archaeology, *Journal of Archaeological Research* 17: 263-95.

McRae, B. H., Dickson, B. R., Keitt, T. H., and Shah, V. B. 2008. Using circuit theory to model connectivity in ecology, evolution, and conservation, *Ecology* 89(10): 2712–24.

Mehrer, M. W., and Wescott, K. L. (Eds.). 2005. *GIS and Site Location Modelling*. Boca Raton, FL: CRC Press.

Menze, B. H., and Ur, J. A. 2012. Mapping patterns of long-term settlement in Northern Mesopotamia at a large scale, *Proceedings of the National Academy of Sciences of the United States of America*. 190: E778–E787.

Parcak, S. H. 2009. *Satellite Remote Sensing for Archaeology*. London: Routledge.

Portugali, J. 2006. Complexity theory as a link between space and place, *Environment and Planning A* 38: 647–64.

Robertson, E. C., Siebert, J. D., Fernandez, D. C., and Zender, M. U. (Eds.). 2006. *Space and Spatial Analysis in Archaeology*. Calgary: University of Calgary Archaeological Association and University of New Mexico Press.

Scharl, A., and Tochtermann, K., (Eds.). 2007. *The Geospatial Web. How Geobrowsers, Social Software, and the Web 2.0 Are Shaping the Network Society*. London: Springer.

Sieber, R. 2006. Public participation Geographic Information Systems: A literature review and framework, *Annals of the Association of American Geographers* 96(3): 491–507.

Wescott, K. L., and Brandon, R. J. (Eds.). 2000. *Practical Applications of GIS for Archaeologists: A Predictive Modelling Kit*. London: Taylor & Francis.

Wheatley, D. 1993. Going over old ground: GIS, archaeological theory and the act of perception, in Andresen, J., Madsen, T., and Scollar, I. (Eds.), *Computing the Past: Computer Applications and Quantitative Methods in Archaeology* 1992, pp. 133–38. Aarhus: Aarhus University Press.

Wheatley, D., and Gillings, M. 2002. *Spatial Technology and Archaeology: The Archaeological Applications of GIS*. New York: Taylor & Francis.

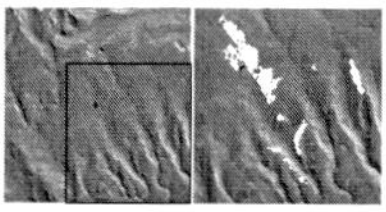

# 2 Intensities, Interactions, and Uncertainties: Some New Approaches to Archaeological Distributions

*Andrew Bevan, Enrico Crema, Xiuzhen Li, and Alessio Palmisano*

Although distribution maps are nearly as old as the discipline of archaeology itself, most archaeologists still rely on personal intuition with regard to their assessment of both the spatial patterns they recover and the environmental processes and human behaviors that might be behind these patterns. To some extent, this general preference for intuitive readings of space in the archaeological record probably reflects several decades of disillusionment with quantitative spatial methods, after a flurry of early interest during the 1970s (for example, Clarke, 1977; Hodder and Orton, 1976), and a continuing wish to prioritize the study of cultural spaces as subjectively experienced and meaningfully constituted by their human inhabitants (such as Gregory and Urry, 1985). Even the enthusiastic uptake of Geographic Information Systems (GIS) from the 1990s onward did little to change this situation with regard to spatial pattern analysis, because most off-the-shelf GIS software was targeted at data management and querying, digital cartography and enhanced visualization, as well as certain focused modeling agendas (for instance, terrain, visibility, and movement). Effectively, the study of distribution maps in archaeology merely carried on as it was, with a healthy dose of expert intuition—and perhaps in slightly richer visual form.

However, while human involvement in the act of interpretation is undeniably a crucial and enduring aspect of archaeological research, there remain good reasons to characterize spatial distributions in more formal, quantitative ways. This chapter focuses on a set of point pattern and process models that, we argue, now puts archaeologists in a position to return to the analysis of spatial pattern and process with renewed ambition, especially with regard to distribution maps. The first section considers current theoretical approaches to point distributions, and subsequent sections then address three case studies that highlight some important conceptual issues and new analytical opportunities.

## Theoretical Perspectives

### Point-Based Simplifications

A dot on a map is usually a considerable simplification. Whether our concern is about the proper 2- or 3-dimensional representation of a real-world entity, its more complicated expression in space-time, or the possible webs of cultural meaning that might envelope it, we certainly risk much by this kind of spatial abstraction. There are also further trade-offs to do with how we record such points, between time and effort on the one hand and any possible archaeological insights we might derive on the other. Do we require great spatial accuracy (such that measured coordinates are close to the actual absolute values), great spatial precision (where measurements of a given location are highly reproducible but not necessarily accurate), or some combination of the two? Furthermore, points can also be thought of as highly simplified events in space-time; as such, they often involve only a fuzzy or very loose sense of duration (how long an event lasts) and equivalence (to what extent an entity in one time-step can still be considered the same entity in the next). Indeed, if we observe such events in traditional blocks of archaeological time (such as periods or phases), this lumping procedure is a further abstraction with its own additional methodological implications and risks.

An important initial stage of spatial (and spatiotemporal) analysis therefore involves deciding what kinds of simplification and trade-off are acceptable for what applications, as well as how best to make use of the information we already have.

### Spatial Randomness, Regularity, and Clustering

Assuming for a moment that point-based abstractions are sometimes justifiable, what do we then want to say about such spatial distributions? The main formal question in the past has been the degree to which a distribution departs or not from what we might expect if we simply scattered points at random across the study area. The latter random, purely stochastic process establishes a theoretical baseline usually referred to as complete spatial randomness, or CSR. Spatial statisticians also tend to assume that the underlying process responsible for generating this random point pattern operates in roughly the same way across the whole study area (that is, it is "homogeneous" and "stationary"[1]) and that if we were to consider the number of points falling in each of a series of similar subunits across this area, we would find that their densities (the more common technical term is point *intensity*, as used hereafter) follow a Poisson distribution and are said to be a realization of a spatial Poisson process. Figure 2.1a depicts an example of a random distribution of points in a rectangular area, generated according to a Poisson process.

In contrast, Figures 2.1b–c depict two alternative patterns in which the point distribution is (b) more "regular" (also often described as "dispersed") or (c) more "clustered" (also often described as "clumped" or "aggregated"). Sometimes such patterns are intuitively obvious, and we could get away without using statistics to consider them, but often our spatial intuition is misleading: for example, some people would suspect slight clustering in Figure 2.1a (in fact, it is purely random), whereas others might not suspect regularity in Figure 2.1b (in fact, there is an arbitrarily imposed minimum distance between points). It has therefore long been acknowledged that the role of quantitative spatial analysis is partly (1) to arbitrate in situations where spatial patterns are uncertain, (2) to characterize such relationships in ways that are useful for explicit comparison, and (3) to offer a formal platform for suggesting possible processes and behaviors behind such spatial patterns. In particular, we often assume that, behind any patterns of regularity or clustering are also some interesting alternative processes, beyond one that is purely stochastic and Poisson. Regular patterns are often thought to be the result of "inhibition" processes. For example, for human settlements we might think of the way in which the existence of one settlement might inhibit the creation of another one immediately next to it (for example, because of competition over resources). For artifact distributions, regularity can be generated by various kinds of post-depositional, taphonomic sorting or be the result of very deliberate human decisions about artifact placement. Clustered patterns, in contrast, are often the result of "attraction" processes. We might think of the movement of people toward larger settlements because of a variety of the advantages such aggregated locations might offer. For artifacts, we can think of processes of discard and subsequent breakage in situ that encourage very clumped scatters of such finds in the archaeological record.

## Spatial Inhomogeneity

So, any point pattern documented across a given study area (of whatever archaeological size, from one observed under the microscope, to one found on a house floor, to one seen across a whole landscape) can be thought of as a realization of one or more underlying processes (see also O'Sullivan and Unwin, 2003: 51–75). In the simplest null case, a single random Poisson process is involved. In other, still simple cases, a nonrandom process is at work, but only one, with effects that are homogeneous across the entire study area (even if

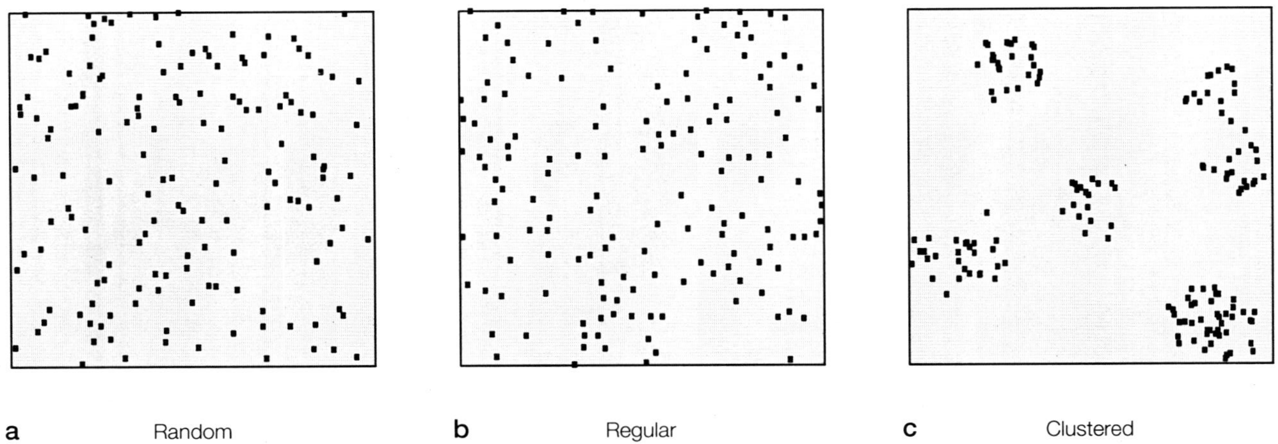


**Figure 2.1** Hypothetical examples of random, regular, and clustered point patterns.

the pattern manifests differently at different spatial scales). However, in many real-world examples, it is likely that multiple processes are at work and/or that they behave differently in different parts of the study area (that is, they can be described as inhomogeneous, heterogeneous, and/or nonstationary; for an archaeological example with aggregated count data rather than point patterns, see Bevan and Conolly, 2009). Given the prevalence of inhomogeneous distributions in real life, it is both theoretically and practically useful to distinguish between the first-order and second-order characteristics of a given point pattern (for instance, Bailey and Gatrell, 1995: 32–35). First-order characteristics are those that describe the average intensity of points across a given region (if this average intensity varies spatially, then the point pattern can be called inhomogeneous), and first-order effects refer to one or more external processes or phenomena that encourage the intensity of points in the study region to vary at different locations. In contrast, the second-order characteristics of a point pattern describe the relative intensity of points as influenced by the spatial configuration of other points in the study area (that is, the pattern's covariance structure) and reflecting different kinds of internal interaction effects among points, such as propensities for attraction or inhibition. A basic lesson from many practical analyses is that it is difficult, and often entirely misleading, to consider second-order effects before properly accounting for first-order effects.

We return to these issues in the first two case studies that follow. For now it is simply worth noting that we can easily build simulations in which to observe what kind of spatial pattern is produced by any single realization of a particular point process of known design. In real-world contexts, however, the process that generated the points is typically unknown, and the challenge becomes the degree to which we can learn about what the first- and second-order effects might be solely via analysis of the resulting pattern. It is ironic to note that, although many archaeologists would be loosely and informally aware of such complex spatial considerations when it comes to their interpretation of the archaeological record, the formal quantitative tools they have so far used have been stuck in some rather idealized and methodologically quarantined boxes. For example, at the scale of landscapes and archaeological sites, "predictive modeling" (for example, Mehrer and Wescott, 2006; Verhagen and Whitley, 2011)[2] has been a commonplace way of assessing first-order properties, demonstrating, for example, correlations between the probability of discovering sites in a particular study area and the distribution of one or more environmental variables (for example, soils, slope steepness, access to water, and so on). Conversely, nearest neighbor tests and quadrat counts have typically been used to assess site spacings, with the implicit assumption that second-order interaction effects are often at work. However, rarely, if ever, are these two methods brought together to treat the issue as an analytically related whole.

One final, complicating factor for archaeologists is the fact that archaeological observations are very partial, imperfect records of past activity. Much of the variability in our observed spatial patterns in archaeology is due to patchy levels of archaeological preservation and investigation. For example, most site distribution maps are the result of historically complex sites and monuments records or unsystematic surveys—many of the perceived clusters of observations concern where people have recently looked, where modern development has recently exposed new archaeology, and so forth. These issues can also be conceived of as kinds of first-order variation in intensity, but ideally we would want to distinguish them from taphonomic and human behavioral effects in the past, ultimately so that we can offer some useful archaeological interpretation.

## Multiscalar and Monte Carlo Approaches

This section moves from general theoretical considerations to explore the relevance of a variety of recent methods for characterizing point patterns and processes. It begins with a hypothetical example to fix some ideas before considering a real, intrasite case study.

### A Hypothetical Example

It may sometimes be very difficult to wholly separate complex first- and second-order effects (some operating in the past, some in the present) in many archaeological datasets, but there remain many advantages to conceptualizing point distributions this way. Alongside the theoretical issues raised above, a whole host of more advanced analytical methods have been discussed in the spatial statistical literature over the last 20–30 years (for a recent overview: Gelfand et al., 2010: 263–423). These have been used in certain applied fields such as astronomy or ecology for a long time but have been slow to percolate into other disciplines, such as archaeology.

Two key methodological advances over the last 35 years (outside of archaeology) have been methods that (1) deliberately seek to address point pattern and process at several different spatial scales and that (2) employ a family of randomization tests known as Monte Carlo simulation (Robert and Casella, 2004), which leverage the speed of modern computational platforms to provide a powerful and flexible way of testing spatial patterns, particularly in cases where the study area is irregular or the underlying effects are complicated. (For other archaeological applications of Monte Carlo simulation, see Crema, Bevan, and Lake, 2010; Drennan and Peterson, 2004; Fisher et al., 1997.) Figure 2.2a, for example, presents a toy example of a point pattern produced by a known process (see Lennard-Jones, 1924) in a notional 10 × 10 m study area that leads to (a) a strong tendency for regular spacing over very short

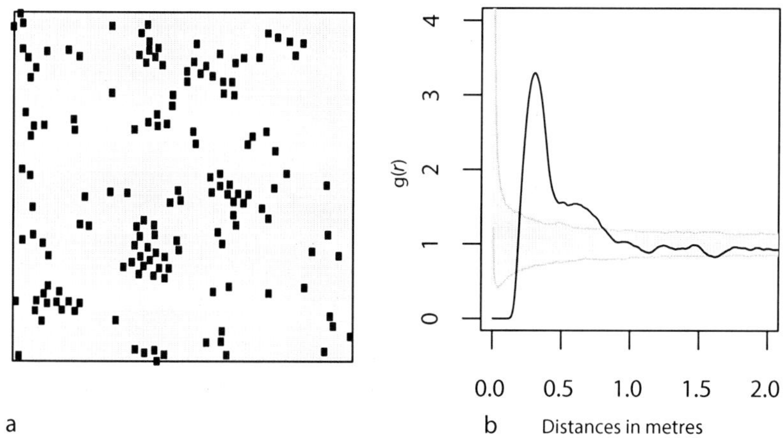


**Figure 2.2** Multiscalar Monte Carlo methods: (a) a point pattern, in a notional 10 × 10 m study area that is both regular and clustered at different distances; (b) a pair correlation function identifying a switch from significant regularity to significant clustering at c. 0.25 m (with the latter slowly tailing off thereafter). The gray area in (b) is a 95% envelope based on 999 random simulations.

distances (up to 0.25 m in this case), but thereafter also to further tendency for clustering at medium distances that gradually tails off to a random pattern at larger ones. In this case, no first-order effects are present, and the process operates in a uniform way across the whole hypothetical study area. To what extent however, can we find methods that correctly identify these different scales of second-order effect based solely on analysis of the resulting site distribution? If, for example, we calculate a traditional nearest neighbor index (Clark and Evans, 1954; Hodder and Orton, 1976: 38–51) that has been, for better or worse, the bread-and-butter of archaeological point pattern analysis for many years, it misleadingly suggests that the pattern is random or only very slightly clustered ($r = 0.91$).

There are, however, more recent spatial statistical methods that consider multiple scales of second-order patterning and explore how likely or unlikely they are to have occurred by chance. (For a technical overview, see Gelfand et al., 2010: 263–423.) Perhaps the most common of these is the K function and its more readable, slightly transformed version, the L function (originally Ripley, 1977; and for some exploratory archaeological uses, Bevan and Conolly, 2006; Crema and Bianchi, In Press; Orton, 2004; Vanzetti et al., 2010). Here we emphasize another related method, the pair correlation function (PCF), which is less well known but arguably more useful in many circumstances (several similar functions go by other names: see Illian et al., 2008: 218–23; Perry, Miller, and Enright, 2006; Wiegand and Maloney, 2004). A PCF measures the intensity of points in donut-shaped rings (annuli) around each point and, as such, is not a cumulative statistic in the same way as a K or an L function (the last two effectively measure the intensity of points in ever-expanding circles that include all previous, smaller ones).

Figure 2.2b shows PCF results for the simulated point pattern. The $x$-axis measures the separation distance between points and the observed results are presented as a black line. This observed result begins well below the theoretically random threshold of $y = 1$, indicating the possible regularity of this pattern at short distances, then climbs well above this threshold, indicating medium distance clustering, before dropping slowly back down toward $y = 1$. For a variety of reasons, however, this theoretical $y = 1$ threshold is often an unreliable baseline, and it is more useful to use Monte Carlo methods that offer an "envelope" of possible values that we might expect under a null model in which the point process generating this pattern is assumed to be wholly random. This envelope is constructed by repeatedly generating sets of an equivalent number of random points and then plotting maximal and minimal PCF values at each distance range. In the case of Figure 2.2b, the gray shaded area marks out not the full range of random PCF values but an envelope enclosing the middle 95% of PCF values from 999 simulation runs.[3] Where the real, observed values are larger than this envelope, the observed pattern can be considered clustered at that distance, whereas where they fall below the envelope, they are more likely to be regularly spaced. In this example, the PCF successfully and accurately documents the shift from significantly regular to clustered effects (the critical feature in this kind of plot is usually the point of inflection at about $y = 1$, for example, here at ca. 0.25 m, rather than the top or bottom of observed humps in the PCF) and thereafter the slow tailing off of this clustering until the pattern becomes wholly random. In fact, the noncumulative nature of the PCF offers certain advantages over K or L functions for analyzing patterns with these kinds of multiple scales and different kinds of interaction, although, in general, such methods offer complementary perspectives.

## Crossbow Triggers and Qin Terracotta Warriors

Of course, real archaeological distributions rarely, if ever, manifest themselves as such completely recovered, simply bounded datasets. The first of our three archaeological case studies therefore explores some of these analytical issues as they arise at the intrasite scale. The tomb complex of the first Chinese emperor, Qin Shihuang (259–210 B.C.E.), is famous, among other things, for its pits of life-sized, terracotta warriors, buried in battle formation, with full military equipment. As an example, we can consider the distribution of bronze crossbow triggers (the trigger being the only part of the crossbow that survives archaeologically) that were found alongside the warriors in the easternmost part of pit 1. A plot of these artifacts against the warriors (Figure 2.3) makes it clear that the overall crossbow trigger distribution itself is strongly clustered in space owing both to the shape of the corridors and the nature of terracotta

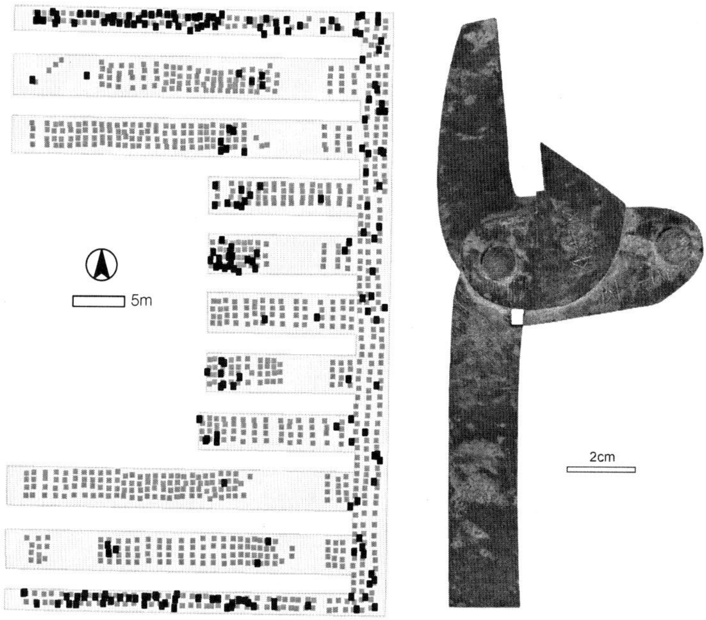


**Figure 2.3** Intrasite spatial distributions: (left) Qin terracotta warriors (gray squares) and bronze crossbow triggers (black circles) in the easternmost parts of pit 1; (right) a photograph of a bronze crossbow trigger.

army's battle formation (with crossbowmen in only certain parts of the army, particularly along the flanks). This situation, in and of itself, is probably not something we need to assess via more complicated statistical treatment, but, in passing, it is worth noting that the corridors represent a very irregular study area that raises some complicated issues to do with "edge effects" (for example, how we handle the fact that the annuli must be truncated to reflect that points cannot fall beyond the corridor area, as well as the possible inaccuracies that arise from our lack of knowledge about areas immediately west of the excavated portion of pit 1).

For our purposes here, however, a key interest is not how to characterize the overall pattern of triggers but how we might modify the methods introduced above to consider patterning among different subgroups of crossbow trigger. More precisely, metrical, typological, and materials analysis of the triggers has been able to distinguish subtle but undeniably different trigger subgroups that suggest the existence of different weapon-casting molds, different metallurgical workshops, and/or different organizational practices. The difficult question therefore becomes: how do we assess the spatial distribution of the trigger subgroups while controlling for the overriding spatial structure of trigger distribution in general?

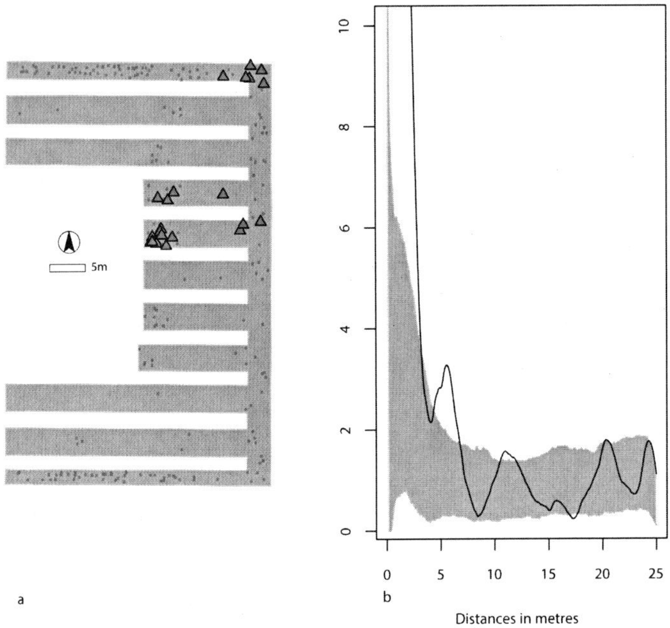


**Figure 2.4** Spatial analysis of trigger groups: (a) group 2 triggers shown as triangles and the others as gray dots; (b) a pair correlation function (observed values in black and 95% critical envelope in gray).

A good example is the trigger subgroup shown in Figure 2.4a. This is a group that, when studied in detail, exhibits small but distinct morphological and typological differences from other triggers. To recap, when we focus on the possible spatial patterning of this subgroup, we clearly want to control for the spatial structure of the triggers as a whole (and by extension the formations of crossbowmen). To do so, we run a Monte Carlo simulation in which the triggers attributed to this particular subgroup are assigned at random among the overall trigger assemblage. In fact, the group 2 triggers in the pit are, themselves, visibly clustered, beyond the pattern imposed by the battle formation (Figure 2.4a), and, again, there may not be a need for a formal method to recognize it in this case. However, this particularly clear-cut example offers a useful proof of concept—and with the knowledge that such standardized evaluation will be far more important in other less obvious cases. Figure 2.4b shows a pair correlation function in which this clustering is very evident in the observed result substantial deviation above the 95% envelope. More precisely, the plot indicates particularly strong clustering of this subgroup up to distances of perhaps 3–4 m radius and then up to 7–8 m, with further possible clustering at much larger distances. There are some interesting processes that are likely to be behind such clustered patterns of trigger subgroups in the pit. For example, they may reflect different workshops producing marginally different crossbow triggers and procedures for the storage and placement of the crossbows in the pit in batches (for example, zones of the pit that were equipped with crossbows in one go). Applied more broadly to other trigger subgroups, other weapon types and other artifacts in the pit, such analyses can begin to map out coherent activity spaces and explore how consistent they were in size, arrangement, and so on (see especially, Li, 2012; Martinón-Torres et al., In Press).

## Inhomogeneous Point Process Models

The preceding case study demonstrates, via a deliberately straightforward example, that such methods can formalize our assessment of spatial patterns at multiple scales, even in the presence of other confounding spatial factors (such as the shape of the corridors and the clustering of crossbow triggers as a whole). In their original form, methods such as K, L, or pair correlation functions were not easily applied to these kinds of inhomogeneous and edge-affected cases, but such problems are now becoming increasingly tractable. The second case study considered here explores the potential of such inhomogeneous approaches for assessments of site location at the landscape scale (for a different application of these methods, see Bevan and Wilson, 2013). It considers some Iron Age I (ca. 12th–11th centuries B.C.E.) settlements documented by fairly systematic surface survey in the central part of the West Bank (modern-day Israel and Palestinian Territories; see Finkelstein and Lederman, 1997; Finkelstein and Magen, 1993). In particular, we focus on an area of hilly dolomite upland of some 766 sq km, across which 99 sites of this period have been documented (Figure 2.5a). This choice of area is deliberate: it reduces the range of complicating factors that need to be considered later in our discussion, both because it was investigated in a fairly even way by a single archaeological project and because it covers an area of generally consistent underlying geology (with some secondary implications for soils, topography, and so on).

A quick visual inspection of Figure 2.5a suggests informally both that there might be a first-order trend toward slightly greater densities of settlement at higher elevations and that there might conceivably be some regular spacing to some of the settlements. We can therefore build some formal point process models to consider whether environmental affordances such as elevation are indeed significant and, above and beyond this, whether there is yet an additional second-order propensity for the location of one settlement to inhibit the location of another nearby. We begin by considering, as examples, four related environmental affordances: elevation, average annual rainfall, ridge-top landforms, and topographic wetness in a local catchment (Figures 2.5a–d).[4] This

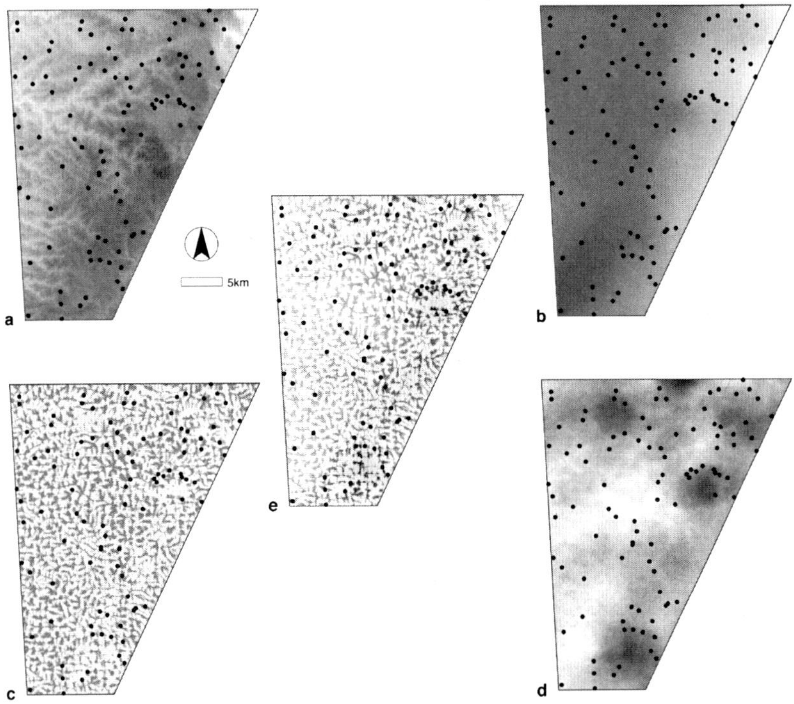


**Figure 2.5** Iron Age I sites in the central West Bank and four possible first-order covariates: (a) elevation (light to dark ranges from 135–1,010 m ASL); (b) average annual rainfall (dark to light ranges from c. 335–720 mm); (c) ridge landforms (darker is more likely to be geomorphometrically classified as a ridge); (d) topographic wetness index summed over a local neighborhood (darker is wetter); and (e) a prediction surface based on the three significant covariates (darker is higher point intensity).

selection is prompted in part by many commentators' informal impressions that rugged topography and hydrology were important factors behind settlement locations in this region and period, for a variety of practical reasons (for instance, Gibson, 2001; Zertal, 1988; in fact, many other possible covariates have been explored but are not considered here). Univariate regression of binned versions of each of these variables against site intensity (Figures 2.6a–d) suggests that rainfall is not a particularly good predictor of the intensity of sites across the landscape but that the other three variables have significant positive correlations ($p < 0.05$ or better). In other words, sites are more common (a) at higher elevations, (b) where the landforms are more ridgelike in shape, and (c) where patches of ground offer better access to surface water or soil moisture. It informally appears that ridgelike locations might be the most influential of these.

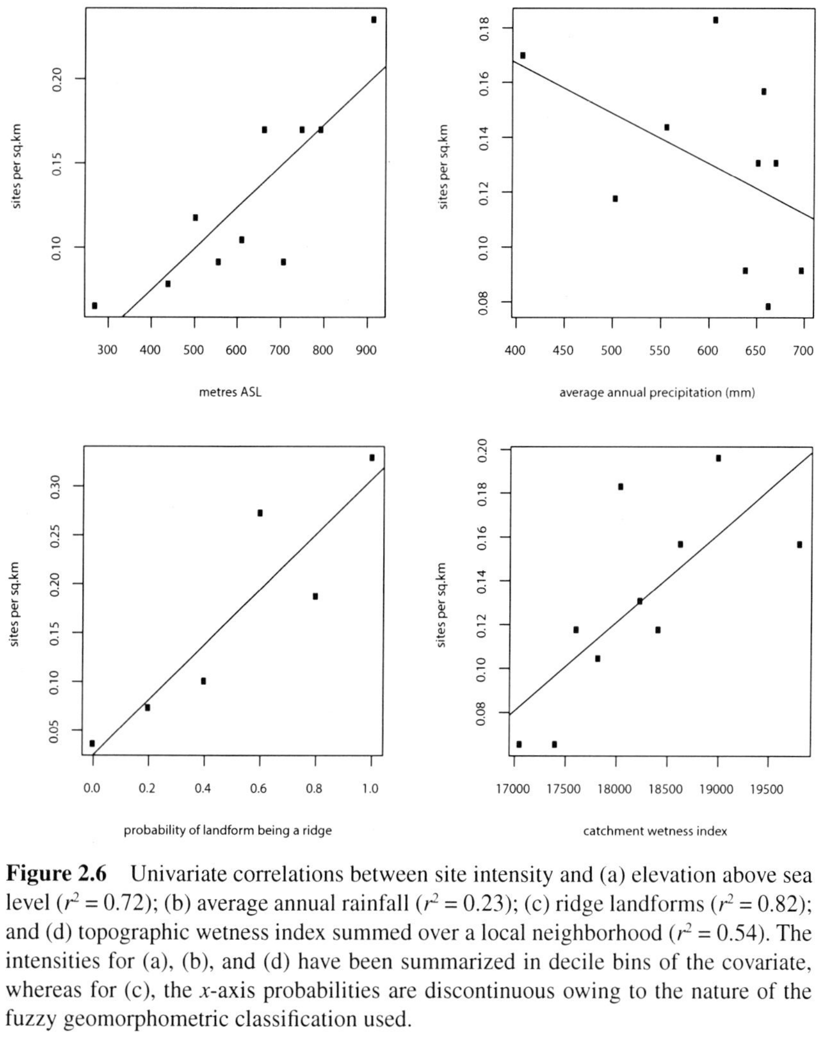


**Figure 2.6** Univariate correlations between site intensity and (a) elevation above sea level ($r^2 = 0.72$); (b) average annual rainfall ($r^2 = 0.23$); (c) ridge landforms ($r^2 = 0.82$); and (d) topographic wetness index summed over a local neighborhood ($r^2 = 0.54$). The intensities for (a), (b), and (d) have been summarized in decile bins of the covariate, whereas for (c), the $x$-axis probabilities are discontinuous owing to the nature of the fuzzy geomorphometric classification used.

These univariate regressions provide a guide to likely relationships between site intensity and various first-order effects (note that we have already removed other possible first-order effects by choosing a study area that is broadly one type of geology and that has been surveyed by one field project with fairly consistent methods). If we now run a multivariate regression and select the best combination of these four variables via stepwise comparison (minimizing an Akaike Information Criterion), we find that rainfall is excluded as we might expect, that the other three variables are all significant ($p < 0.05$ or better), and that this new model with a first-order trend is substantially more effective than a null, random hypothesis. We can then create a predicted first-order intensity surface (Figure 2.5e) that can be used to return to the question of second-order interactions in a more complete way.

First, as a point of contrast, it is worth considering a pair correlation function of the settlements sites, along with an envelope of wholly random Monte Carlo simulations (Figure 2.7a). The observed PCF shows something very close to regularity at shorter distances of up to ca.1 km, although the Monte Carlo envelope suggests that this might be of marginal significance. If we then look at a simple histogram of nearest neighbor distances (Figure 2.7b), we can see a spike at just over 1,000–1,250 m, and a Monte Carlo 95% envelope

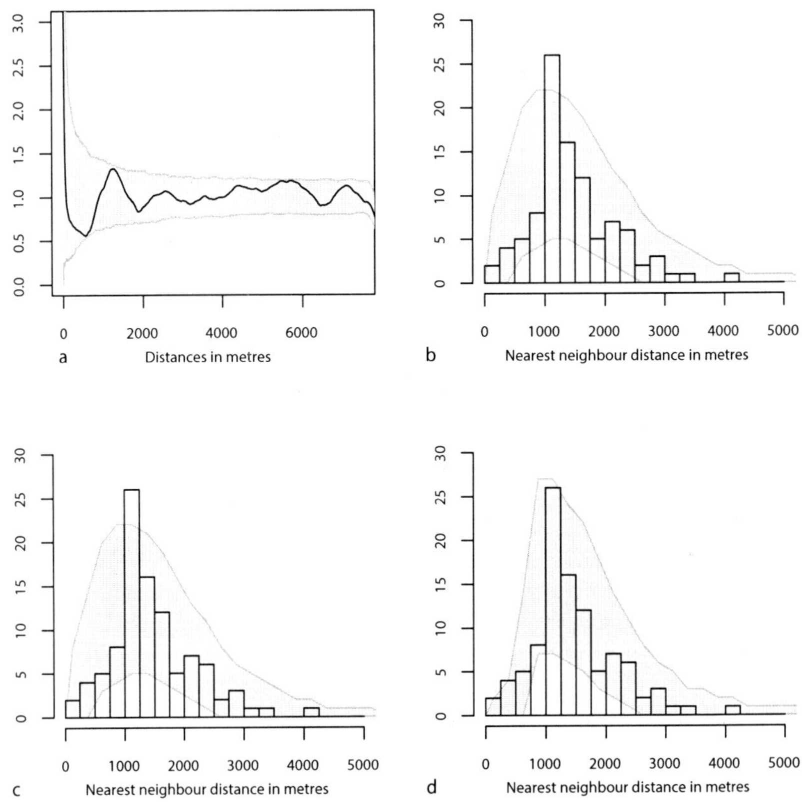


**Figure 2.7** Point process models and goodness-of-fit: (a) a pair correlation function of the observed sites with a 95% envelope from wholly random Poisson process; (b) a histogram of nearest neighbor distances with a 95% envelope from wholly random Poisson process; (c) a histogram of nearest neighbor distances with a 95% envelope also conditioned on the first-order covariates model; (d) a histogram of nearest neighbor distances with a 95% envelope also conditioned on both the first-order covariates and a second-order, area-interaction model.

suggests that this pattern is unlikely to occur by chance. (For this method with histograms of nearest neighbor distances, see also Wilson and Melnick, 1990.) In other words, the nearest neighbor histogram provides a slightly more discerning picture of very short distance patterning that confirms the evidence for regular spacing that was initially visible in the PCF. (The same observation is valid for other multiscalar functions such as K and L as well.) At this stage, however, we cannot be sure whether such regular spacing has been induced by the spatial structure of some important external influence on site location (for instance, evenly spaced ridgetops) or is due to internal processes that inhibited settlements being located close to one another (for example, competition over resources).

Figure 2.7c seeks to tease apart the relative contribution of first- and second-order effects by showing the same nearest neighbor histogram for the observed values, but this time with a simulated envelope conditioned on the spatially inhomogeneous intensities predicted by our first-order covariates. Put plainly, the Monte Carlo sets of points are now forced to respect the spatial inhomogeneity modeled by the predicted intensity surface (Figure 2.5e). As mentioned, we might conceivably anticipate that one or more of the first-order variables, such as evenly spaced ridgelines, might have accounted for some of the short distance regularity in settlement, but this is not the case. Of course, it is possible that this continuing regularity simply means that some important environmental covariate has not been considered, but, nevertheless, we are now at least moving closer toward ruling this possibility out.

Finally, therefore, we can explore the goodness-of-fit (via informal visual comparison here, although more formal statistical treatment is also possible) of an explicit model of what might be causing these second-order point interactions. Perhaps the most relevant one is Baddeley and van Lieshout's "area-interaction model" (1995), which generates patterns of inhibition and clustering with reference to a defined circular neighborhood around each point. The implicit idea of this model—that points have notional territories of influence around them—is obviously attractive given our understanding of how many human settlements work. We can draw on our knowledge of the observed spacing between settlements and set the parameters to suggest a radius for the interaction neighborhood of just over 655 m radius (half the median nearest neighbor distance) and inhibitive effects that are very strong but not absolute within this zone. These parameters lead to sites spacings that are often twice the neighborhood radius and that often suggest formal or informal village catchments of less than 135 ha, with such a scale being not unreasonable, given evidence for fairly small Iron I community sizes of dozens to no more than a couple of hundred people in this area. Figure 2.7d demonstrates that the histogram of observed nearest neighbor distances now falls within the Monte Carlo envelope. As we discuss shortly, other explanatory models might conceivably offer better or equivalent fits, but by narrowing the range of possibilities in this formal manner, we clarify our thinking about what might be plausible kinds of causal phenomena.

## Models with Temporal Uncertainty

Our final case study is again one focused on settlements and landscapes, but with a greater emphasis on diachronic comparison in the presence of uncertain dating. Temporal uncertainty is an elephant in the room of much archaeological interpretation. It is a near ubiquitous feature of archaeological datasets, whether these are radiocarbon dates, geoarchaeological deposits, or individual artifacts. There is insufficient space here to discuss this topic at length, but for our purposes, one primary risk in the spatial analysis of point distributions is that they might reflect a chronological palimpsest that thwarts our ability to unpick single-period, contemporary point patterns. This is also a topic to which many archaeologists have discretely turned a blind eye: for example, it is no more than a convenient analytical assumption that all 99 of the Iron I settlements in the central West Bank study area discussed previously were inhabited at exactly the same time during that phase, which spans a couple of centuries. (Indeed, a few of the unusually clumped sites in the eastern part of the study area might be seasonally occupied.) In other cases, the chronological range of the sites under investigation is even broader, and the risk of drawing misleading conclusions is correspondingly greater. This is especially true with regard to the assessment of second-order effects and the processes that lead to them. Michael Barton (this volume) nicely outlines an example of a regularly spaced pattern of small sites in north-central Arizona that might be due to patterns of shifting clearance, short-term cultivation, and abandonment, in which many of the sites involved belonged to the same broad period, but might not be strictly contemporary. The implied processes of interaction are, in this case, quite different, and the discussion at the end returns to the well-known problems of equifinal models that this example raises.

### Aoristic Methods

One way to engage more effectively with temporal uncertainty is for us to make the best of all our available temporal information, however fuzzy. Occasionally, we can define an explicit probability distribution that suggests how likely it is for an event to have occurred at a certain stage in time based on a range of sources of knowledge (for example, diagnostic artifacts, clear stratigraphic relationships, absolute radiocarbon dates, and so on). Even so, such results rarely produce a normal distribution that can be conveniently summarized by a single summary value and confidence interval. Instead, we are more likely to have the kinds of irregular probability distribution often produced, for example, via Bayesian modeling of calibrated radiocarbon dates and associated soft information. More important, such information-rich cases are rare: in most instances, we can suggest only a very approximately bounded time span within which the event is likely to have occurred and, within this, assume a simple, uniform probability distribution (that is, implying that an event has a similar chance of having occurred at any stage within the time span). Aoristic analysis is an approach that provides a way of quantifying these temporal uncertainties and incorporating them into subsequent analysis (particularly in the case of simple time spans where we assume uniform probabilities). It was initially developed in criminology (Ratcliffe, 2000) and has subsequently been adopted by some archaeologists for looking at both individual artifacts and larger archaeological sites (Crema, 2012; Crema, Bevan, and Lake, 2010; Johnson, 2004; Pentedeka et al., 2010).

More precisely, given a specific set of points with their temporal probability distributions, we will have a limited number of *possible* spatiotemporal patterns that might actually have arisen. Instead of ignoring this uncertainty and producing a single, but misleading, spatial analysis, we can generate different possible spatial patterns based on these temporal probabilities and then obtain a distribution of the more and less likely results. Figure 2.8 is a schematic representation of both the problem and the possible solution: take, in this case, six point events (A–F) that each occurred in one of five time-steps (t1–t5) but can often be ascribed archaeologically only to wider time spans (Figure 2.8, left). Each of the actual scenarios in the middle and right-hand-panels of this figure are possible realizations of the pattern, among many others, given the state of our temporal knowledge. The only way to explore what possible spatial patterns might really have been present is therefore to analyze a whole host of possible realizations and explore if there are any first- or second-order spatial properties that persistently crop up. If for instance, 90% of all the possible point patterns are spatially clustered, we will have a relatively high confidence that the observed pattern was indeed clustered.

However, although from a theoretical standpoint, it might be tempting to consider each and every possible spatiotemporal configuration, in practical terms this is computationally prohibitive, because the number of possible scenarios is often intractable. The alternative, however, is simply to sample a finite number of possible realizations via Monte Carlo simulation and calculate the frequency of certain spatial patterns in the results. (See Crema, Bevan, and Lake, 2010, for further details; Izquierdo et al., 2009, for a similar perspective.)

## Middle Jomon Settlement

Our final case study combines aoristic and Monte Carlo methods to consider settlement patterns among the Jomon hunter-gatherers of central Japan. In a

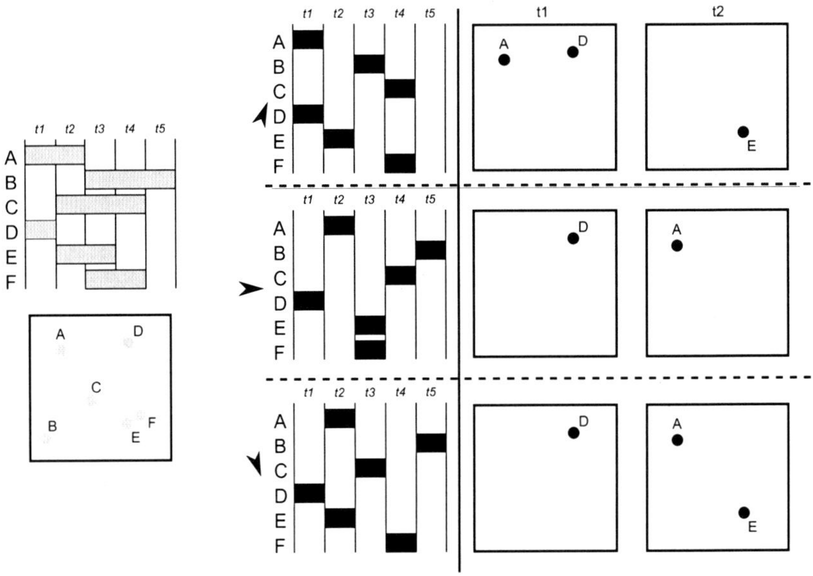

**Figure 2.8** Temporal uncertainty in point patterns. The lower-left panel depicts a simple hypothetical point pattern and (above this) our degree of temporal knowledge about each of the six point events (A–F) over five time-blocks (t1–t5). The gray horizontal bars represent the time span of each event, showing that point event D has low uncertainty (the point event belongs exclusively to time-block t1), and event B has higher uncertainty (the point event belongs somewhere between time-blocks t3 and t5). The three panels in the middle show three possible realizations of the actual temporal (middle panel) and spatiotemporal patterns (right panel, for time-blocks t1 and t2).

sense, it offers an ideal case for tackling the issue of temporal uncertainty because, although careful pottery study and an amazingly dense number of emergency excavations (Habu, 2004; Kobayashi, 2008) provide some of the most detailed spatial distributions and relative chronologies known for any prehistoric complex hunter-gather groups worldwide, it remains true that some Jomon pithouses and broader settlements can be ascribed to only fairly broad chronological ranges, whereas others can be dated far more accurately.

Moreover, Jomon settlement exhibits some interesting possible patterns that may relate to changing demography, social practices, and subsistence strategies. For example, several studies (for example, Imamura, 1996) have indicated a sudden rise in the overall number of Jomon residential units during the first part of the Middle Jomon period (ca. 3530–2470 cal BC) followed by a rapid collapse after few centuries. Some authors explain such dynamics as due to increasingly intensive use of certain plant resources during the Early Middle Jomon that made it possible to maintain higher population densities but that became more problematic during a subsequent climatic cooling phase in Late Middle Jomon, which may have led to a reduction in the overall availability of these resources (Habu, 2008; Imamura, 2002). In terms of spatial patterning, most commentators agree that there were larger, more nucleated settlements before the proposed population collapse and more dispersed, smaller settlements after it. Although such a broad dichotomy seems plausible for the Middle Jomon, considering tempos of change over smaller timescales or comparing these processes with those in earlier or later periods of Japanese prehistory remains difficult. Aoristic analysis and Monte Carlo simulation can, however, provide a good analytical framework for such research.

The case study area for the analysis that follows is located on the western side of Tokyo Bay, near the modern city of Chiba. We have chosen an arbitrary square-shaped area of 15 × 15 km within which 120 separate open-area excavations have documented some 1,418 Jomon pithouses that can each be attributed to somewhere between the Early and Late Jomon period (ca. 5050–1270 cal BC; Figure 2.9a). For each of these pithouses, we can define a more precise time span of existence from the description of pottery and associated artifacts available from excavation reports. We can then spit the whole 5050–1250 cal BC time frame into arbitrary chronological blocks of 100 years each and calculate the probability that a given pithouse actually exists during that block. For our purposes, we compare two distinct time-blocks: 2950–2850 cal BC during a phase of population increase and 2750–2650 cal BC at the observed

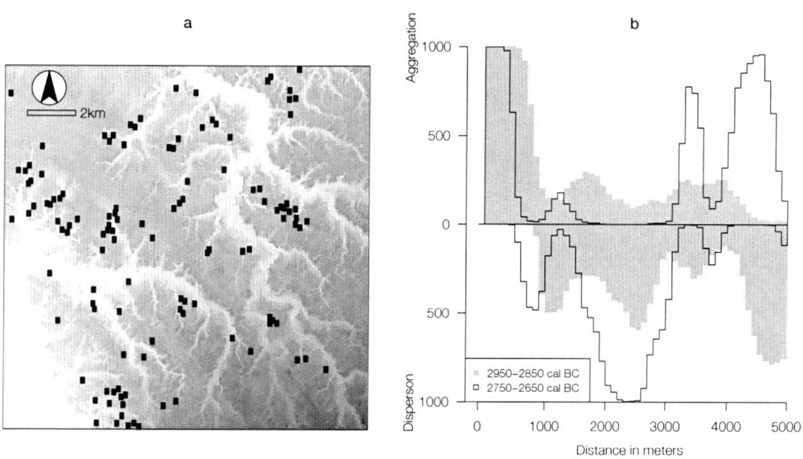


**Figure 2.9** (a) Distribution of excavation units containing at least one pithouse attributed to Early to Late Jomon (filled and hollow circles) and locations where at least one pithouse with a probability of existence higher zero at one of the two temporal blocks examined is present (filled circles); (b) number of PCFs with significant clustering (above the horizontal line) and significant dispersion (below the horizontal line) for time-blocks 2950–2850 cal BC (gray shaded bars) and 2750–2650 cal BC (hollow bars) with 1,000 simulated spatiotemporal patterns.

peak in population. For each of these two, we then generate 1,000 Monte Carlo simulations of possible spatiotemporal patterns. In these simulations, and again for the purposes of this example, we designate a new settlement to exist at a certain location, if an excavated area at that spot is allocated at least one pithouse for that chronological block.

The resulting simulations provide a series of realized settlement patterns; in order to explore differences in the respective spatial distributions for each period, we computed a series of PCFs. Just as in our earlier case studies, each observed PCF can be compared to an envelope generated from 999 Monte Carlo sets, each with an identical number of points and each allowed to exist only within the 120 parent locations where Early to Late Jomon pithouses have been actually been recovered. This method of constrained randomization is similar to the one adopted for the case study on bronze crossbow triggers and allows us to account both for the patchy nature of modern excavation and for some general first-order, locational choices the Jomon may have had over the long term (the latter being of great general interest of course, but not for our analysis here).

Figure 2.9b shows the frequency of instances among the 1,000 simulated spatiotemporal patterns where the observed results depart from the 95% envelope of random values. During the earlier phase of population increase (2950–2850 cal BC) settlements appear to be aggregated to varying degrees over distances up to 600 or even 1,000 m radius. (And indeed, whether we see this as an observation about intersettlement clustering or simply large extensive settlement areas is partly just a question of semantics.) At much larger spatial scales, there are a few instances of dispersed patterns, but these are fairly rare, with an exception at ca. 4,700–4,900 m, where about 70% of the simulated patterns showed dispersion. Also notice how, at some spatial scales (for example, at ca. 1,500 m), the number of instances of clustering and dispersion are roughly equal, implying that our levels of available information are insufficient to draw any robust conclusions. A few hundred years later, when the population size reached its peak (2750–2650 cal BC), the settlement pattern is notably different. First, the short-distance clustering is still present but now occurs only of a much smaller range (< 300 m radius). This result indicates that clusters of nucleated settlements may have started to decrease in their sizes, an idea that is supported also by a small peak of dispersed patterns at ca. 1,000 m. Second, there is strong evidence for patterns of dispersion (that is, regular spacing) at separation distances of c. 2,500 m. Third, at ca. 4,900 there is a relatively high number of simulations (> 95%) that lead to aggregated patterns, the opposite result to the one seen in the previous time-block. Clearly, the patterns for the two different time periods are not the same, with the former being characterized by a greater nucleation and the latter possibly by greater dispersion at medium distances and aggregation at higher distances (that is, broad clumps of settlement activity with intervening spaces of some

5 km between these). Although we still need to treat such results cautiously, the analysis suggests that the beginnings of the dispersed patterning known to be unequivocally present by the mid-3rd millennium B.C.E. is already visible during the period of peak settlement, reflecting possible early instances of group fission that have been plausibly argued as being driven by the diminishing availability of local food resources.

## Discussion

The theoretical discussion and three case studies above should convey the degree to which it is worth reinvesting in the formal modeling of point patterns and processes in archaeology. In any case, from this discussion we can make several practical conclusions:

1. Careful definition of a study area is important.
2. Histograms of the distribution of nearest neighbor distances offer a fairly robust way of exploring short-range regularities in point spacing and can be made more robust as a confirmatory method via Monte Carlo simulation. Traditional Clark and Evans tests are far less discerning.
3. Multiscalar methods such as K, L, and pair correlation functions are potentially useful for understanding second-order interactions but are inappropriate on their own if there are grounds for thinking that the patterns exhibit spatial inhomogeneity.
4. We can use multivariate regression models (in a manner similar to established practices in archaeological predictive modeling) to provide first-order measures of the varying intensity of points across a study area as influenced by a range of external variables. These can then offer a platform from which to consider second-order interactions via the preceding multiscalar methods even where spatial inhomogeneity is present.
5. The kinds of temporal uncertainty present in most archaeological datasets can be successfully addressed in spatial analysis by adopting a probabilistic and Monte Carlo framework.

To end this chapter we revisit one well-known criticism of formal approaches to spatial analysis and modeling in archaeology. A common suggestion is that such efforts are, at best, frustrating and, at worst, have little interpretative value (for example, Hodder, 1977) because (1) several different modeled processes can sometimes be shown to produce the same or similar outcomes (that is, they are equifinal or convergent), and (2) the same model can sometimes be shown to lead to quite different outcomes depending on its exact starting conditions or the role played by random chance (that is, it is multifinal or divergent). However, although these challenges should encourage us to avoid statements that imply a cast-iron certainty about causal relationships, they should not dissuade us from trying to model them at all: a smaller set of equally plausible models is still better than a situation in which anything goes (see also Premo, 2010). In this sense, the preceding discussion has led us from a traditional quantitative emphasis on simply rejecting null hypotheses toward one in which greater emphasis is placed on comparing the fit of a series of potentially plausible explanations (that is, in general sympathy with an information criterion approach). In other words, Monte Carlo methods allow us to embrace equifinality by weighting alternative hypotheses in probabilistic terms. This approach should be welcomed as more in tune with interpretative archaeologies of landscape that often look positively on the coexistence (temporary or otherwise) of rival explanations; there is no reason that a similarly healthy lack of certainty should not be welcomed for quantitative approaches as well.

## Acknowledgments

The crossbow trigger case study derives from a much wider collaboration between UCL Institute of Archaeology and the Museum of the Terracotta Army, and our particular thanks go to Marcos Martinón-Torres and Thilo Rehren and the Terracotta Army Museum staff. The Iron I settlement survey dataset is one provided by the West Bank and East Jerusalem Archaeological Database (digitallibrary.usc.edu/wbarc/), with slight modifications by Palmisano. We would like to thank Adi Keinan (UCL Institute of Archaeology) for her help with this and related West Bank data, as well as Adi Ben-Nun (GIS Center, Hebrew University of Jerusalem) for allowing us to make use of the rainfall contour dataset. The dataset for the Jomon case study at Chiba has been obtained from the excavation reports kindly made available by the Cultural Properties Centre of Chiba Prefecture Education Foundation. The analysis discussed here was conducted primarily in R (R Development Core Team, 2011), with some preprocessing in GRASS GIS (GRASS Development Team, 2008) and Landserf (Wood, 2009). Our thanks to Adrian Baddeley for advice on point process models and the use of the spatstat package (Baddeley and Turner, 2005), as well as to Mark Lake and two anonymous reviewers for comments on a chapter draft.

### Notes

1. In certain contexts, there are differences between these two terms, but for economy in subsequent discussion, they are used interchangeably, as are their opposites (inhomogeneity, heterogeneity, and nonstationarity). One related aspect of spatial data that does not receive any attention here, but that implies a limited form of nonstationarity, is anisotropy (that is, situations in which points are found more frequently in certain prevailing directions; see Bevan et al., 2013, Markofsky and Bevan, 2011; for archaeological discussion).

2. In archaeology, this term has developed an unnecessarily narrow meaning, related to cultural resource management and models of site location probabilities. However, beyond archaeology, it is just a general term for any kind of model that leads to explicit predictions of one kind or another.
3. For simplicity and consistency in each of the analyses developed in this chapter, we have run 999 Monte Carlo simulations, have combined these with the observed values, and have then taken the 25th and 975th ranked values to define the borders of the envelope depicted in each plot. At first glance, it might seem as if these envelopes could be treated as also defining a 0.05 significance level, but in fact this is potentially misleading for tests that consider multiple critical values simultaneously. Alternative envelope calculations that do produce exact significance envelopes are feasible (for example, see envelope in the R spatstat package) but are more complicated to implement consistently across the different methods used here, so they have not been included.
4. The digital elevation model (DEM) used here is NASA's 90 m SRTM dataset (Jarvis et al., 2008). The rainfall data has interpolated from 50 mm contours of average annual precipitation (the original contours are courtesy of the GIS Center, Hebrew University of Jerusalem). Ridgelike landforms were defined from the DEM via a fuzzy feature classification across focal filter scales from 3 × 3 to 11 × 11 cells (Fisher, Wood, and Cheng 2004). Catchment-based topographic wetness was calculated via focal filtering of a standard topographic wetness index surface (itself derived from the DEM) in a way that summed all values within a circular neighborhood of 2.5 km radius (about half an hour's walk and a common threshold for daily travel budgets).

## References


Baddeley, A. J., and Turner, R. 2005. spatstat: An R package for analyzing spatial point patterns, *Journal of Statistical Software* 12(6): 1–41.

Baddeley, A. J., and van Lieshout, M. N. M. 1995. Area-interaction point processes, *Annals of the Institute of Statistical Mathematics* 47(4): 601–19.

Bailey, T., and Gatrell, T. 1995. *Interactive Spatial Data Analysis*. Harlow: Longman.

Bevan, A., and Conolly, J. 2006. Multi-scalar approaches to settlement pattern analysis, in Lock, G., and Molyneaux, B. (Eds.), *Confronting Scale in Archaeology: Issues of Theory and Practice*, pp. 217–34. New York: Springer.

________. 2009. Modelling spatial heterogeneity and nonstationarity in artifact-rich landscapes, *Journal of Archaeological Science* 36(4): 956–64.

Bevan, A., Jobbová, E., Helmke, C., and Awe, J. 2013. Directional layouts in central lowland Maya settlement, *Journal of Archaeological Science* 40(5): 2373–83.

Bevan, A., and Wilson, A. 2013. Models of settlement hierarchy based on partial evidence, *Journal of Archaeological Science* 40(5): 2415–27.

Clarke, D. (Ed.). 1977. *Spatial Archaeology*. Boston: Academic Press.

Clarke, P. J., and Evans, F.C. 1954. Distance to nearest neighbour as a measure of spatial relationships in a population, *Ecology* 35: 445–53.

Crema, E. R. 2012. Modelling temporal uncertainty in archaeological analysis, *Journal of Archaeological Method and Theory* 19(3): 440–61.

Crema, E. R., Bevan, A., and Lake, M. 2010. A probabilistic framework for assessing spatio-temporal point patterns in the archaeological record, *Journal of Archaeological Science* 37(5): 1118–30.

Crema, E. R., and Bianchi, E. In Press. Looking for patterns in the noise: Non-site spatial analysis at Sebkha Kalbia (Tunisia), in Mulazzani, S. (Ed.), *Le Capsien de hergla (Tunisie). Culture, environnement et économie*. Frankfurt: Africa Magna Verlag.

Drennan, R. D., and Peterson, C. E. 2004. Comparing archaeological settlement systems with rank-size graphs: A measure of shape and statistical confidence, *Journal of Archaeological Science* 31(5): 533–49.

Finkelstein, I., and Lederman, Z. 1997. *Highlands of Many Cultures: The Southern Samaria Survey*. Jerusalem: Graphit Press.

Finkelstein, I., and Magen, Y. 1993. *Archaeological Survey of the Hill Country of Benjamin*. Jerusalem: Israel Antiquities Authority.

Fisher, P., Farrelly, C., Maddocks, A., and Ruggles, C. 1997. Spatial analysis of visible areas from the Bronze Age cairns of Mull, *Journal of Archaeological Science* 24(7): 581–92.

Fisher, P., Wood, J., and Cheng, T. 2004. Where is Helvellyn? Fuzziness of multi-scale landscape morphology, *Transactions of the Institute of British Geographers* 29(1): 106–28.

Gelfand, A. E., Diggle, P. J., Fuentes, M., and Guttorp, P. 2010. *Handbook of Spatial Statistics*. London: CRC/Taylor and Francis.

Gibson, S. 2001. Agricultural terraces and settlement expansion in the highlands of early Iron Age Palestine: Is there any correlation between the two? in Mazar, A. (Ed.), *Studies in the Archaeology of the Iron Age in Israel and Jordan*. Sheffield: Sheffield Academic Press.

GRASS Development Team. 2008. *Geographic Resources Analysis Support System (GRASS) Software*. Open Source Geospatial Foundation Project, http://grass.osgeo.org.

Gregory, D., and Urry, J. (Eds.). 1985. *Social Relations and Spatial Structures*. London: Macmillan.

Habu, J. 2004. *Ancient Jomon of Japan*. Cambridge: Cambridge University Press.

________. 2008. Growth and decline in complex hunter-gatherer societies: A case study from the Jomon period Sannai Maruyama site, Japan, *Antiquity* 82: 571–84.

Hodder, I. 1977. Spatial studies in archaeology, *Progress in Human Geography* 1: 33–64.

Hodder, I., and Orton, C. 1976. *Spatial Analysis in Archaeology*. Cambridge: Cambridge University Press.

Illian, J., Penttinen, A., Stoyan, H., and Stoyan, D. 2008. *Statistical Analysis and Modelling of Spatial Point Patterns*. New York: Wiley.

Imamura, K. 1996. *Prehistoric Japan: New Perspectives on Insular East Asia*. London: UCL Press.

________. 2002. *Jomon no yutakasa to genkai*. Tokyo: Yamakawa.

Izquierdo, L. R., Izquierdo, S. S., Galán, J. M., and Santos, J. I. 2009. Techniques to understand computer simulations: Markov Chain Analysis, *Journal of Artificial Societies and Social Simulations* 12, http://jasss.soc.surrey.ac.uk/12/1/6.html.

Jarvis, A., Reuter, H. I., Nelson, A., and Guevara, E. 2008. *Hole-Filled SRTM for the Globe Version 4*, available from the CGIAR-CSI SRTM 90m Database, http://srtm.csi.cgiar.org.

Johnson, I. 2004. Aoristic analysis: Seeds of a new approach to mapping archaeological distributions through time, in Stadtarchäologie Wien, Magistrat der Stadt Wien

(Ed.), *Enter the Past: The E-Way into the Four Dimensions of Cultural Heritage (CAA 2003)*, pp. 448–52. Oxford: Archaeopress.

Kobayashi, K. 2008. Jomonjidai no rekinendai, in Kosugi, Y., Taniguchi, Y., Nishida, Y., Mizunoe, W., and Yano, K. (Eds.), *Rekishi no monosashi: Jomon jidai kenkyu no hennen taikei*, pp. 257–69. Tokyo: Douseisha.

Lennard-Jones, J. E. 1924. On the determination of molecular fields, *Proceedings of the Royal Society of London A* 106: 463–77.

Li, X. 2012. Standardisation and Labour Organisation of the Bronze Weapons for the Qin Terracotta Warriors, China (Ph.D. thesis, University College London).

Markofsky, S., and Bevan, A. 2011. Directional analysis of surface artefact distributions: A case study from the Murghab Delta, Turkmenistan, *Journal of Archaeological Science* 39(2): 428–39.

Martinón-Torres, M., Li, X., Bevan, A., Xia, Y., Zhao, K., and Rehren, T. In Press. Forty thousand arms for a single emperor: From chemical data to the labour organization behind the bronze arrows of the Terracotta Army, *Journal of Archaeological Method and Theory*.

Mehrer, M. W., and Wescott, K. (Eds.). 2006. *GIS and Archaeological Predictive Modeling*, London: CRC/Taylor and Francis.

Orton, C. 2004. Point pattern analysis revisited, *Archeologia e Calcolatori* 15: 299–315.

O'Sullivan, D., and Unwin, D. 2003. *Geographic Information Analysis*. Hoboken, NJ: Wiley.

Pentedeka, A., Kiriatzi, E., Spencer, L., Bevan, A., and Conolly, J. 2010. From fabrics to island connections: Macroscopic and microscopic approaches to the prehistoric pottery of Antikythera, *Annual of the British School at Athens* 105: 1–81.

Perry, G. L. W., Miller, B. P., and Enright, N. J. 2006. A comparison of methods for the statistical analysis of spatial point patterns in plant ecology, *Plant Ecology* 187(1): 59–82.

Premo, L. 2010. Equifinality and explanation: The role of agent-based modeling in postpositivist archaeology, in Costopoulos, A., and Lake, M. (Eds.), *Simulating Change: Archaeology into the Twenty-First Century*, pp. 28–37. Salt Lake City: University of Utah Press.

R Development Core Team. 2011. *R: A Language and Environment for Statistical Computing*. Vienna: R Foundation for Statistical Computing, www.R-project.org/.

Ratcliffe, J. H. 2000. Aoristic analysis: The spatial interpretation of unspecified temporal events, *International Journal of Geographical Information Science* 14(7): 669–79.

Ripley, B. D. 1977. Modelling Spatial Patterns, *Journal of the Royal Statistical Society B* 39: 172–212.

Robert, C. P., and Casella, G. 2004. *Monte Carlo Statistical Methods* (2nd ed.). New York: Springer.

Vanzetti, A., Vidale, M., Gallinaro, M., Frayer, D. W., and Bondioli, L. 2010. The iceman as a burial, *Antiquity* 84: 681–92.

Verhagen, P., and Whitley, T. G. 2011. Integrating archaeological theory and predictive modeling: A live report from the scene, *Journal of Archaeological Method and Theory* (online first copy).

Wiegand, T., and Moloney, K. A. 2004. Rings, circles, and null-models for point pattern analysis in ecology, Oikos 104: 209–29.

Wilson, S. M., and Melnick, D. J. 1990. Modelling randomness in locational archaeology, *Journal of Archaeological Science* 17(4): 403–12.
Wood, J. 2009. *LandSerf: A Geographic Information System for the Visualization and Analysis of Surfaces*. London City University, www.landserf.org/.
Zertal, A. 1988. The water factor during the Israelite Settlement Process in Canaan, in Heltzer, M., and Lipinski, E. (Eds.), *Society and Economy in the Eastern Mediterranean* (c. 1500–1000 BC), pp. 341–52. Leuven: Peeters.

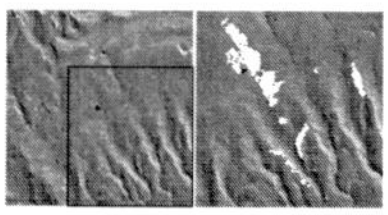

# 3 An Examination of Automated Archaeological Feature Recognition in Remotely Sensed Imagery

*Kenneth Kvamme*

World populations continue to burgeon and with them associated developments and landscape modifications that impinge on the archaeological record, whether through increasing areas of urbanization or such practices as deep plowing and landscape leveling. Hence, as evidence of the human experience continues to decline around the world it is increasingly apparent that renewed efforts must be made to locate and document archaeological resources before they are forever lost. Although this task may seem daunting, given that archaeological sites cover the globe and number in the millions, certain new technologies offer promise for aiding in their discovery. Satellite and aerial imagery cover the planet, and much occurs at high spatial resolutions sufficient for resolving archaeological manifestations, as evidenced by the frequent visibility of archaeological features in such offerings as Google Earth™ (Ur, 2006) (Figure 3.1). Elements of archaeological sites can be visualized in such imagery in the form of standing ruins, through shadows cast by subtle terrain variations covering buried settlements, or by changes in plant cover. The last includes the "crop mark" phenomenon, whereby plants growing over ruins may be stunted or show yellowing owing to disadvantageous growing environments, while those occurring over former ditches or middens may illustrate advanced growth or greening from favorable subsurface moisture or soil nutrients (Wilson, 2000) (Figure 3.1).

*Computational Approaches to Archaeological Spaces* by Andrew Bevan and Mark Lake, Eds., 53–68 

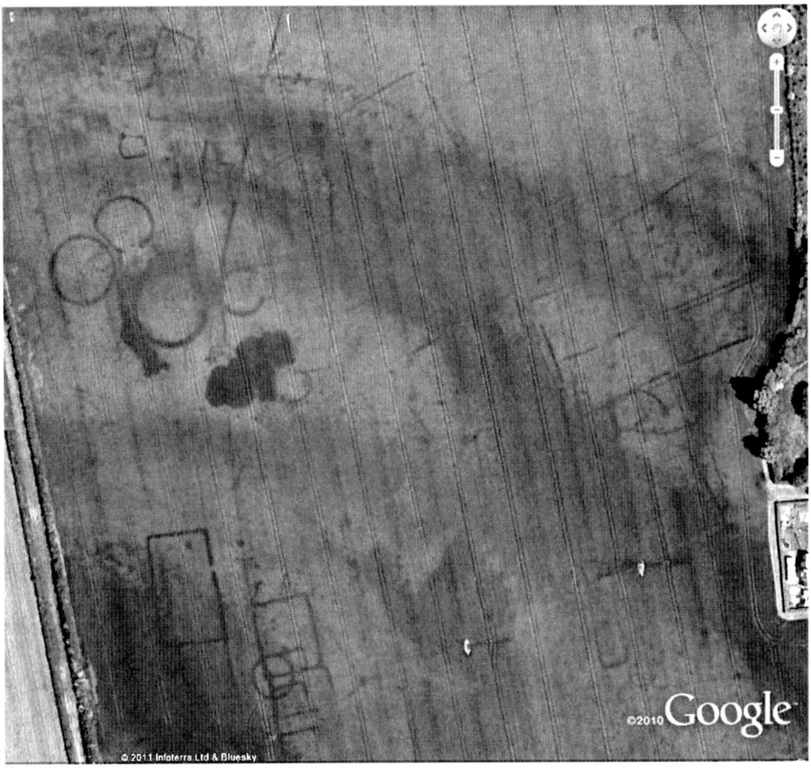


**Figure 3.1** Circles, straight lines, and rectangles pointing to ancient human constructions made visible by crop marks in an agricultural field near Oxford, United Kingdom (2010 Google Earth™ image).

Archaeologists have employed aerial imagery since the early 20th century, and it is probable that more archaeological sites have been discovered through aerial means than by all other prospecting methods combined. The recent transition to high-resolution satellite imagery for the same purpose is a natural evolution (De Laet, Paulissen, and Waelkens, 2007; Lasaponara and Masini, 2007), although at present only a limited number of satellites offer imagery with meter or submeter spatial resolutions, and such data are typically confined to a single panchromatic band. Data acquired from aerial platforms consistently offer the highest spatial resolutions yielding greater detail, the ability to resolve smaller archaeological features, and formats ranging from normal color to multispectral and thermal infrared (Barnes, 2003; Bewley, 2000; Dabas and Tabbagh, 2000). Regardless of data source, discoveries are typically made by direct visual inspection of imagery by individuals skilled in aerial or satellite image interpretation. This is a laborious process that requires hours of "eyeball time" scanning imagery kilometer by kilometer for likely archaeological manifestations. In our recent Computer Age this process has been aided by a variety of digital image transformation techniques such as high-pass filters, contrast and brightness enhancements, and color palette manipulations, which help to reveal subtle archaeological evidence not normally visible (Lasaponara and Masini, 2007; Ninfo et al., 2009), but these tools have also magnified the time required for such visual inspections.

Recent advances in computer technology also offer prospects for the automatic detection of archaeological features in remotely sensed imagery. Such detection seems highly desirable given the labor and time requirements of traditional methods and the vast archaeologically unexplored regions that face urbanization or agricultural development. In the following sections an automated approach for the rapid scanning of aerial or satellite imagery for specific types of archaeological features is investigated. This research closely follows the pioneering work in archaeology by Trier, Larsen, and Solberg (2009) and research in other sciences confronted with similar problems. The approach taken focuses on pattern recognition using shape characteristics and template-matching. It makes heavy use of GIS and image-processing methods and is largely experimental in nature, although a real-world case study is also offered. The reader is referred to such texts as Conolly and Lake (2006) and Wheatley and Gillings (2002) for details about GIS in archaeology.

## Pattern Recognition Based on Shape

A central principle in aerial image interpretation is that regular geometric shapes are generally of human origin. In other words, circles, squares, rectangles, or straight lines in remotely sensed imagery are likely the outcome of human activities. These forms are generated much less frequently by natural processes, although some natural features do possess distinctive shapes, such as alluvial fans, floodplain meanders, and volcanic cones (Avery and Berlin, 1992: 52). An apparent rectangle in remotely sensed imagery likely represents a human construction that, depending on scale, could represent the perimeter of a house, a larger administrative building or monument, a plaza or courtyard, or even agricultural field boundaries. This principle is probably the most important one in the domain of pattern recognition and frequently permits classification of landscape features to particular archaeological types (see Kvamme, 2008, for an overview of pattern recognition in archaeology). Examples of archaeological features exhibiting regular geometric shapes are not hard to find. The most recognized examples are probably the Great Pyramids at Giza, which are instantly identified from space, even in the very coarse spatial resolutions of imagery available in the 1970s (Quann and Bevan, 1977). Figure 3.1 illustrates several regular geometric shapes made visible by crop marks in an agricultural field in the United Kingdom. Such indications are common over wide areas of the globe, but can a computer be trained to recognize such regular features?

## Automated Feature Detection through Pattern Recognition

Pattern recognition represents a large area in computing where such common undertakings as Optical Character Recognition (OCR) form a strong and growing software domain that is fast becoming commonplace. It also figures prominently in the field of robotics, where complex systems must recognize outside world elements in order to interact with them. The central idea behind OCR is pattern-matching to recognize characters in an image of a printed page. Templates are commonly employed for this task, whereby various forms of the letter "a," for example, are matched against pixels in a scan to determine whether they might represent that letter. Other templates are employed for other letters or symbols to translate a scanned image into text that may be manipulated by word processors or other software. This sort of approach has been employed in other applications and disciplines to locate features of interest in imagery, such as the military for locating airplanes and tanks or lunar scientists for identifying and counting craters (see Russ, 1995: 341).

As digital archaeology and remote sensing continue to grow, automatic recognition systems will become increasingly important for the identification of archaeological features in the world's landscapes. Little prior work has yet been undertaken in archaeology, however. An early approach by Lemmens, Stančič, and Verwaal (1993), which was moderately successful, developed an algorithm that looked for abrupt brightness transitions at the same radius around any given locus to identify circular crop marks in scanned aerial photography. More recently, Trier, Larsen, and Solberg (2009) attempted to identify circular archaeological mounds in Ikonos satellite imagery in Norway, with good results in a wide number of images using a template-matching algorithm. Similar template-matching has also been applied to high-resolution lidar elevation data to identify archaeological pit features, also with good success (Trier and Pilo, 2012). Template-matching happens to be a common approach used by space scientists for identifying and counting craters on planetary bodies, although many variations to the basic algorithm exist to allow for ellipses on curved surfaces and various forms of shadowing (Kim et al., 2005). In this chapter a basic template-matching technique is investigated for the identification of circular features in imagery that is easy to implement using common GIS tools.

The foregoing studies focused on circular archaeological features, an advantage because they are invariant with image orientation. In other words, a circular template will match circles in an image regardless of how it is rotated, a circumstance untrue of such other shapes as squares, rectangles, and triangles (Figure 3.2a). To match patterns for other shapes, rotated versions

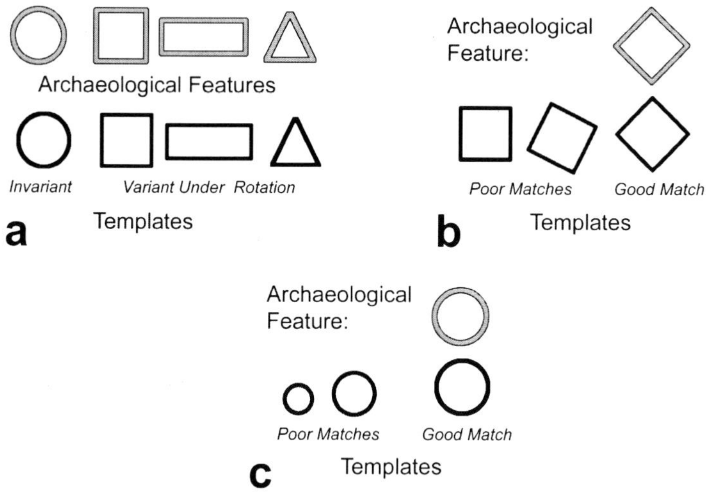


**Figure 3.2** Issues in the use of pattern templates for shape recognition: (a) circles are invariant under image rotations unlike other shapes; (b) template rotation for pattern-matching; (c) matching of template size.

of templates must also be employed (Figure 3.2b), a much more laborious undertaking requiring multiple passes through an image, each with a different template. Size variations must also be considered regardless of shape, in which case a range of templates must be employed simply because a too-small or too-large template will not suitably match features in the imagery (Figure 3.2c). In the following sections, template-matching is examined with two experimental applications followed by a real-world example from the Northern Great Plains, United States.

## Experimental Application 1: Finding Circles

An artificial archaeological raster image of a 300 × 400 m region, without real-world "noise," was developed containing circular archaeological features in the form of "rings" approximately 20 m in diameter (Figure 3.3a). The spatial resolution was set at .5 m. In this image some of the rings were made to overlap, and all are coded with 1s and 0s elsewhere for a binary map. Because crisp features such as these are unlikely in air or space imagery a second "blurred" image was also developed, wherein the rings of the initial image were subjected to strong low-pass or smoothing filters to make them less distinct or fuzzy, with values ranging from just over 0 to about 0.1 in magnitude (Figure 3.3b). Insets in Figure 3.3a–b illustrate the respective details of the crisp and blurred rings.

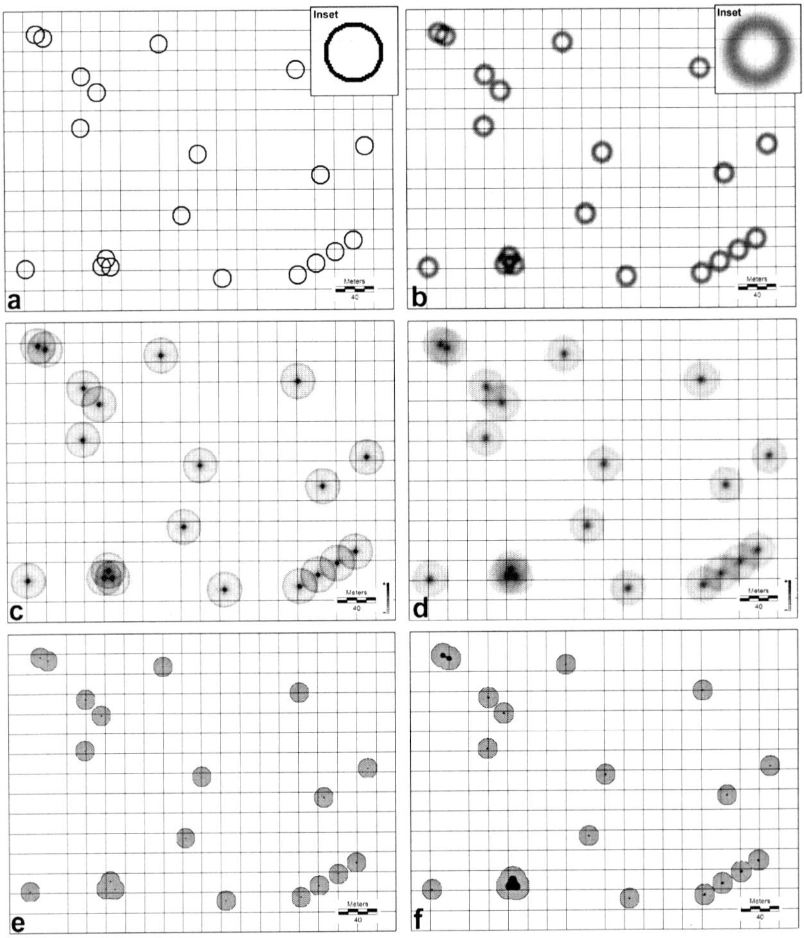


**Figure 3.3** Experiment 1 in locating circles: (a) map of crisp circles; (b) map of blurred circles; (c) outcome of template-matching on crisp circles; (d) outcome of template-matching on blurred circles; (e) derived circle center points (black) and reconstructed circles (gray) from crisp data; (f) derived circle center points (black) and reconstructed circles (gray) from blurred data. In c–d dark-grays-to-blacks indicate close template matches with higher sums.

A model or template of a crisp ring was developed with the same diameter and thickness of the crisp rings in Figure 3.3a. The 39 × 39 pixel template contains 1,521 elements in a binary mask containing a 1 in the 356 pixels making up the ring and a 0 in the remaining 1,165 pixels (Figure 3.4a–b). In application this template is centered on each pixel of a raster image, and its elements are cross-multiplied against corresponding elements in the raster and summed. Obviously the sum can accumulate only when a 1 in the template is multiplied against a non-0 element in the raster. A maximum value will occur whenever the 1s of the template closely match many non-0 elements in the input raster.

The template was applied to the crisp experimental map (Figure 3.3a), yielding results with distinctive "hot-spots" containing maxima of 356 whenever the template was exactly coincident with a circle and lesser sums (shown in gray) at loci of only partial overlap (Figure 3.3c). The form of this outcome is seen more clearly in Figure 3.4c, in a 3-dimensional portrayal of the vicinity of a single circle that clearly reveals the maximum sum with rapid fall-off to very small values when the template is even slightly mismatched but overlapping.

It is significant that less distinct, but nevertheless good, results occur when the template is applied to the blurred raster (Figure 3.3d). The local maxima in either case are easily defined by their larger magnitudes. GIS reclassification operations may then be employed to isolate these maxima, which correspond to circle center points (black points, Figures 3.3e–f). Surrounding circles of appropriate diameters may be reconstructed through GIS buffering operations (gray areas, Figures 3.3e–f).

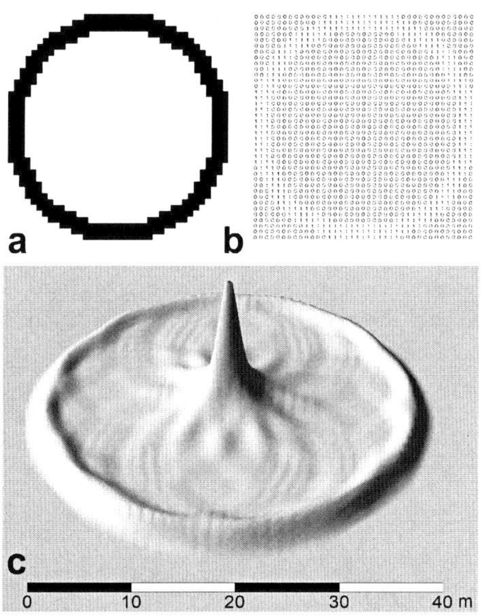


**Figure 3.4** Details of templates and template-matching: (a) graphic view of template for a circular ring 39 pixels in diameter and 3 pixels thick; (b) binary coding scheme for the same template; (c) 3-dimensional representation of the template-matching output sums in the vicinity of a single crisp circle showing the large central sums associated with close matches.

## Experimental Application 2: Circles of Different Sizes

A number of large circles about 40 m in diameter were added to the previous mix, including overlapping ones and a partial circle (Figure 3.5a). The small-circle template of Figure 3.4a was applied to these data, which defined the center points of the small circles, as before, but it was unable to generate center points for the larger circles, although tell-tale "rings" of elevated sums associated with these circles can be seen in the data (Figure 3.5b). A second

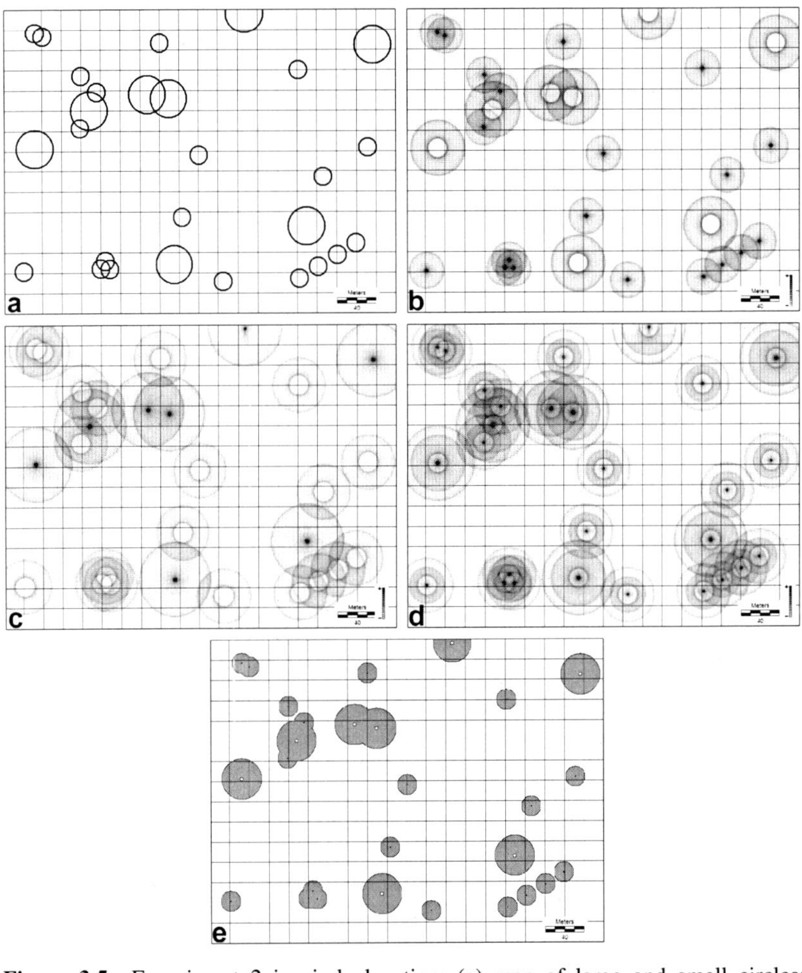


**Figure 3.5** Experiment 2 in circle location: (a) map of large and small circles; (b) outcome of template-matching for small circles; (c) outcome of template-matching for big circles; (d) sum of template matches (b + c) for big and small circles; (e) derived circle center points (black for small and white for big) and reconstructed circles (gray). In b–d dark-grays-to-blacks indicate close template matches with higher sums.

template was therefore developed that matched the thickness and 77 pixel diameter of the large circles. Application of this template to the data produced center points for the big circles, including the partial one, but it was unable to indicate center points for the small circles owing to its mismatched size (Figure 3.5c). These results demonstrate that distinct templates may be necessary to define features of different sizes (Figure 3.2c), although some lines of research suggest that a wider or "thicker" template will capture circles over a broader range of sizes (Atherton and Kerbyson, 1999), a tactic not employed here. To reconstruct the full range of circles GIS map algebra was employed to sum the two results (Figure 3.5d), generating an image with center points for all circles large and small. Small circles are associated with a maximum sum of 356 (the number of 1s in the template), whereas big circles indicate a maximum of 717. GIS reclassification procedures were applied to define center points by targeting these values. Incidentally, the large half-circle could be predicted to have a maximum sum of slightly more than 360 (that is, half of 717), and it was found to equal about 400.

In a real-world remote sensing application the idiosyncrasies of nature will cause circular anomalies to be somewhat irregular and even incomplete, causing templates to not exactly fit them. A broad range of possible outcomes (that is, sums in the processed imagery) will likely be associated with circular features requiring a range of possible maxima to be investigated in the definition of circle centers. The results of the blurred-circle experiment (Figure 3.3d) suggest the approach has power even with fuzzy data, however. A larger problem for automation lies with partially imaged features, such as the half-circle in Figure 3.5, or with incompletely revealed features that are so common in archaeological remote sensing, as when only part of a subsurface structure expresses itself. In these cases the sum resulting from a template match will be far lower than values associated with complete features of the same size and shape. Such edge effect problems are common in archaeological spatial analyses. One solution might be to include a buffer zone of some distance around an area of interest to insure only complete features are imaged or to employ some sort of edge-weighting scheme. In any case, if enough of a feature is included in an image a local maximum may result (as in the partial circle of Figure 3.5), so these too could be investigated as possible candidates for feature reconstruction.

## Archaeological Application: Aerial Imagery

To further examine the potential benefits and pitfalls of the circle-seeking approach of previous sections a real-world example is briefly explored. The focus is an aerial image acquired from a low- and slow-flying powered parachute (Hailey, 2005) over the Fort Clark State Historic Site, in central North Dakota, which contains the remains of a 19th-century Native American Mandan and

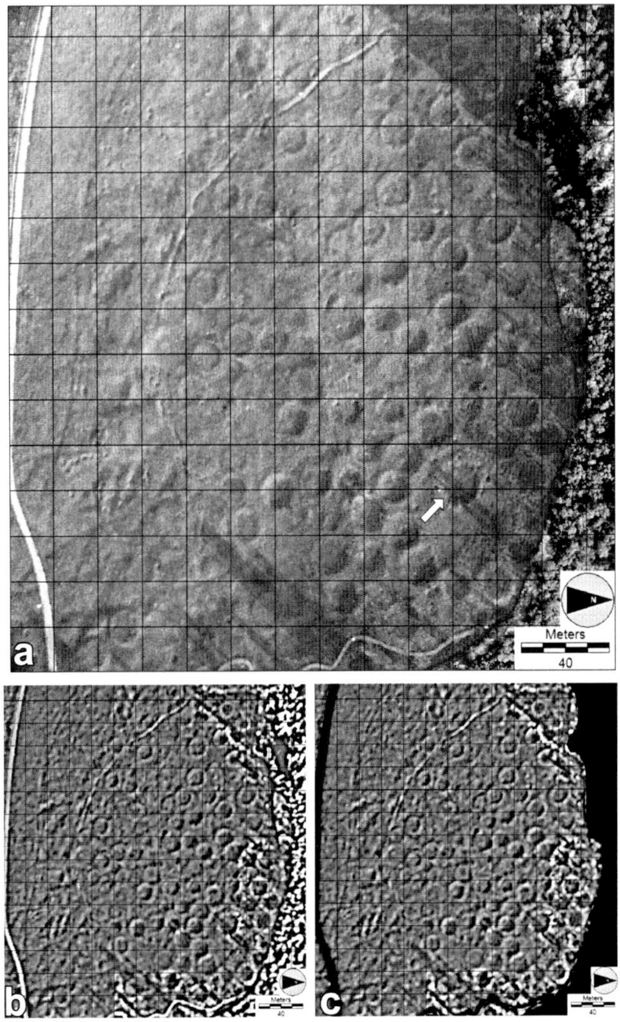


**Figure 3.6** Aerial imagery of the Mandan-Arikara village at the Fort Clark State Historic Site, North Dakota: (a) raw grayscale view (arrow points to larger ceremonial lodge); (b) result after application of low- and high-pass filters and contrast stretch; (c) final preprocessed data after masking out the road, trail, and forest.

later Arikara village (Figure 3.6a). The image was acquired in 2004 with my colleague, Dr. Tommy Ike Hailey, in an early morning flight to enhance shadowing from an altitude of about 100 m with a Minolta Dimage A2, 8 megapixel color digital camera. The region illustrates numerous circles associated with collapsed earthlodges (a type of dwelling) from the ca. 1822–1861 occupation

(Wood, 1993). The earthlodge in the 19th century was a circular structure built of a wooden frame in the shape of a hemisphere and covered with about a third of a meter of soil. When a lodge was abandoned it would eventually collapse, and the soil covering it would form a raised or mounded ring about its perimeter. These rings and the associated lodge floor depressions within appear much like craters on planetary bodies. The spatial resolution of the original image is approximately 12 cm/pixel, and the scene measures about 290 × 270 m. It also shows part of a surrounding fortification ditch to the west and south, an access road to the south, a tourist pathway to the east, and floodplain forest to the north (Figure 3.6a). The apparent lodge or house locations range in size from about 11 m to 17.5 m in diameter, with a single larger structure of 20 m diameter adjacent to an open plaza known from historical records to be associated with Arikara ceremonial activities (arrow, Figure 3.6a).

In an 8-bit grayscale image of Figure 3.6a bright pixels hold high values (with white = 255), while dark pixels are linked with low values (black = 0). The low mounds of soil surrounding each lodge tend to be brightly lit, holding higher numeric values. When a circular pattern template of appropriate size is multiplied against this image, the template should generally yield high sums when it is centered over a lodge. The image presents a number of problems, however. First, shadows from trees to the east and north are cast across some of the houses, altering image contrast and the brightness of the apparent circular rings. Second, the forest to the east exhibits the brightest pixels, which could cause unusually large sums to accrue in this area. The same is also true of the southern road and eastern trail, which also show bright reflectance. It was therefore decided to preprocess the image to enhance chances for pattern recognition.

Preprocessing included five operations.

1. The image was de-sampled, reducing the spatial resolution to .24 m. This was done to speed up processing time but also to consolidate or smooth image noise.
2. A low-pass or smoothing filter was applied to further consolidate anomalies (in effect, to "melt" together the pixels making up the circular rings).
3. A high-pass filter was applied to reduce the effects of the shadowed areas on feature contrast. (This filter computes the difference between each pixel and the average of its local background, thereby reducing background variations.)
4. A contrast stretch was performed to exaggerate brightness differences between the rings and the background (Figure 3.6b).
5. A mask was created that replaced the road, trail, and forest with 0s (shown in black, Figure 3.6c) to eliminate the influence they would introduce to the algorithm owing to their extreme brightness.

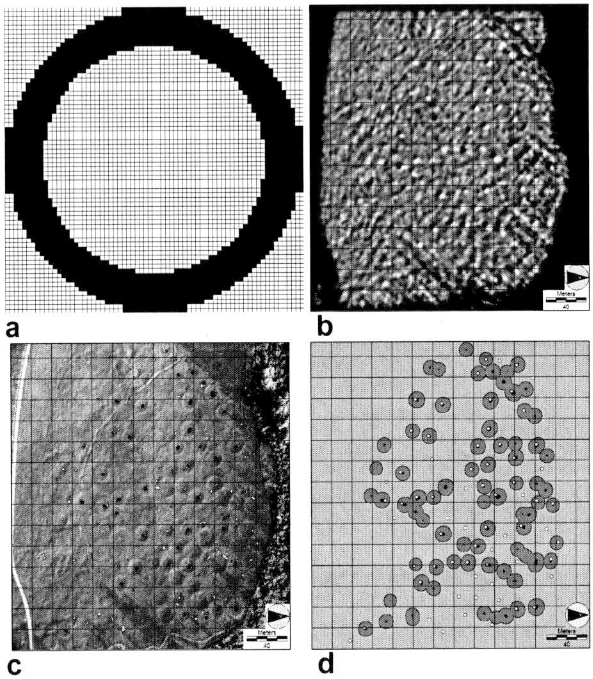


**Figure 3.7** Application and findings of template-matching algorithm: (a) the 69 × 69 pixel template; (b) outcome of template applied to preprocessed data; (c) liberal (white) and conservative (black) maxima from the algorithm superimposed on the initial image; (d) circular houses (dark gray) grown from the conservative predicted center points (black) with small white circles indicating "actual" house centers derived from a simple visual count.

With most of the houses 13–17 m in diameter, a template was developed that was centered on a circle with an average diameter of 14.6 m (61 pixels). The wall of the circles was made 2.2 m "thick" (9 pixels) to allow for variations in house diameter. This means that the smallest ring diameter that would fit into the template is 12.7 m (53 pixels), whereas the largest is 16.6 m (69 pixels), thereby allowing potential capture of a wide range of house sizes in the image (Figure 3.7a). The 69 × 69 template contains 1,608 1s.

The template was applied to the preprocessed image of Figure 3.6c, with results illustrated in Figure 3.7b. Several interesting outcomes are apparent in the result. First, as might be expected, it is extremely busy, because real-world imagery contains enormous variation. With the image characterized by gray values in the range 0–255, and with the 69 × 69 template containing 1,608 1s, enormous sums and a large range of sums are possible. For instance, a hypothetical broad white space (represented by 255) larger than the template will produce a maximum of 410,040, whereas on the average the image should yield a sum of about 206,000 (because middle gray is represented by a value of 128). The 0s forced onto the image by the preprocessing operation that masked out the road, trail, and forest produce the dark perimeter. Despite the high-pass filter and contrast enhancements the effects of the image shadows may be seen in the diagonal striping (Figure 3.7b).

What is significant for our purpose are the positive "data spikes" or very large sums indicated in white (about 2–4 m in diameter). Such peaks in the generated surface point to loci in the input image that are surrounded by bright circles of the correct diameter (that is, lodges). GIS reclassification methods were employed to isolate these peaks with a liberal threshold of 230,000 and a second more conservative one of 240,000. They are illustrated in Figure 3.7c superimposed in white and black, respectively, over the initial image. It is immediately apparent that the template-matching was generally successful in the identification of circles in this image, with many apparent successes. The liberal threshold, while capturing a few more apparent houses than the conservative result, indicates many loci that are clearly in error ("false-positives") associated with irregular bright regions in the landscape.

The conservative set of points was employed to "grow" 14-m-diameter house circles, illustrated in Figure 3.7d. The conservative outcome does a fair job in locating most of the house circles, but many obvious ones are missed by the application (that is, "false-negatives"), particularly in the northeastern portion of the village (compare Figure 3.7c–d). By my count, which is admittedly subjective and may vary with different viewings, there are 68 likely house circles in the image, some of which are quite subtle. Their center points are identified with small white circles in Figure 3.7d. Of them, 42 (62%) are correctly specified by the algorithm, while 26 (38%) are missed, which points to only fair performance. The template-matching algorithm identified 76 possible houses, however, leaving 34 false-positives (45%) after the 42 correct specifications by my count.

There are several explanations for the errors. First, the houses in the northeast tend to be larger because, historically, they were populated by important families located near the heart of the village close to the plaza and ceremonial lodge. Nearly all these lodges are 17 m or more in diameter (including the 20-m-diameter ceremonial lodge) and therefore surpass the diameter of the template employed. Second, this area is plagued by tree shadows, making it darker with less contrast and therefore reducing the size of the sums generated by the template. Finally, these larger houses are associated with deeper depressions in the landscape, so they each contain dark shadows that may affect the algorithm. Obviously, this experiment may be improved through consideration of other circular template sizes, templates that permit consideration of light and dark shadows that may be associated with rings imaged under a particular lighting condition or time of day (Kim et al., 2005), and by better preprocessing of the image to reduce the effects of tree shadows and other sources of noise.

## Conclusions

The results considered here represent an attempt to automatically detect archaeological features in air or space imagery based on their regular shape characteristics. Procedurally, this work closely follows steps outlined by Trier, Larsen, and Solberg (2009). This research was undertaken to further explore and demonstrate the utility of these methods as well as shortcomings and problems associated with them. It was also employed as an exercise in an Advanced Raster GIS class at the University of Arkansas to investigate the utility of ready-made GIS software—in this case Idrisi (16) Taiga—to undertake such operations. It is apparent that considerable research must yet be undertaken to improve these methods and to assess their performance using different types of imagery, in multiple environments, and by examining difficulties associated with various archaeological feature shapes.

The detection of archaeological features in remotely sensed imagery is likely to become increasingly important for locating and discovering archaeological sites in large, unexplored regions or anywhere on the globe where archaeological sites are under threat. Efficient means must therefore be found to facilitate this process. Rapid scanning of imagery through such automated approaches could permit potential sites of archaeological interest to be quickly located. In practice, a series of different template shapes, sizes, and orientations (under various rotations) could be passed sequentially over imagery. The outcome of this process, perhaps color-coded to indicate circular versus square features under various size categories, could be regarded as a "first screening." Such results could then be passed to trained image interpreters, pointing them to regions of potential archaeological interest and offering a means to save time and labor associated with this endeavor. In this way the scale of archaeological image analysis could potentially be augmented, because even partial automation would greatly increase the size of regions commonly investigated. This conceivably could lead to a scale change in archaeological spatial analyses, because multiple cultural landscapes, perhaps widely separated, could potentially be identified across vast regions, perhaps leading to new questions associated with their interactions and placement.

One thing that air and space remote sensing and ground-based geophysics has taught archaeology in the past few decades is that entire buried cultural landscapes exist around the globe that contain tremendous densities and enormous variation in archaeological content. Techniques such as those presented here can contribute to the revelation of such landscapes in the daunting body of high-resolution air and space imagery that is rapidly being produced at an exponential rate. With improvements in software and user-interfaces, techniques such as these could easily be automated and distributed to the archaeological community to facilitate discoveries around the globe.

## Acknowledgments

I wish to thank Andrew Bevan and Mark Lake for their invitation to participate in this volume and the former, in particular, for excellent review comments and suggestions. I also wish to thank several anonymous reviewers for their very helpful comments. This research was ultimately made possible by the State Historical Society of North Dakota, and particularly by Fern Swenson, Director of the Archaeology and Historic Preservation Division, who have promoted several years of research at the Fort Clark State Historic Site. Tommy Ike Hailey of Northwestern State University, Louisiana, deserves enormous credit for letting me talk him into over-flying Fort Clark to acquire imagery one cold morning while we were working on another project more than 100 km distant, as does Jo Ann C. Kvamme for copiloting and acquiring some of that imagery.

## References

Atherton, T. J., and Kerbyson, D. J. 1999. Size invariant circle detection, *Image and Vision Computing* 17: 795–803.

Avery, T. E., and Berlin, G. L. 1992. *Fundamentals of Remote Sensing and Airphoto Interpretation* (5th Ed.). New York: Macmillan.

Barnes, I. 2003. Aerial remote-sensing techniques used in the management of archaeological monuments on the British Army's Salisbury Plain training area, Wiltshire, *Archaeological Prospection* 10: 83–90.

Bewley, R. H. 2000. Aerial photography for archaeology, in Ellis, L. (Ed.), *Archaeological Method and Theory: An Encyclopedia*, pp. 3–10. New York: Garland Publishing.

Conolly, J., and Lake, M. 2006. *Geographical Information Systems in Archaeology*. Cambridge: Cambridge University Press.

Dabas, M., and Tabbagh, A. 2000. Thermal prospecting, in Ellis, L. (Ed.), *Archaeological Method and Theory: An Encyclopedia*, pp. 626–30. New York: Garland Publishing.

De Laet, V., Paulissen, E., and Waelkens, M. 2007. Methods for the extraction of archaeological features from very high-resolution Ikonos-2 remote sensing imagery, Hisar (southwest Turkey), *Journal of Archaeological Science* 34: 830–41.

Hailey, T. I. 2005. The powered parachute as an archaeological aerial reconnaissance vehicle, *Archaeological Prospection* 12: 69–78.

Kim, J. R., Muller, J.-P., van Gasselt, S., Morley, J. G., Neukum, G., and the HRSC Col Team. 2005. Automated crater detection, a new tool for Mars cartography and chronology, *Photogrammetric Engineering and Remote Sensing* 71(10): 1205–17.

Kvamme, K. L. 2008. Remote sensing: Archaeological reasoning through physical principles and pattern recognition, in Sullivan III, A. P. (Ed.), *Archaeological Concepts for the Study of the Cultural Past*, pp. 65–84. Salt Lake City: University of Utah Press.

Lasaponara, R., and Masini, N. 2007. Detection of archaeological crop marks by using satellite Quickbird multispectral imagery, *Journal of Archaeological Science* 34: 214–21.

Lemmens, J. P. M. M., Stančič, Z., and Verwaal, R. G. 1993. Automated archaeological feature extraction from digital aerial photographs, in Andresen, J., Madsen, T., and Scollar, I. (Eds.), *Computing the Past CAA 92: Computer Applications and Quantitative Methods in Archaeology*. Aarhus: Aarhus University Press.

Ninfo, A., Fontana, A., Mozzi, P., and Ferrarese, F. 2009. The map of Altinum, ancestor of Venice, *Science* 325: 577.

Parcek, S. H. 2009. *Satellite Remote Sensing for Archaeology.* London: Routledge.

Quann, J., and Bevan, B. 1977. The pyramids from 900 kilometers, *MASCA Newsletter* 13: 12–14.

Russ, J. C. 1995. *The Image Processing Handbook* (2nd ed.). Boca Raton, FL: CRC Press.

Trier, O. D., Larsen, S. O., and Solberg, R. 2009. Automatic detection of circular structures in high-resolution satellite images of agricultural land, *Archaeological Prospection* 16: 1–15.

Trier, O. D., and Pilo, L. H. 2012. Automatic detection of pit structures in airborne laser scanning data, *Archaeological Prospection* 19: 103–21.

Ur, J. 2006. Google Earth and archaeology, *SAA Archaeological Record* 6(3): 35–38.

Wheatley, D., and Gillings, M. 2002. *Spatial Technology and Archaeology: The Archaeological Applications of GIS*. New York: Taylor and Francis.

Wilson, D. R. 2000. *Air Photo Interpretation for Archaeologists*. Charleston, SC: Arcadia Publishing.

Wood, W. R. 1993. Integrating ethnohistory and archaeology at Fort Clark State Historic Site, North Dakota, *American Antiquity* 58: 544–59.

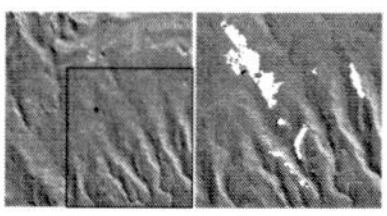

# 4 An Introduction to Integrative Distance Analysis (IDA)

*Terence Clark*

Within the natural and social sciences there is currently a wave of renewed interest in the realm of spatial statistics (for example, Clark, 2010, 2013; Hart and Engelbrecht, 2011; Lansing et al., 2007; LeSage et al., 2009; Miller, Franklin, and Aspinall, 2007; Pettorelli, Dray, and Maillard, 2005; Schillaci, Meijaard, and Clark, 2009; Soininen, Paavola, and Muotka, 2004; Spielman and Yoo, 2009). From archaeology to zoology, there is a growing sentiment that the proximity of data points is meaningful and structures the results we measure. The field of ecology is at the forefront in measuring and describing proximity-based variation (for example, Borcard, Legendre, and Drapeau, 1992; Clarke and Ainsworth, 1993; Cottenie, 2002; Dray, Chessel, and Thioulouse, 2003; Jackson, 1995; Legendre and Legendre, 1998; Peres-Neto and Jackson, 2001; Perry, Miller, and Enright, 2006). Ecological variables, like archaeological ones, are spatially patterned. The composition of environmental communities is structured by a series of environmental variables—not least among them, the distance to neighbor communities. Numerous ecological articles have focused on accounting for and removing the effect of spatial autocorrelation of variables (for instance, Diniz, Bini, and Hawkins, 2003; Fortin, 1999; Koenig, 1999; Legendre, 1993; Lichstein et al., 2002; Tobin, 2004).

Geographic Information Systems (GIS) are in widespread use in archaeology, and spatial autocorrelation is well known to archaeologists (Kvamme, 1990; Neiman, 1997; Premo, 2004; Williams, 1993). This renewed interest in proximity, however, goes beyond merely the use of GIS; a number of new

*Computational Approaches to Archaeological Spaces* by Andrew Bevan and Mark Lake, Eds., 69–98 

or rediscovered statistical procedures are also at the forefront (for example, Anselin and Getis, 1992; Bentley and Shennan, 2003; Bevan and Conolly, 2009, 2011; Clark, 2010, 2013; Hart and Engelbrecht, 2011; Jackson, 1995; Jordan and Shennan, 2003; Lansing et al., 2007; Legendre, 1993; Peres-Neto and Jackson, 2001; Soininen, Paavola, and Muotka, 2004; Zhao, Shuter, and Jackson, 2008). Within this milieu of proximity-focused analysis, I present a methodological approach to large datasets that not only accounts for geographic structure affecting data composition but also allows for much more complex structural forms to be identified and described.

## Proximity-Based Models

According to Tobler's First Law of Geography, "everything is related to everything else, but near things are more related than distant things" (1970: 236), a sentiment echoed by Milke's (1935, 1949) Theory of Cultural Similarity, Hägerstrand's (1967) Innovation Diffusion as a Spatial Process, and Hudson's (1972) Geographical Diffusion Theory. Similar work on the spatial patterning of diffusion was the foundation for a series of synthetic texts in ethnology, particularly arising from the Culture Element Distribution series of Kroeber and his students (Barnett, 1939; Driver and Coffin, 1975; Driver and Massey, 1957; Driver et al., 1972; Jorgensen, 1966, 1969, 1980; Kroeber, 1925, 1939).

In its simplest form, shared cultural traits are seen as a product of historical interaction between cultures. More shared traits means more interaction or a shorter history of separation, an idea furthered in Darwinian Archaeology with cultural differences being partially explained as a product of time since divergence from a common ancestor and cumulative cultural drift (Hart and Terrell, 2002; Mace, Holden, and Shennan, 2005; Shennan, 2002).

This principle forms the basis for gravity models (for example, Haynes and Fotheringham, 1984; Hodder, 1974; Johnson, 1977) and the distance decay model (such as Hägerstrand, 1952; Renfrew, 1977; Zeitlin, 1982), which have seen widespread use in archaeology. Geography alone does not explain the patterns of variation and similarity seen cross-culturally. Cultural adaptationists believe that technology is a response to changing ecological conditions (Binford, 1962, 1972, 1989; Clarke, 1968; Kirch, 1980; White, 1959). Changes in the resource base cause adaptation of technology in an attempt to create an optimal system for resource exploitation. Reduced to its simplest form, tools are the means by which we get food from the environment.

The resources people extract and process must determine at least some of the tools people use. By the Principle of Limited Possibilities (Goldenweiser, 1913), there are only so many cultural outcomes for a given problem. There are only so many ways to catch a fish or skin a bison. The necessity to extract and process resources for food, shelter, and clothing in all areas of the globe means that the tools that perform these functions should have some similarity of form independent of geographic proximity. Bifacial scrapers do occur in various regions partly as a result of convergent evolution, not necessarily diffusion alone.

What is clear is that artifacts are to some degree the products of both history and function, a combination of choice and need. IDA does not attempt to disentangle this protracted argument, but it offers a methodology to isolate variation between data classes that may shed some light on the debate.

Analysis of proximity should be of great importance to archaeologists. The relationships of a single data point to other data points and to other data classes are at the heart of archaeological inquiry. Stated simply, this concept is context. Although archaeologists devote a great deal of effort to recording and understanding context during excavation, the same cannot be said for much of the analysis that presently occurs. The physical relationship of artifacts and fauna rarely are expressed more scientifically than with the use of simple density diagrams. Proximity-based analysis, which could equally be deemed statistical analysis of context, allows for the same scientific rigor to be applied to context during analysis as during excavation.

## Integrative Distance Analysis (IDA)

IDA was designed to understand and account for proximity, both spatial and nonspatial, within complex datasets. It is accomplished with the use of a series of statistical procedures that measure the congruence of data classes. It is best described as a comprehensive multiscalar analysis of proximity. Taken together, I refer to these procedures as Integrative Distance Analysis (IDA).

### Founding Assumptions

IDA operates on the following basic assumptions.

1. Datasets have structure. The entirety of the data, likely separated into data classes, has some meaningful structure to it. Data are not random. Whether it be a series of pot sherds on the surface of a field or the decorated motifs that appear on them, nonrandom variation in space, time, and form exists and is the basis of archaeological classification.
2. Structure can be measured and described. Variability is not nebulous but rather can be measured through quantification, classification, or by some other means. For the purposes of IDA, the form of these measures need not be equivalent across data classes; all that matters is that they are internally consistent within each data class. Since IDA converts all variation into distance matrices, the form of raw data is not significant. Presence/absence counts can be compared to proportional and/or volumetric data.

3. Data classes are not entirely independent. Each data class, whether it is geography, technology, language, or genes, does not exist in a vacuum. At least some of the artifact assemblage exists to perform tasks. Knives cut, choppers chop, and baskets store. Subsistence technology is not independent of economy. Changes in resource use will cause changes in the types of artifacts that are used to extract and process those resources. Examining only a single data class will lead to higher correlation values than truly exist (Type I errors) (Cottenie, 2002: 116). The level of integration is variable but can be very high (Clark, 2008, 2010, 2013).
4. Complex processes occur across multiple data classes. Processes such as intensification, domestication, population replacement, and identity maintenance are complex. They operate across several data classes simultaneously. Intensification can be seen as increasing the efficiency of exploitation; it is multifaceted with broad effects and material culture signatures. Changes toward more efficient technology (nets, plows, weirs, and so on), more efficient organization of labor, resource ownership, irrigation, cooperative labor, social hierarchies, and/or worldview; or even changes in the efficiency of the plant or animal itself (domestication) amount to intensification (Boserup, 1965). By this definition, multiple data classes must be affected when complex processes occur.
5. No single data class can then explain complex processes. Following this line of logic, it becomes impossible to understand the nature of complex processes by simply looking at one data class. Technology alone cannot describe culture change, and faunal analysis alone cannot describe intensification. Furthermore, understanding how other data classes are affected during these processes will tell us more about the processes themselves.

## Getting Started

IDA works by linking multiple data classes together. This is achieved by comparing the relative position of data objects between data classes in multivariate space. It can function at a number of spatial scales depending on the data and the research question involved. At the regional scale, where data objects are archaeological sites or components, questions about community-level processes may be asked. At the site scale, where data objects may be individual layers within excavation units, questions about site use and occupation history may be broached.

*Data Preparation* One of the strengths of IDA is that any form of data may be compared to all others. It will work equally well on ratio, interval, ordinal, or nominal data. The only requirement is that each data class be recorded consistently. Since IDA looks at the relative relationship of site objects, to function it merely needs an expression of object relationships. The most common way to express data object relationships is with the use of a distance matrix. Distance matrices provide a measure of the dissimilarity of each object to all others in multivariate space.

Data are prepared for analysis by converting each class into a distance matrix. The selection of distance measure is dependent only on user preference and computer program limitation. All the examples given in this chapter express distances in Euclidean (or straight line and symmetric) distance as defined here (Legendre and Legendre, 1998: 277):

$$d_{ij}^2 = \sum_{k=1}^{p} (y_{ik} - y_{jk})^2 = (y_i - y_j)' (y_i - y_j)$$

Formula for Euclidean distance

where *i*, *j*, and *k* are points in *p*-dimensional space.

The resultant matrices provide measures of distance between each data object, in this case site components, for the data class in question. The distance between two data points is an expression of their dissimilarity in multivariate space (Gower, 2006; Pruzansky, 2006). This pairwise comparison of all objects creates a triangular matrix that depicts the entirety of the variation seen within that particular data class. This concept is analogous to a triangular travel distance matrix at the bottom of a road map, where all distances shown reflect the driving distance between city locations and the matrix as a whole represents the sum total of geographic distance variation between cities on the map.

## Statistical Procedures

IDA uses a series of statistical procedures to measure, isolate, and describe variation within and between data classes. Not a single test, but several, it is better thought of as a camera with a series of statistical lenses, where each step in the procedure focuses more tightly on a different aspect of variation. The end product then is not simply a measure of correlation of variation but a nuanced description of the character of variation seen within a dataset.

For this analysis the computer programs NCSS 2007, PAST 2.08, PROTEST, and ArcGIS 9.3 were used. Other equivalent programs could be easily substituted.

## Case Study

As a means of illustration, results from Clark's (2008, 2010) Integrative Distance Analysis of ethnographic variables recorded by Jorgensen (1980) is presented in this section. This dataset provides an ideal case study, because there are multiple data classes that have a high degree of colinearity. Jorgensen (1980) recorded 292 ethnographic variables for 172 aboriginal groups in western United States and Canada. These variables were grouped into seven classes that he termed *technology*, *subsistence economy*, *economic organization*, *social organization*, *political organization*, *sodalities*, and *warfare*, *ceremonialism*, and *spirits and shamanism*. Clark (2008) focused solely on the 27 Northwest Coast groups ranging from the Northern Tlingit to the Lower Chinook (see Table 4.1 and Figure 4.1). Data for technology, subsistence economy, economic organization, and social organization were compared to geographic location using IDA.

The ethnographic groups presented in this study serve as analogs for archaeological culture data. The use of *archaeological culture*, or similar terms such as *phase*, *culture*, or *culture type*, underscores a widespread issue in culture historical studies. The artifactual remains available to archaeologists, long

**Table 4.1** Northwest Coast Aboriginal Groups

| *Group* | *Region* | *Abbreviation* |
|---|---|---|
| Northern Tlingit | North Coast | NT |
| Southern Tlingit | North Coast | ST |
| Massett Haida | North Coast | MH |
| Skidegate Haida | North Coast | SH |
| Tsimshian | North Coast | Ts |
| Gitksan | North Coast | Gt |
| Haisla | North Coast | Hsl |
| Haihais | Wakashan | Hai |
| Bella Bella | Wakashan | BB |
| Fort Rupert | Wakashan | FR |
| Bella Coola | Wakashan | BC |
| Clayoquot | Wakashan | Cl |
| Makah | Wakashan | Ma |
| Klahuse | Central Coast Salish | Kh |
| Pentlatch | Central Coast Salish | Pe |
| Squamish | Central Coast Salish | Sq |
| Cowichan | Central Coast Salish | Co |
| West Sanetch | Central Coast Salish | WS |
| Upper Stalo | Central Coast Salish | US |
| Lower Fraser | Central Coast Salish | LF |
| Lummi | Central Coast Salish | Lu |
| Klallam | Central Coast Salish | Km |
| Twana | Southern Coast Salish | Tw |
| Quinault | Southern Coast Salish | Qn |
| Puyallup | Southern Coast Salish | Pu |
| Quileute | Southern Coast Salish | Ql |
| Lower Chinook | Southern Coast Salish | LC |

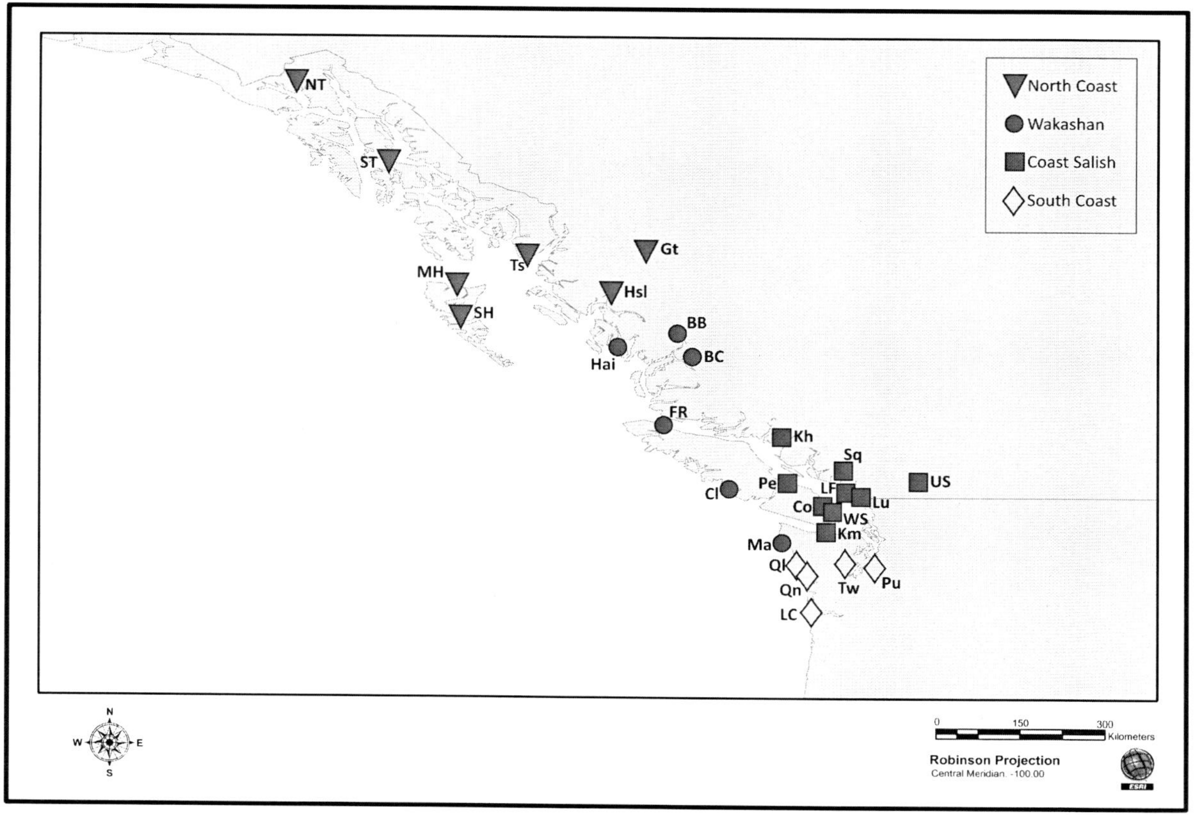


**Figure 4.1** Map of Northwest Coast aboriginal groups

held as the sole determinant of culture history, are but a small segment of material culture, let alone a direct corollary of past culture as a whole. Archaeologists can recover only those tools durable enough to preserve until they are rediscovered. (For a cautionary tale, see the site of Ozette in the U.S. state of Washington [Croes, 1977; Samuels, 1991], where perishable artifacts account for over 95% of the total artifact assemblage.) Even in the unlikely event of complete artifactual recovery, artifacts are not a direct reflection of culture as a whole (Clark, 2008). Artifacts are but one aspect of a larger cultural system. This study shows that although artifacts are significantly correlated with cultural units, aspects of social and economic organization proved far more powerful in predicting cultural affiliation. The use of artifacts as a stand-alone proxy for past cultures is therefore problematic. The original culture historical approach was in fact not the creation of histories of past cultures but merely the creation of seriated technological categories whose link to past cultures was not determined.

The use of the term *phase* is in widespread usage in North America and perhaps exemplifies the issue at hand. As Willey and Phillips (1958: 22) stated: "an archaeological unit possessing traits sufficiently characteristic to distinguish it from all other units similarly conceived, whether of the same or other cultures or civilizations, spatially limited to the order of magnitude of a locality or region and chronologically limited to a relatively brief interval of time." As used in North America, phases are simply groups of artifacts that covary in space and time. Phases can be inductively reasoned and statistically tested. As a classificatory device, phases have great utility, but, more often than not, they are stretched beyond their meager limitations to make proclamations about past cultures. Links between a small portion of a past toolkit and some form of past culture or society need to be demonstrated, not assumed. This point was not lost on Abbott (1972), who argued that seasonally distinct toolkits could easily be mistaken for separate phases within a single culture. Distinct summer and winter assemblages could be interpreted as two contemporaneous phases if seasonal mobility provided no overlap in site distribution.

Moreover, Abbott (1972) noted that separate cultures with similar economic foci may choose similar tools for practicality rather than ethnic affiliation. Independent cultural groups may choose to adopt efficient tools of their neighbors, and this could be interpreted as evidence for a larger, more homogenous culture that was present historically.

Croes and Hackenberger (1988) continued the point, suggesting that phases are simply economic plateaus that have no distinct or discernible cultural affiliation. For them phases are adaptive plateaus that serve the economic realities of past peoples. They represent a response to ecological characteristics of a particular place and time. They state specifically that this adaptation is independent of ethnicity or culture.

*Phase* as it is generally used does not take into account these issues. Willey and Phillips (1958: 49) believed that "the equivalent of phase, then, ought to be 'society.'" They saw a clear link between phase and past culture, a notion rarely challenged or demonstrated in culture historical studies.

Archaeological cultures or phases should be seen as an imperfect amalgam of influences and processes, which were lucky enough to stand the test of time and preserve, and not as a mirror on past society. From this seemingly intractable morass, IDA offers a statistical method to help disentangle multiple influences on single streams of data and also individual factors that may influence multiple data streams. With the use of IDA a comprehensive understanding of archaeological cultures may be possible.

## Step 1: Analyzing Overall Variation

Following the conversion of all data classes into distance matrices, the matrices are then compared using the Mantel test (Mantel, 1967). The Mantel test is the simplest and most widely used method for comparing distance matrices (Cottenie, 2002). The Mantel statistic compares two such distance matrices for correlation. It is widely used in scientific disciplines (Borcard, Legendre, and Drapeau, 1992; Fortin and Gurevitch, 1993; Leduc et al., 1992; Legendre and Fortin, 1989; Legendre and Troussellier, 1988; Peres-Neto and Jackson, 2001; Smouse, Long, and Sokal, 1986). In addition, a series of iterations are run on randomly constructed replicate matrices to determine if the measured result differs significantly from random. The replicates are populated with random values but have column and row sums equivalent to the original matrix. The Mantel technique is particularly powerful as it can operate on any form of data (ratio, interval, ordinal, or nominal) (Gower, 2006; Urban, 2006). The Mantel formula is (Urban, 2006: 1)

$$r = \sum_{i=1}^{n} \sum_{j=1}^{n} w_{ij} y_{ij}$$

Formula for the Mantel test

where $w_{ij}$ and $y_{ij}$ are measures of the proximity of locations $i$ and $j$.

At this stage of analysis, each set of two data classes are compared. This pairwise comparison ensures that each data class is compared to all others. The results of this test provide a correlation value, corr, which varies between 0 and 1. Classes that have no correlation have a score of 0, whereas data classes that correlate perfectly have a score of 1. The $p$ value represents the probability that the results seen in the corr value are a product of random chance derived from a user-defined set of iterations, in this case 10,000. Following standard scientific protocol only $p$ values of less than 0.05 (less than a 5% chance of occurring randomly) are accepted as significant.

Results of the Mantel tests are shown in Table 4.2. All correlations were very highly significant with $p$ values less than 0.0005. The majority of the

**Table 4.2** Mantel Test Results

| | | | | |
|---|---|---|---|---|
| Technology | 0.486 | | | |
| | *0* | | | |
| Subsistence | 0.504 | 0.361 | | |
| | *0.0001* | *0* | | |
| Economic | 0.562 | 0.650 | 0.435 | |
| Organization | *0* | *0* | *0* | |
| Social | 0.805 | 0.398 | 0.447 | 0.555 |
| Organization | *0* | *0* | *0.0005* | *0* |
| Corr. Value | Geographic | Technology | Subsistence | Economic |
| *p* Value | Location | | | Organization |

results returned *p* values of 0 indicating not a single one of the 10,000 iterations corresponded with the measured result. Geographic location proved an excellent predictor of all data classes, with the strongest association being with social organization at 0.805 or 80.5% of the variation in social organization predicted by geographic location. However, it is apparent that all classes under study are highly intercorrelated. It is interesting that the weakest, yet still highly significant, correlation is between technology and subsistence at 0.361. The weakest predictor of the toolkit is economic tasks those tools, at least in some measure, perform.

*Object Deviation Linkage Analysis* The overall correlation is not necessarily uniformly distributed across sites within the dataset. The independent data class—in this case, geographic proximity of site locations—gives a prediction for the outcome of the dependent data class. The discrepancy between real and predicted values for each data point is the object deviation; it is a measure of how each site in the dependent data class was influenced by the independent data class. By looking at object deviation we can see which sites are positively or negatively affecting the overall measured outcome; it allows a quick way to see which objects are not conforming to overall trends. To measure object deviation the normalized independent distance matrix is subtracted from the normalized dependent distance matrix to account for differences in scale. This process provides a measure of deviation for each data location from the predicted value. Table 4.3 shows the output of linkage analysis of social organization and geographic proximity for this study. To better illustrate the pattern, present positive linkages are shown in Figure 4.2 and negative linkages are shown in Figure 4.3. The positive linkage lines correspond to values greater than one and two standard deviations above the mean. Positive linkages are when points are closer in the measured data class that would be predicted by geographic proximity. This result shows that, although the North Coast cultures are highly clustered and removed from all others, difference is less than

**Table 4.3** Linkage Analysis Northwest Coast Geography and Social Organization

| | | | | | | | | | | | | | | | | | | | | | | | | | | | |
|---|---|---|---|---|---|---|---|---|---|---|---|---|---|---|---|---|---|---|---|---|---|---|---|---|---|---|---|
| NT | 0 | | | | | | | | | | | | | | | | | | | | | | | | | | |
| ST | +1 | 0 | | | | | | | | | | | | | | | | | | | | | | | | | |
| MH | 0 | 0 | 0 | | | | | | | | | | | | | | | | | | | | | | | | |
| SH | +1 | 0 | 0 | 0 | | | | | | | | | | | | | | | | | | | | | | | |
| Ts | +2 | 0 | −1 | −1 | 0 | | | | | | | | | | | | | | | | | | | | | | |
| Gt | +1 | 0 | 0 | 0 | 0 | 0 | | | | | | | | | | | | | | | | | | | | | |
| Hsl | 0 | 0 | 0 | 0 | 0 | 0 | 0 | | | | | | | | | | | | | | | | | | | | |
| Hai | 0 | 0 | −1 | −1 | −1 | −1 | −2 | 0 | | | | | | | | | | | | | | | | | | | |
| BB | +1 | 0 | −1 | −1 | −1 | −1 | −2 | 0 | 0 | | | | | | | | | | | | | | | | | | |
| FR | +1 | 0 | −1 | −1 | −1 | −1 | −2 | 0 | 0 | 0 | | | | | | | | | | | | | | | | | |
| BC | 0 | 0 | −1 | −1 | −1 | −2 | −2 | 0 | −1 | 0 | 0 | | | | | | | | | | | | | | | | |
| Cl | +1 | 0 | 0 | 0 | 0 | −1 | −1 | 0 | 0 | 0 | 0 | 0 | | | | | | | | | | | | | | | |
| Ma | +2 | +1 | 0 | 0 | 0 | 0 | 0 | 0 | 0 | 0 | +1 | +1 | 0 | | | | | | | | | | | | | | |
| Kh | +1 | 0 | 0 | 0 | 0 | −1 | −1 | 0 | −1 | 0 | 0 | 0 | 0 | 0 | | | | | | | | | | | | | |
| Pe | +1 | 0 | 0 | 0 | 0 | −1 | −1 | 0 | 0 | 0 | 0 | 0 | 0 | 0 | 0 | | | | | | | | | | | | |
| Sq | +2 | +1 | 0 | 0 | 0 | −1 | 0 | 0 | 0 | 0 | 0 | 0 | 0 | +1 | 0 | 0 | | | | | | | | | | | |
| Co | +2 | +1 | 0 | 0 | 0 | 0 | 0 | 0 | 0 | 0 | 0 | 0 | 0 | 0 | 0 | 0 | 0 | | | | | | | | | | |
| WS | +2 | +1 | 0 | 0 | 0 | 0 | 0 | 0 | 0 | 0 | 0 | 0 | 0 | 0 | 0 | +1 | 0 | 0 | | | | | | | | | |
| US | +2 | +1 | 0 | 0 | 0 | −1 | 0 | 0 | 0 | 0 | 0 | 0 | 0 | 0 | 0 | 0 | 0 | 0 | 0 | | | | | | | | |
| LF | +2 | +1 | 0 | 0 | 0 | 0 | 0 | 0 | 0 | 0 | 0 | +1 | 0 | 0 | 0 | 0 | 0 | 0 | 0 | 0 | | | | | | | |
| Lu | +2 | +1 | 0 | 0 | 0 | 0 | 0 | 0 | 0 | +1 | 0 | +1 | 0 | 0 | 0 | 0 | 0 | 0 | 0 | 0 | 0 | | | | | | |
| Km | +2 | +1 | 0 | 0 | 0 | −1 | −1 | 0 | 0 | 0 | 0 | 0 | −1 | −1 | −1 | −1 | −1 | −1 | 0 | −1 | −1 | 0 | | | | | |
| Tw | +2 | +1 | +1 | 0 | 0 | 0 | 0 | +1 | 0 | 0 | +1 | 0 | 0 | 0 | 0 | 0 | 0 | 0 | 0 | 0 | 0 | −1 | 0 | | | | |
| Qn | +2 | +1 | 0 | 0 | 0 | 0 | 0 | 0 | 0 | 0 | +1 | 0 | 0 | 0 | 0 | 0 | 0 | 0 | 0 | 0 | 0 | −1 | 0 | 0 | | | |
| Pu | +2 | +1 | 0 | 0 | 0 | 0 | 0 | 0 | 0 | 0 | 0 | 0 | −1 | −1 | 0 | −1 | −1 | −1 | 0 | −1 | −1 | 0 | −2 | −1 | 0 | | |
| Ql | +2 | +1 | 0 | 0 | 0 | 0 | 0 | 0 | 0 | 0 | 0 | 0 | 0 | 0 | 0 | 0 | 0 | 0 | 0 | 0 | 0 | 0 | 0 | 0 | 0 | 0 | |
| LC | +2 | +1 | 0 | 0 | 0 | 0 | 0 | 0 | 0 | 0 | +1 | 0 | 0 | 0 | 0 | 0 | 0 | 0 | 0 | 0 | 0 | 0 | 0 | 0 | −1 | 0 | 0 |
| | NT | ST | MH | SH | Ts | Gt | Hsl | Hai | BB | FR | BC | Cl | Ma | Kh | Pe | Sq | Co | WS | US | LF | Lu | Km | Tw | Qn | Pu | Ql | LC |

their geographic isolation would predict. Given the ubiquity of large, seagoing canoes, the vast distances that isolate the northern groups could be overcome.

Negative linkage lines correspond to measures one and two standard deviations below the mean. Negative linkages provide a visualization method to see discontinuities in the data. Natural and cultural boundaries are often highlighted in this manner. In Figure 4.3 there are strong negative linkages between the Wakashan and North Coast cultures. Although these cultures are geographically close there are robust differences in social organization that are highlighted in this figure.

The Mantel test showed that social organization is highly correlated with geographic location, but linkage analysis showed that that pattern is not consistent along the north-south axis. There is a cultural break in social organization between the North Coast and Wakashan groups.

The Mantel test is ideal for garnering a general sense of the influence of one data class on another. However, the overall variation within a data class can be complex and difficult to interpret. Fortunately, statisticians have a number of methods for compressing the overall variation into small, more easily interpreted units called *dimensions* or *factors* and eliminating statistical noise.

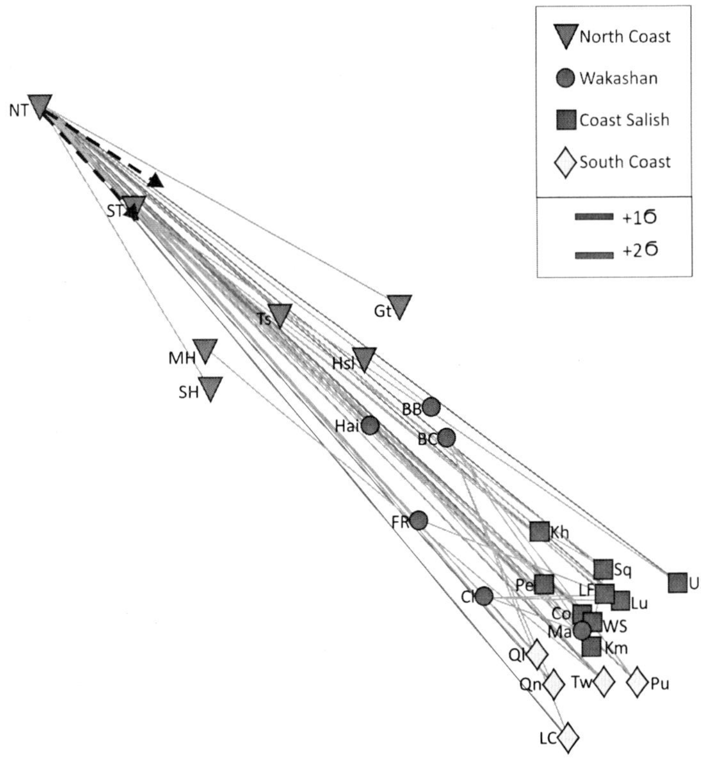


**Figure 4.2** Positive linkages for Northwest Coast geography and social organization

## Step 2: Compress Overall Variation

There are a number of related statistical procedures that can all be used to compress overall variation, which is done by reducing the dimensionality of the dataset. Classic or metric Multidimensional Scaling (MDS), also known as Principle Coordinates Analysis (PCoA), Principle Components Analysis (PCA), and Correspondence Analysis (CA), are all scaling procedures that serve this function (Baxter, 1994; James and McCulloch, 1990; Legendre and Legendre, 1998; Manly, 1994). Generally all these statistics provide similar results, with the main differences being that MDS retains original distance associations in Euclidean space, whereas PCA does not (Legendre and Legendre, 1998; Young, 2006). In other words, if the user chooses a non-Euclidean input distance measure (that is, Bray-Curtis, chord, genetic), MDS maintains those distances through to output. PCA converts all distances to Euclidean throughout the analysis (Flury, 2006; Legendre and Legendre, 1998). In cases where data are metric, Euclidean distances are used, and PCA is conducted

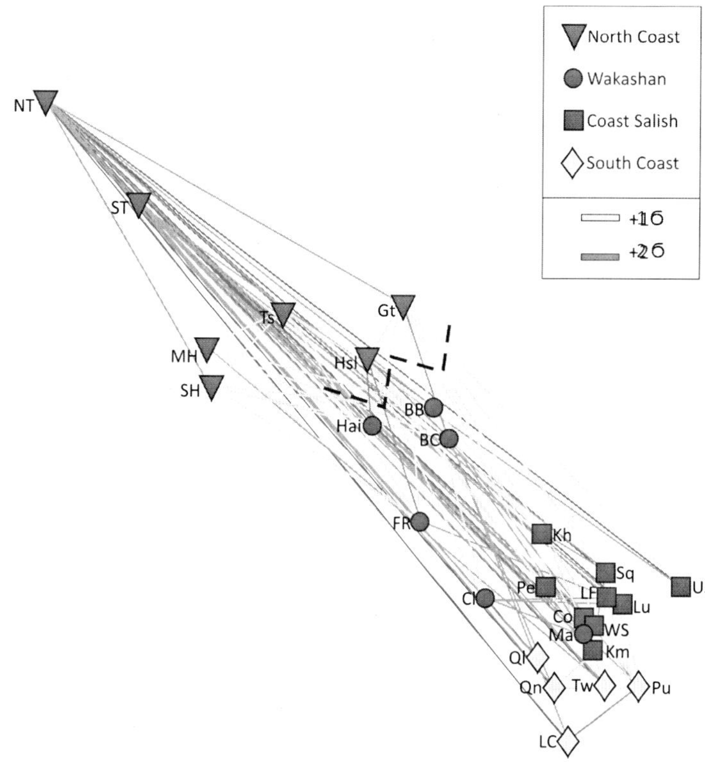


**Figure 4.3** Negative linkages for Northwest Coast geography and social organization

on a correlation matrix of attributes, the results will be the same (James and McCulloch, 1990: 143).

CA retains the chi-squared distances among points rather than strict Euclidean distances (Borcard, Legendre, and Drapeau, 1992). It is better suited when one expects variation to be continuous and unimodal and is often limited to count data (Borcard, Legendre, and Drapeau, 1992).

MDS was chosen for inclusion in IDA for two reasons. First, its use in archaeological studies is widespread (Bollong et al., 1997; Burley, 1980; Clark, 2000; Drennan, 1976; Kruskal, 1971; Matson, Ludowicz, and Boyd, 1980; Matson and True, 1974; Thom, 1992). Second, because MDS can be run directly on the already constructed distance matrices, there is an intuitive and procedural elegance following from the Mantel test through to MDS. Both PCA and CA are run on the original variable table and compute distances internally. At best this means replicating a step, the creation of distance matrices, and at worst it may allow error into the study if the distances are not identical to those previously constructed.

In all these procedures, overall variation seen in the distance matrix is compressed into a series of dimensions or factors. These dimensions or factors each express a percentage of the overall variation. Each dimension or factor, or sometimes a set of two together, can be interpreted separately. This dissection of variation is very useful, because each data class is likely shaped by a number of other data classes as well as time. By isolating these dimensions or factors it becomes much easier to assign an interpretation to each one or set of two. The formula for metric MDS is presented below (NCSS, 2004: 432–35).

$$d_{ij} = \sqrt{\sum_{k=1}^{p} (x_{ik} - x_{jk})^2}$$

Formula for metric multidimensional scaling

where $d_{ij}$ is the distance and $x_{ik}$ is the value of *i*th row and *k*th column.

In MDS the total number of dimensions recovered from a data matrix is $(n - 1)$, which means that in a dataset composed of 10 objects, a total of 9 dimensions will be present. MDS procedures assign a strength measurement to each of the dimensions that express the proportion of overall variation explained. Dimension 1 contributes the most, followed by dimension 2, and so on. The sum variation explained by all dimensions is 100%. The goal of MDS is to explain as much of the variation as possible in the lowest number of dimensions.

During interpretation it can be difficult to know which of the possibly many dimensions are significant and worthy of discussion. Arbitrary determinations such as including only those dimensions accounting for greater than 10% of the variation have been used (Matson and True, 1974; Orchard and Clark, 2005, 2006). Measuring the stress or goodness-of-fit is a non-arbitrary means of making this decision. The stress equation is listed here; it is a measure of the goodness-of-fit of the actual measured distances and predicted ones based on the number of cases and dimensions present, as shown below (NCSS, 2004: 433–35).

$$S = \sqrt{\frac{\sum (d_{ij} - \hat{d}_{ij})^2}{\sum d_{ij}^{2}}}$$

Formula for stress

where $d_{ij}$ is the measured distance and $\hat{d}_{ij}$ is the predicted distance.

As a general rule Kruskal (1964) (cf. NCSS, 2004: 435–43) suggested the guidelines given in Table 4.4 for determining the use of stress in dimensional cutoff.

As the number of dimensions used increases, the stress decreases. The goal of MDS is to explain as much of the variation as possible in the lowest number

**Table 4.4** Kruskal's Guidelines for Determining the Use of Stress in Dimensional Cutoff

| *Stress* | *Goodness-of-Fit* |
|---|---|
| 0.200 | poor |
| 0.100 | fair |
| 0.050 | good |
| 0.025 | excellent |
| 0.000 | perfect |

of dimensions, with an acceptable level of stress. This reduces the greatest amount of uninterpretable statistical noise.

Continuing the example from Clark (2008), the MDS plot for social organization is shown in Figure 4.4. In this case dimension 1 accounted for 42.96% of the total variation and dimension 2 another 10.40%. Of the 26 dimensions, the first 5 were used to achieve a stress level of 0.049, which is considered good. These 5 dimensions described 73.21% of the variation and the remaining 26.79% is deemed statistical noise.

The figure shows clustering based on geographic subarea with a strong separation of the North Coast cultures in dimension 1. The interpretation of

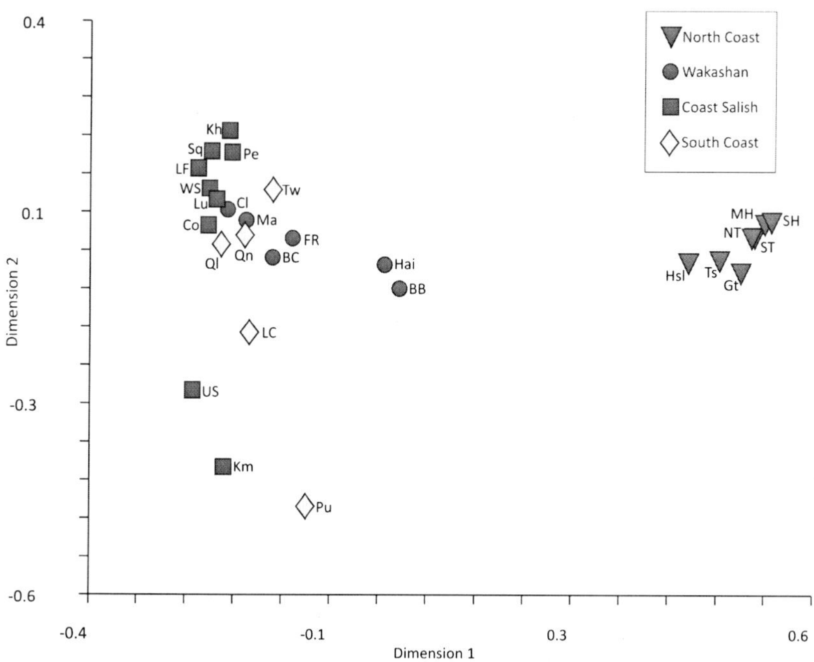


**Figure 4.4** MDS dimension 1 and 2 of Northwest Coast social organization

this MDS plot is that it depicts a gradient of rigidity of social organization, right to left or north to south, a pattern previously described elsewhere (Coupland, Clark, and Palmer, 2009).

Multivariate statistical analysis of this nature is very common in archaeology (Clark, 2000; Drennan, 1976; Kruskal, 1971; Matson, Ludowicz, and Boyd, 1980; Matson and True, 1974). At this stage, however, data classes are independent, and interclass proximity has not yet been analyzed. The next two steps add proximity analysis to the MDS results.

## Step 3: Analyze Compressed Variation

The analysis of compressed variation takes two forms. The first looks at individual dimensions and the second looks at combinations of two or three dimensions. Individual dimensions are assessed using Moran's $I$ (Moran, 1950). The Moran's $I$ test is commonly used to measure spatial autocorrelation in GIS applications (Conolly and Lake, 2006; Kvamme, 1990; Neiman, 1997; Premo, 2004; Wheatley and Gillings, 2002; Williams, 1993). It allows a single dimension under study to be compared to 2- or 3-dimensional data for a reference data class. Commonly, the two or three dimensions of the reference data class are the longitude, latitude, and sometimes altitude of object locations. Thus Moran's $I$ is used to measure the predictive strength of geographic proximity in shaping the values of the tested dimensions (Legendre, 1993; Waldhör, 2006). The formula for Moran's $I$ (Waldhör, 2006: 1):

$$I = \frac{n}{S_0} \frac{\sum_{i=1} \sum_{j=1}^{n} w_{ij}(x_i - \overline{x})(x_j - \overline{x})}{\sum_{j=1}^{n} (x_i - \overline{x})^2}$$

Formula for Moran's $I$

where $x_i$ and $x_j$ are the values observed at locations $i$ and $j$, and $w_{ij}$ is the proximity measure between locations $i$ and $j$, and $S_0$ is a scaling factor equivalent to

$$\sum_{i=1}^{n} \sum_{j=1}^{n} w_{ij}.$$

This procedure measures the correspondence of the geographic structure with the structure of the data factor in question. It can be repeated for all the dimensions of all the data classes. As each of these dimensions can have a separate interpretation, Moran's $I$ is ideal in determining which of them are related to physical proximity and which are independent of proximity. Results are presented in correlation or corr values ranging from $-1$ to 1. Positive values indicate direct relationships while negative ones indicate inverse relationships. As with the Mantel test, 10,000 iterations of randomly constructed samples are compared to the observed values to provide a $p$ value.

**Table 4.5** Spatial Autocorrelation Results for Northwest Coast: Social Organization

| *Dimension* | *Moran's* I | p *Value* |
|---|---|---|
| 1 | 0.52 | highly clustered < 0.01 |
| 2 | 0.01 | random |
| 3 | 0.031 | highly clustered < 0.01 |
| 4 | 0.008 | clustered 0.05–0.10 |
| 5 | 0 | random |

Spatial autocorrelation results for the Clark (2008) example are shown in Table 4.5. The dimensions of social organization have been tested against geographic proximity. In this case dimensions 1 and 3 are highly clustered, and $p$ values are lower than 0.01. Fifty-two percent of the variation in social organization expressed in dimension 1 can be predicted by geographic proximity.

The testing of single dimensions is a valuable exercise, because it indicates where within the realm of overall variation geographic proximity is influential. For the most part, however, the output of MDS analysis is visualized and interpreted in two dimensions. Indeed, the MDS plots are referred to as *maps of variation* (Baxter, 1994). The next test allows for the analysis and graphical depiction of any 2- or 3-dimensional data to any other 2- or 3-dimensional data by overlaying different data class maps on each other.

This test is achieved with the use of the Procrustes superimposition method (Gower, 1971). Procrustes uses rotation, translation, reflection, and dilation to fit one set of data on another (Gower, 1971). Analagous to the fitting of map layers in GIS, Procrustes orients, rotates, and scales a target data class to a reference data class. All data have been rotated and scaled to physical location. Physical proximity acts as the universal reference data class, and all other classes are fit to it. This procedure allows for viewing of multiple data classes as map overlays are superimposed upon a geographic map. The method is depicted in Figure 4.5.

The original configuration of data class $X$, depicted as ABC, and data class $Y$, depicted as abc, is plotted in (i). In box (ii) abc has been standardized and scaled. The size and center point have been adjusted to fit with ABC. If required, abc is reflected (iii). Last, abc is rotated to provide the least sum of squared differences of homologous observations (locations that exist in both data classes) with ABC (iv).

Computationally, Procrustes proceeds with the following steps (Peres-Neto and Jackson, 2001: 170–71):

$$X_{scl} = \frac{(I-P)X}{\sqrt{tr[(I-P)X'(I-P)]}}$$

Centering and scaling

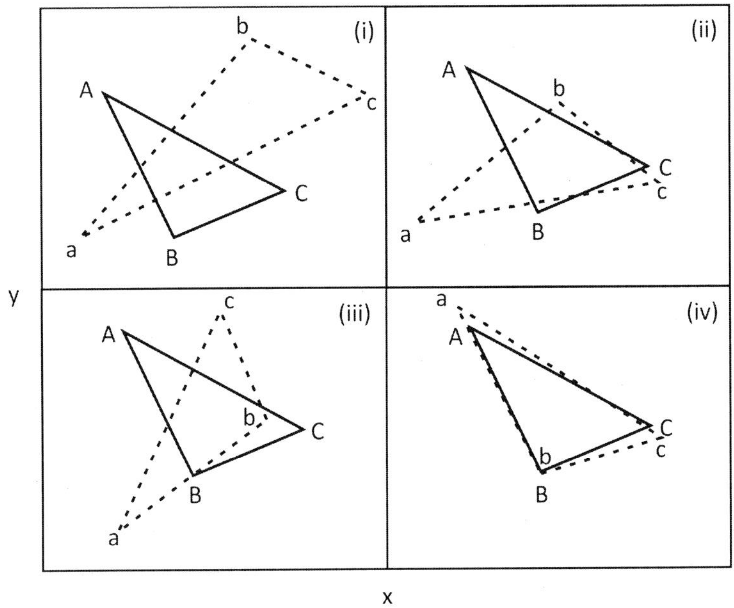


**Figure 4.5** Steps of Procrustes analysis: (i) original configuration; (ii) standardization/scaling; (iii) reflection; (iv) rotation (after Peres-Neto and Jackson, 2001: 171)

where $I$ is an $(nxn)$ identity matrix and $P$ is an $(nxn)$ matrix with all elements $= 1/n$.

The process is then repeated for $Y$.

$$m_{xy}^2 = \Delta^2(X_i', Y_i')$$

Reflection and rotation

where $\Delta X_i', Y_i'$ is the distance between points $X_i'$ and $Y_i'$

The best fit is calculated by the least-squares superimposition of one configuration to a reference configuration $X$ (Peres-Neto and Jackson, 2001:170).

Traditional Procrustes analysis (Gower, 1971) does not measure point to point concordance or offer a test of significance. IDA, however, uses the iterative Procrustes Randomization Test (or PROTEST) (Jackson, 1995), which includes a measure of congruence and a $p$ value. After two data classes have been superimposed, a value of correlation $m^2$ and a $p$ value are calculated based on 10,000 randomly constructed samples. Replicate samples have random values but an equal sum of the original configuration (Jackson, 1995). Unlike corr, which defines perfect congruence as 1, $m^2$ is the sum of residual error; therefore, values of 0 are perfectly correlated, whereas values near 1 show no correlation.

PROTEST has been used on several ecological samples to show community structure and evolutionary relationships (Burgess, Osborne, and Caley, 2010; Jackson, 1995; Mykra, Heino, and Muotka, 2008; Peres-Neto and Jackson, 2001; Schneider and Borlund, 2007; Venturelli and Tonn, 2005; Zhao, Shuter, and Jackson, 2008). The benefit of this analysis is that the resultant output is a map that allows for the easy visual identification of concordance of individual cases (Peres-Neto and Jackson, 2001). IDA uses maps for each data class to visually inspect patterns that occur across classes.

To illustrate PROTEST we return to the Northwest Coast ethnographic example. Table 4.6 shows the pairwise PROTEST results for all data classes. Once again all $p$ values are very highly significant. The strongest correlation is physical distance and social organization at 0.368; in other words, 63.2% of the variation in social organization can be accounted for by the actual map locations of sites. The PROTEST plot for geography and social organization is presented in Figure 4.6. Arrows of deflection are shown that express the direction and extent of deviation from the reference data class to the rotated one. The arrows begin at the geographic location of sites and terminate at the rotated position of the social organization plot. Since both the start and end points are significant, the arrow itself is a useful interpretative tool.

If a priori groupings are present in the data then this information can be used to garner a sense of how these groupings fit into the overall results. Within the Northwest Coast sample of aboriginal cultures, there are subareal groupings of the North Coast, Wakashan, Central Coast Salish, and Southern Coast Salish. By centering the deviation arrows associated with these groups, we can see patterned variation. This centering is achieved by giving all the deviation arrows a common starting point. To examine subgroup difference two measures are used: deflection and dispersion.

Deflection is simply the maximum within-group range of deviated direction expressed in degrees. Deflection measures give a general sense of the direction of the arrows of deviation. A smaller range means a more uniform

**Table 4.6** PROTEST Results

| | | | | |
|---|---|---|---|---|
| Technology | 0.529<br>*0.0001* | | | |
| Subsistence | 0.693<br>*0.0008* | 0.790<br>*0.0062* | | |
| Economic Organization | 0.411<br>*0.0001* | 0.433<br>*0.0001* | 0.676<br>*0.0002* | |
| Social Organization | 0.368<br>*0.0001* | 0.533<br>*0.0001* | 0.688<br>*0.0008* | 0.385<br>*0.0001* |
| $m^2$ value<br>*p* value | Geographic Location | Technology | Subsistence | Economic Organization |

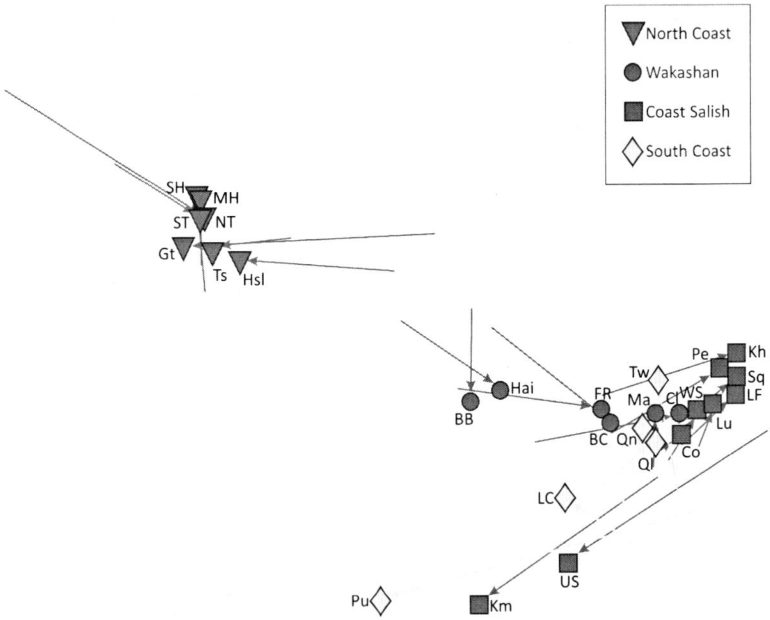


**Figure 4.6** PROTEST results for Northwest Coast geography and social organization

pattern of in-group variation across the two data classes. Dispersion is a measure of the spread of subgroup points in both the independent and dependent data classes. It is calculated by averaging the within-group Euclidean distance. The coefficient of dispersion or *D* is calculated as

$$D = \frac{X}{Y}$$

Formula for coefficient of dispersion

where *X* is the dependent dispersion and *Y* is the independent dispersion.

The subgroup deflection and dispersion values for geographic proximity and social organization are presented in Table 4.7. The Southern Coast Salish have the most uniform response across the measured data classes at 155°, which is the smallest maximum angle of deflection. The dispersion scores show that although the North Coast and Wakashan subareas have much greater geographic dispersion they have similar social organization values. The coefficient of dispersion shows that geography has less predictive power over social organization in the Northern Coast and Wakashan subareas, whereas the Central and Southern Coast Salish exhibit a more direct relationship between the two data classes. A depiction of the deflection and dispersion is presented in Figure 4.7.

**Table 4.7** Northwest Coast Geographic Location and Social Organization Subgroup Deflection and Dispersion Values

| | | *Dispersion* | | |
|---|---|---|---|---|
| *Subarea* | *Deflection* | *Geographic Location* | *Social Organization* | D *Coefficient* |
| North Coast | 214° | 3.153 | 0.531 | 5.938 |
| Wakashan | 179° | 2.284 | 0.470 | 4.860 |
| Central Coast Salish | 193° | 1.034 | 0.416 | 2.486 |
| Southern Coast Salish | 155° | 1.020 | 0.538 | 1.896 |

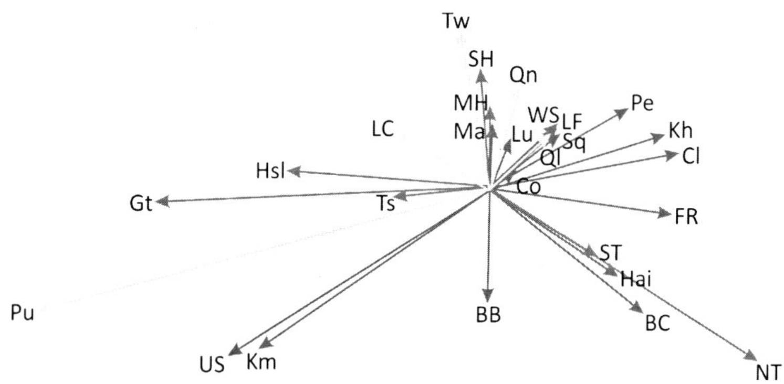


**Figure 4.7** Centered PROTEST results for Northwest Coast geography and social organization

Deflection and dispersion measures are particularly useful in studies that involve change over time. The influence of geography in structuring material culture has been demonstrated to lessen with the increase of social complexity and regional cultural homogenization (Clark, 2008, 2010).

## Step 4: Complex Patterning

Comparing two data classes at a time illuminates a great deal about the overall and specific variation found within a dataset. The last step in IDA is to isolate the variation further to look at more complex relationships between data classes. As Tables 4.2 and 4.6 indicate, there is a high degree of intercorrelation or integration between data classes. Geography significantly shapes all other measured classes, and comparing any two others without addressing geography will lead to inflated correlation values. The last step in IDA removes the influence of one dataset so that all others may be compared in a pairwise manner. This comparison can be done with the use of the partial Mantel and partial PROTEST procedures, for overall and compressed variation, respectively.

Partial statistics hold the influence of one data class *X* constant while viewing the relationship between two others, *J* and *K*. To achieve this analysis, all variables of either the dependent distance matrix or the rotated PROTEST scores are subject to multiple regression to the entire reference or independent data class. This process in effect, removes the influence of the reference data class on the dependent class. The residuals produced from multiple regression for each dependent data class are then subject to the Mantel test for distance matrices and PROTEST for compressed dimensional scores.

The standard formula for multiple regression is (NCSS, 2004: 301–05)

$$J = \beta_0 + \beta_1 X_1 + \beta_2 X_2 + \cdots + \beta_p X_p + \varepsilon$$

Formula for multiple regression

where *J* is the dependent variable, *X* is the independent variable, and $\varepsilon$ represents the residual error for each observation.

To return to the ethnographic Northwest Coast example, Table 4.8 shows partial Mantel test scores for technology, subsistence, economic organization, and social organization. The influence of geographic proximity has been removed. Social organization is not significantly correlated with technology, subsistence, or economic organization at the 0.05 *p* value. The strong measures that had previously been associated with social organization were a product of geographic intercorrelation (see Table 4.2). The strongest correlation is between technology and economic organization at 0.410, or 41% of the variation in one explained by variation in the other.

Table 4.9 provides the PROTEST results with the influence of geography excluded. When we look at geographically isolated compressed variation, the strongest correlation remains between social and economic organization. Only the relationships of subsistence and technology, and subsistence and social organization, fail to meet the significance cutoff.

It is interesting that the overall comparison of technology and subsistence shows a significant relationship with a partial correlation of 0.211 (Table 4.8), but the partial PROTEST results show no significant relationship (Table 4.9).

**Table 4.8** Partial Mantel Results, Geographic Location Removed

| | | | |
|---|---|---|---|
| Subsistence | 0.211<br>*0.0136* | | |
| Economic Organization | 0.410<br>*0* | 0.169<br>*0.0244* | |
| Social Organization | 0.075<br>*0.1961* | 0.038<br>*0.3203* | 0.132<br>*0.07121* |
| corr. value<br>*p* value | Technology | Subsistence | Economic Organization |

**Table 4.9** Partial PROTEST Results, Geographic Location Removed

| | | | |
|---|---|---|---|
| Subsistence | 0.935<br>*0.2989* | | |
| Economic Organization | 0.749<br>*0.003* | 0.812<br>*0.0126* | |
| Social Organization | 0.789<br>*0.0058* | 0.927<br>*0.2734* | 0.635<br>*0.0001* |
| $m^2$ value<br>*p* value | Technology | Subsistence | Economic Organization |

This means that, in overall terms, at least some tools are adaptive and respond to the changing needs of resource acquisition. However, when we focus only on the two most important dimensions represented with PROTEST instead of the overall variation the pattern is no longer present. Thus, in a general sense, function shapes the some part of the toolkit, but it is not the most important factor.

If we remove the influence of geography from the overall and compressed analyses the relationships between other classes can be investigated. The process of isolating variation can be repeated by removing the influence of economy, social organization, and the other data classes.

## Strengths and Weaknesses of IDA

IDA reduces the complexity of large datasets by the use of distances matrices. This technique allows for the overall structure of the dataset to be analyzed. However, there is loss of data when this occurs. Individual relationships are replaced with overall summaries of data structure. Therefore, it should be used in consort with, not replace, other statistics that look at simpler aspects of variation.

The strength of IDA lies with its flexibility and its ability to isolate background and foreground variation. Overall variation, seen in the distance matrices and measured with the Mantel test, gives an excellent indication of the broad structure of a dataset. Overall variation is much more revealing when compared to foreground variation of the compressed data. Foreground variation is the one or two most important dimensions under study; it can be thought of as de-trended variation. The interpretations of background and foreground structure may differ and examining both can reveal more complicated structure in the data.

Because complex datasets represent a conflation of several, and sometimes disparate, structural stimuli, using IDA allows for the isolation of these influences.

## Methodological Summary

IDA is a powerful tool in measuring and characterizing variation in large datasets. It is designed to look at variability within and across data classes and, through a series of steps, zoom in on particular aspects of variation so they can be better understood. Beginning with overall variation and moving inward to the compressed and more interpretable results of MDS, the importance of proximity is demonstrated. Object deviation linkage analysis allows for the delineation of natural or cultural breaks or discontinuities. Overlaying multiple linkage maps will show strong patterns associated with identity boundaries and other complex processes. Subgroup variation can be examined with measures of deflection and dispersion. The last step is to isolate variation by the use of partial statistics. By removing the influence of one data class when comparing two others issues of colinearity can be resolved.

Archaeologists must constantly struggle with complex datasets. Different aspects of material culture, environmental variables, proximity, time, and many other factors may structure measured results. IDA offers a toolkit that allows variation seen across complex datasets to be understood.

Complex datasets are not limited to the field of archaeology, and exploring other fields with IDA would be very useful. Habitat choice for ungulates, the composition marsh ecosystems, voting patterns among minorities, and many other topics are similarly structured datasets. Data are recorded in several classes that are not wholly independent. Understanding the complex relationships between types of data can help us to better understand the effects of at times competing influences.

### References

Abbott, D. N. 1972. The utility of the concept of phase in the archaeology of the southern northwest coast, *Syesis* 5: 267–78.

Anderberg, M. P. 1973. *Cluster Analysis for Applications*. New York: Academic Press.

Anselin, L., and Getis, A. 1992. Spatial statistical-analysis and geographic information systems, *Annals of Regional Science* 26(1): 19–33.

Barnett, H. G. 1939. Culture element distributions, IX: Gulf of Georgia Salish, *University of California Anthropological Records* 1(5): 221–95.

Baxter, M. 1994. *Exploratory Multivariate Analysis in Archaeology*. Edinburgh: Edinburgh University Press.

Bentley, R. A., and Shennan, S. J. 2003. Cultural transmission and stochastic network growth, *American Antiquity* 68(3): 459–85.

Bevan, A., and Conolly, J. 2009. Modelling spatial heterogeneity and nonstationarity in artifact-rich landscapes, *Journal of Archaeological Science* 36(4): 956–64.

________. 2011. Terraced fields and Mediterranean landscape structure: An analytical case study from Antikythera, Greece, *Ecological Modelling* 222: 1303–14.

Binford, L. R. 1962. Archaeology as anthropology, *American Antiquity* 28(2): 217–25.

———. 1972. Contemporary model building: Paradigms and the current state of paleolithic research, in Clarke, D. (Ed.), *Models in Archaeology*, pp. 109–66. London: Methuen.

———. 1989. *Debating Archaeology*. San Diego: Academic Press.

Bollong, C. A., Jacobson, L., Peisach, M., Pineda, C. A., and Sampson, C. G. 1997. Ordination versus clustering of elemental data from PIXE analysis of herder-hunter pottery: A comparison, *Journal of Archaeological Science* 24(4): 319–27.

Borcard, D., Legendre, P., and Drapeau, P. 1992. Partialling out the spatial component of ecological variation, *Ecology* 73(3): 1045–55.

Boserup, E. 1965. *The Conditions of Agricultural Growth: The Economics of Agrarian Change under Population Pressure*. London: Allen & Unwin.

Burgess, S. C., Osborne, K., Caley, M. J. 2010. Similar regional effects among local habitats on the structure of tropical reef fish and coral communities, *Global Ecology and Biogeography* 19: 363–75.

Burley, D. V. 1980. *Marpole: Anthropological Reconstructions of a Prehistoric Northwest Coast Culture Type*. Burnaby: Simon Fraser University, Department of Archaeology, Publication no. 8.

Clark, T. N. 2000. Prehistoric Culture Change on Southern Vancouver Island: The Applicability of Current Explanations of the Marpole Transition. Unpublished Master's thesis. Department of Anthropology, University of Victoria.

———. 2008. Integrative Archaeology: A New Method for Complex Data. Paper presented at the 2008 Canadian Archaeological Association Conference, Peterborough.

———. 2010. *Rewriting Marpole: The Path to Cultural Complexity in the Gulf of Georgia*. Unpublished Ph.D. dissertation, Department of Anthropology, University of Toronto.

———. 2013. *Rewriting Marpole: The Path to Cultural Complexity in the Gulf of Georgia. Mercury Series*. Ottawa: Canadian Museum of Civilization/University of Ottawa Press.

Clarke, D. L. 1968. *Analytical Archaeology*. London: Methuen.

Clarke, K. R., and Ainsworth, M. 1993. A method for linking multivariate community structure to environmental variables, *Marine Ecology Progress Series* 92: 205–19.

Conolly, J., and Lake, M. 2006. *Geographical Information Systems in Archaeology*. Cambridge: Cambridge University Press.

Cottenie, K. 2002. *Local and Regional Processes in a Zooplankton Metacommunity*. Unpublished Ph.D. dissertation. Department of Biology, Katholieke Universiteit Leuven.

Coupland, G., Clark, T. N., and Palmer, A. 2009. Hierarchy, communalism, and the spatial order of Northwest Coast plank houses: A comparative study, *American Antiquity* 74(1): 77–106.

Croes, D. R. 1977. *Basketry from the Ozette Village Archaeological Site: A Technological, Functional and Comparative Study*. Unpublished Ph.D. dissertation, Department of Anthropology, University of Washington.

Croes, D. R., and Hackenberger, S. 1988. Hoko River archaeological complex: Modeling prehistoric northwest coast economic evolution in prehistoric economies of the Pacific northwest coast, in Isaac, B. (Ed.), *Research in Economic Anthropology*, Suppl. 3, pp. 19–85. Greenwich, CT: JAI Press.

Diniz, J. A. F., Bini, L. M., and Hawkins, B. A. 2003. Spatial autocorrelation and red herrings in geographical ecology, *Global Ecology and Biogeography* 12(1): 53–64.

Dray, S., Chessel, D., and Thioulouse, J. 2003. Procrustean co-inertia analysis for the linking of multivariate datasets, *Ecoscience* 10(1): 110–19.

Drennan, R. D. 1976. Refinement of chronological seriation using nonmetric multidimensional-scaling, *American Antiquity* 41(3): 290–302.

Driver, H. E., and Coffin, J. L. 1975. *Classification and Development of North American Indian Cultures: A Statistical Analysis of the Driver-Massey Sample*. Philadelphia: American Philosophical Society.

Driver, H. E., Kenny, J., Hudson, H., and Engle, O. M. 1972. Statistical classification of North-American Indian ethnic units, *Ethnology* 11(3): 311–39.

Driver, H. E., and Massey, W. C. 1957. *Comparative Studies of North American Indians*. Philadelphia: American Philosophical Society.

Drucker, P. 1943. Archaeological survey of the northern Northwest Coast, Anthropological Papers 20, *Bureau of American Ethnology Bulletin* 133.

________. 1955. *Indians of the Northwest Coast*. New York: Natural History Press.

________. 1958. Sources of northwest coast culture, in Meggers, B. (Ed.), *New Interpretation of Aboriginal American Culture History*, pp. 59–81. Washington, D.C.: Anthropological Society of Washington.

ESRI. 2008. ArcGIS 9.3. Computer program. ESRI Inc.

Flury, B. D. 2006. Principal components, in *Encyclopaedia of Statistical Sciences Online*, John Wiley and Sons. URL: http://onlinelibrary.wiley.com/book/10.1002/0471667196

Fortin, M.-J. 1999. Spatial statistics in landscape ecology, in Klopatek, J. M., and Gardner, R. H. (Eds.), *Landscape Ecological Analysis*, pp. 253–79. New York: Springer.

Fortin, M.-J., and Gurevitch, J. 1993. Mantel tests: Spatial structure in field experiments, in Scheiner, S. M., and Gurevitch, J. (Eds.), *Design and Analysis of Ecological Experiments*, pp. 342–59. New York: Chapman & Hall.

Goldenweiser, A. A. 1913. The principle of limited possibilities in the development of culture, *The Journal of American Folklore* 26(101): 259–90.

Gower, J. C. 1971. Statistical methods of comparing different multivariate analyses of the same data, in Hodson, F. R., Kendall, D. G., and Tautu, P. (Eds.), *Mathematics in the Archaeological & Historical Sciences*, pp. 138–49. Edinburgh: Edinburgh University Press.

________. 2006. Measures of similarity, dissimilarity, and distance, in *Encyclopaedia of Statistical Sciences Online*, John Wiley and Sons. URL: http://onlinelibrary.wiley.com/book/10.1002/0471667196

Hägerstrand, T. 1967. *Innovation Diffusion as a Spatial Process*. Chicago: University of Chicago Press.

Hart, J. P., and Engelbrecht, W. 2011. Northern Iroquoian ethnic evolution: A social network analysis, *Journal of Archaeological Method and Theory* 19(2): 322–49.

Hart, J. P., and Terrell, J. 2002. *Darwin and Archaeology: A Handbook of Key Concepts*. Westport, CT: Bergin & Garvey.

Haynes, K. E., and Fotheringham, A. S. 1984. *Gravity and Spatial Interaction Models*. Beverly Hills: Sage.

Hodder, I. 1974. Regression analysis of some trade and marketing patterns, *World Archaeology* 6(2): 172–89.

Hodson, F. R., Sneath, P., and Doran, J. 1966. Some experiments in the numerical analysis of archaeological data, *Biometrika* 53: 311–24.

Hudson, J. C. 1972. *Geographical Diffusion Theory*. Evanston, IL: Northwestern University.

Jackson, D. 1995. PROTEST: A Procrustean Randomization Test of community environment concordance, *Ecoscience* 2: 297–303.

James, F. C., and McCulloch, C. E. 1990. Multivariate-analysis in ecology and systematics: Panacea or Pandora's Box? *Annual Review of Ecology and Systematics* 21: 129–66.

Johnson, G. A. 1977. Aspects of regional analysis in archaeology, *Annual Review of Anthropology* 6: 479–508.

Jordan, P., and Shennan, S. 2003. Cultural transmission, language, and basketry traditions amongst the California Indians, *Journal of Anthropological Archaeology* 22(1): 42–74.

Jorgensen, J. 1969. *Salish Language and Culture: A Statistical Relationship of Internal Relationships, History and Evolution*, Language Science Monographs 3. Bloomington: Indiana University.

———. 1980. *Western Indians*. San Francisco: W. H. Freeman and Co.

Kirch, P. V. 1980. Polynesian prehistory: Cultural adaptation in island ecosystems, *American Scientist* 68(1): 39–48.

Koenig, W. D. 1999. Spatial autocorrelation of ecological phenomena, *Trends in Ecology & Evolution* 14(1): 22–26.

Kroeber, A. L. 1925. *Handbook of the Indians of California*. Washington, D.C.: Dover Publications.

———. 1939. *Cultural and Natural Areas of North America*. Berkeley: University of California Publications in American Archaeology and Ethnology no. 38.

Kruskal, J. B. 1964. Multidimensional scaling by optimizing goodness of fit to a non-metric hypothesis, *Psychometrika* 29: 1–27.

Kruskal, J. B., and Wish, M. 1978. *Multidimensional Scaling*. Beverly Hills: Sage.

———. 1971. Multidimensional scaling in archaeology: Time is not the only dimension, in Hodson, F. R., Kendall, D. G., & Tautu, P. (Ed.), *Mathematics in the Archaeological and Historical Sciences*, pp. 119–32. Edinburgh: Edinburgh University Press.

Kvamme, K. L. 1990. Spatial autocorrelation and the classic Maya collapse revisited: Refined techniques and new conclusions, *Journal of Archaeological Science* 17(2): 197–207.

Lansing, J. S., Cox, M. P., Downey, S. S., Gabler, B. M., Hallmark, B., Karafet, T. M., Norquest, P., Schoenfelder, J. W., Sudoyo, H., Watkins, J. C., and Hammer, M. F. 2007. Co-evolution of languages and genes on the island of Sumba, eastern Indonesia, *Proceedings of the National Academy of Sciences of the United States of America* 104(41): 16022–26.

Leduc, A., Drapeau, P., Bergeron, Y., and Legendre, P. 1992. Study of spatial components of forest cover using partial Mantel tests and path-analysis, *Journal of Vegetation Science* 3(1): 69–78.

Legendre, P. 1993. Spatial autocorrelation: Trouble or new paradigm? *Ecology* 74(6): 1659–73.

Legendre, P., and Fortin, M. J. 1989. Spatial pattern and ecological analysis, *Vegetatio* 80(2): 107–38.

Legendre, P., and Legendre, L. 1998. *Numerical Ecology*. New York: Elsevier.

Legendre, P., and Troussellier, M. 1988. Aquatic heterotrophic bacteria: Modeling in the presence of spatial auto-correlation, *Limnology and Oceanography* 33(5): 1055–67.

LeSage, J., Banerjee, S., Fischer, M. M., and Congdon, P. 2009. Spatial statistics: Methods, models, and computation, *Computational Statistics and Data Analysis* 53(8): 2781–85.

Lichstein, J. W., Simon, T. R., Shriner, S. A., and Franzreb, K. E. 2002. Spatial autocorrelation and autoregressive models in ecology, *Ecological Monographs* 72(3): 445–63.

Mace, R., Holden, C. J., and Shennan, S. 2005. *The Evolution of Cultural Diversity: A Phylogenetic Approach*. London: UCL Press.

Manly, B. F. J. 1994. *Multivariate Statistical Methods: A Primer*. Boca Raton, FL: Chapman & Hall/CRC.

Mantel, N. 1967. Detection of disease clustering and a generalized regression approach, *Cancer Research* 27(1): 209–20.

Matson, R. G. 1974. Clustering and scaling of Gulf of Georgia Sites, *Syesis* 7: 101–14.

Matson, R. G., Ludowicz, D., and Boyd, W. 1980. Excavations at Beach Grove in 1980. Report on file, Culture Library, Ministry of Tourism, Culture and the Arts, Victoria.

Matson, R. G., Pratt, H., and Rankin, L. 1991, 1989 and 1990. Crescent Beach Excavations, Final Report. Report on file, Culture Library, Ministry of Tourism, Culture and the Arts, Victoria.

Matson, R. G., and True, D. L. 1974. Site relationships at Quebrada Tarapaca, Chile: A comparison of clustering and scaling techniques, *American Antiquity* 39(1): 51–73.

Milke, W. 1935. *Südostmelanesien: Eine Ethnostatistische Analyse*. Würzburg.

———. 1949. The quantitative distribution of cultural similarities and their cartographic representation, *American Anthropologist, New Series* 51(2): 237–52.

Miller, J., Franklin, J., and Aspinall, R. 2007. Incorporating spatial dependence in predictive vegetation models, *Ecological Modelling* 202(3–4): 225–42.

Moran, P. A. P. 1950. Notes on continuous stochastic phenomena, *Biometrika* 37 (1): 17–23.

Mykra, H., Heino, J., and Muotka, T. 2008. Concordance of stream macroinvertebrate assemblage classifications: How general are patterns from single-year surveys? *Biological Conservation* 141(5): 1218–23.

NCSS. 2004. Computer Program. NCSS Inc.

Neiman, F. D. 1997. Conspicuous consumption as wasteful advertising: A Darwinian perspective on spatial patterns in classic Maya terminal monument dates, in Barton, C., and Clark, G. (Eds.), *Rediscovering Darwin: Evolutionary Theory and Archaeological Explanation*, pp. 267–90. Arlington, VA: Archaeological Papers of the American Anthropological Association, 7, American Anthropological Association.

Newman, S. 1974. Linguistic retention and diffusion in Bella Coola, *Language and Society* 3: 201–14.

Orchard, T. J., and Clark, T. N. 2005. Multidimensional scaling of northwest coast faunal assemblages: A case study from southern Haida Gwaii, British Columbia, *Canadian Journal of Archaeology* 29(1): 88–112.

———. 2006. Was Salmon Specialization a Northwest Coast Universal? Intensification and Generalization of Prehistoric Northwest Coast Economies. Poster presented at the 39th Annual Meeting of the Canadian Archaeological Association, Toronto.

Peres-Neto, P. R., and Jackson, D. A. 2001. How well do multivariate data sets match? The advantages of a Procrustean superimposition approach over the Mantel test, *Oecologia* 129(2): 169–78.

Perry, G. L. W., Miller, B. P., and Enright, N. J. 2006. A comparison of methods for the statistical analysis of spatial point patterns in plant ecology, *Plant Ecology* 187(1): 59–82.

Pettorelli, N., Dray, S., and Maillard, D. 2005. Coupling principal component analysis and GIS to map deer habitats, *Wildlife Biology* 11(4): 363–70.

Premo, L. S. 2004. Local spatial autocorrelation statistics quantify multi-scale patterns distributional data: An example from the Maya Lowlands, *Journal of Archaeological Science* 31(7): 855–66.

Price, T. D. 2000. *Europe's First Farmers*. New York: Cambridge University Press.

Pruzansky, S. 2006. Proximity data, in *Encyclopaedia of Statistical Sciences Online*. John Wiley and Sons.

Renfrew, C. 1977. Alternative models for exchange and spatial distribution, in Earle, T. K., and Ericson, J. E. (Eds.), *Exchange Systems in Prehistory*, New York: Academic Press.

Roberts, J. M., Moore, C. C., and Romney, A. K. 1995. Predicting similarity in material culture among New-Guinea villages from propinquity and language: A log-linear approach, *Current Anthropology* 36(5): 769–88.

Samuels, S. R. 1991. House structure and floor midden, *Ozette Archaeological Project Research Report*, Volume I. Washington State University Reports of Investigation 63. National Park Service, Northwest Regional Office. Pullman: Washington State University.

Schillaci, M. T., Meijaard, E., and Clark, T. N. 2009. The effect of island area on body size in a primate species from the Sunda Shelf Islands, *Journal of Biogeography* 36(2): 362–71.

Schneider, J. W., and Borlund, P. 2007. Matrix comparison, Part 2: Measuring the resemblance between proximity measures or ordination results by use of the Mantel and Procrustes statistics, *Journal of the American Society for Information Science and Technology* 58(11): 1596–609.

Shennan, S. 2002. *Genes, Memes, and Human History: Darwinian Archaeology and Cultural Evolution*. London: Thames & Hudson.

Smith, E. A., and Winterhalder, B. 1992. *Evolutionary Ecology and Human Behavior*. New York: Aldine de Gruyter.

Smouse, P. E., Long, J. C., and Sokal, R. R. 1986. Multiple regression and correlation extensions of the Mantel test of matrix correspondence, *Systematic Zoology* 35(4): 627–32.

Sneath, P., and Sokal, R. 1973. *Numerical Taxonomy: The Principles and Practice of Numerical Classification*. San Francisco: W. H. Freeman and Co.

Soininen, J., Paavola, R., and Muotka, T. 2004. Benthic diatom communities in boreal streams: Community structure in relation to environmental and spatial gradients, *Ecography* 27(3): 330–42.

Spielman, S. E., and Yoo, E. H. 2009. The spatial dimensions of neighborhood effects, *Social Science & Medicine* 68(6): 1098–105.

Thom, B. 1992. An investigation of inter-assemblage variability within the Gulf of Georgia phase, *Canadian Journal of Archaeology* 16: 24–31.

Tobin, P. C. 2004. Estimation of the spatial autocorrelation function: Consequences of sampling dynamic populations in space and time, *Ecography*: 767–75.

Tobler W. 1970. A computer movie simulating urban growth in the Detroit region, *Economic Geography*, 46(2): 234–40.

True, D. L., and Matson, R. G. 1970. Cluster analysis and multidimensional scaling of archaeological sites in northern Chile, *Science* 169: 1201–03.

Urban, D. L. 2006. The Mantel test, in *Encyclopaedia of Statistical Sciences Online*. John Wiley and Sons.

Venturelli, P. A., and Tonn, W. M. 2005. Invertivory by northern pike (*Esox lucius*) structures communities of littoral macroinvertebrates in small boreal lakes, *Journal of the North American Benthological Society* 24(4): 904–18.

Waldhör, T. 2006. Spatial autocorrelation coefficient, Moran's, in *Encyclopaedia of Statistical Sciences Online*. John Wiley and Sons.

Wheatley, D., and Gillings, M. 2002. *Spatial Technology and Archaeology: The Archaeological Applications of GIS*. New York: Taylor & Francis.

White, L. A. 1959. *The Evolution of Culture: The Development of Civilization to the Fall of Rome*. New York: McGraw-Hill.

Willey, G., and Phillips, P. 1958. *Method and Theory in American Archaeology*. Chicago: University of Chicago Press.

Williams, J. T. 1993. Spatial autocorrelation and the classic Maya collapse: One technique, one conclusion, *Journal of Archaeological Science* 20(6): 705–09.

Young, F. W. 2006. Multidimensional scaling, in *Encyclopaedia of Statistical Sciences Online*. John Wiley and Sons.

Zeitlin, R. N. 1982. Toward a more comprehensive model of inter-regional commodity distribution: Political variables and prehistoric obsidian procurement in Mesoamerica, *American Antiquity* 47(2): 260–75.

Zhao, Y., Shuter, B. J., and Jackson, D. A. 2008. Life history variation parallels phylogeographical patterns in North American walleye (*Sander vitreus*) populations, *Canadian Journal of Aquatic Sciences* 65: 198–211.

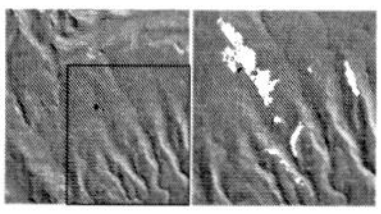

# 5 Network Models and Archaeological Spaces

*Ray Rivers, Carl Knappett, and Tim S. Evans*

A fundamental ingredient of archaeological analysis is an understanding of spatial relationships. We concur with Renfrew (Renfrew, 1981: 268) that "the very activity of examining the spatial correlates of early social structure has the useful effect of posing important problems in general and simple terms." Thus it is not altogether surprising that different authors have provided almost as many answers as there are questions, and to shape the discussion that follows we have chosen to put a heavy emphasis on the Bronze Age Aegean, the period for which our own models (Evans, Knappett, and Rivers, 2009; Knappett, Evans, and Rivers, 2008, 2011) of maritime trade and exchange routes were devised. The act of constructing and developing them has illuminated problems common to many approaches.

The original focus of our work was the southern Aegean in the Middle (MBA) and early Late Bronze Age. This is well bounded in time, concluding with the widespread fire destruction of settlements across Crete, sometime after the eruption of Thera. The archaeological record poses some interesting questions. On the one hand, we wanted to understand why some sites, such as Knossos on Crete, grew to be so large and influential. On the other hand, we wanted to understand how the volcanic eruption of Thera seemed to have such little immediate effect on the exchange between Minoan Crete and the surrounding regions despite destroying Akrotiri, which might have been anticipated as the main gateway from Knossos into the Cyclades, Peloponnese, and the Greek mainland.

*Computational Approaches to Archaeological Spaces* by Andrew Bevan and Mark Lake, Eds., 99–126 

Although both address spatial issues, superficially these concerns seem very different. Thus, the emergence and characteristics of such sites as Knossos are partially explained in local terms, with these local conditions only then facilitating exchange with other sites. The global stability of the Minoan exchange network, meanwhile, might be thought of as explicable in terms of Aegean geography and maritime technology, since the appearance of the sail from about 2000 B.C.E. facilitated new levels of interregional interaction. However, to separate local power from global organization too strongly is to set up a false dichotomy, since the centers of local power are also likely to be key players in global exchange. The same routes that enable these centers to maintain their influence are also conduits for the entwined cultural and economic connections between them. As a result, their growth and importance will be conditioned by the characteristics of the larger networks to which they belong, leading to what Batty terms an "archaeology of relations" (Batty, 2005: 149).

In our Aegean work we have argued that network models can provide the natural framework to transcend this dualism between physical and relational space. Networks are, in essence, no more than a set of nodes or vertices connected by links or edges. Most simply, in archaeology we identify nodes with archaeological sites and the communities that inhabited these sites, and links with the exchange between them. Networks thus have the potential, in an almost self-evident way, of encoding spatial correlates by incorporating both the local attributes of sites together with their global interactions and showing the reciprocal effect of the one on the other. It is no surprise that island archipelagos lend themselves to network analysis. With a dominant means of interaction (sea travel) and, typically, a dominant sailing technology (canoe or sailing vessels) they allow for simple connectivity.

By way of contrast the spatial arrangements of points may be studied using statistical techniques such as those discussed in Chapter 2 (Bevan, Crema, and Palmisano, 2013). However, the bilateral relationships between sites are not made explicit in such methods. Agent Based Models, such as those mentioned in Chapters 6 and 7 (Barton, 2013; Premo, 2013), can define explicit links, but such methods are complementary to the approaches discussed here.

In this article we are primarily interested in how networks of relationships are conditioned by what, in shorthand, we might term *geographical space*. Extensions to other types of space can be achieved using the approach discussed in Chapter 4 (Clark, 2013). In the next section we give an overview of some of the issues that arise in constructing network models for archaeology that rely heavily on the basic attributes of physical geography: distance and topography. Null models are useful to this end, in helping to provide simple benchmarks. This prologue prefigures the most substantive part of the chapter, an analysis of how optimization models, which add agency to geography, might be useful in Aegean prehistory. Although such modeling has obvious limitations, in its stress on the material circumstances of human-environment interaction on social change, it allows for a "restricted set of spatial variables to play a role in shaping historical transformations" (Smith, 2003: 53). As a result, in a step-by-step way, we can construct tractable quantitative models amenable to simulation.

In trying to decide whether and how such models work, we see that it is hard to improve on the sentiment in Rihll and Wilson that

> the purpose of a good model is to formulate simple concepts and hypotheses concerning them, and to demonstrate that, despite their simplicity, they give approximate accounts of otherwise complex behaviour of phenomena. If a model "works" (faithfully represents the known evidence) then it shows that the assumptions and hypotheses built into the model contribute to an explanation of the phenomena. (1991: 62)

Unfortunately, although the evidence, the archaeological record, may be good for some eras, for the prehistoric era of our models it is sufficiently patchy in both space and time as to be ambiguous. If the datasets were good, whether the model could be made to "work" would become apparent quite quickly. Unfortunately, with the data as incomplete as they are, we might expect to find so many choices of model to be commensurate with the data that we would not be sure how to proceed. In practice, this is not the case. A simple touchstone is that the models should produce a range of site occupancies and a range of site activity that go beyond the range of potential resources. Despite this seeming generality we stress that there is no universal behavior implied by network models, and this fact alone turns out to be a surprisingly good discriminant. Different models give different patterns of exchange, and we need to choose the model type appropriately, with even a poor record in mind, before we look more closely at the detailed behavior that they can suggest. Specifically, we require assumptions about (maritime) technology and social organization specific to the period in question. There are yet additional issues about determinacy, and here we do our best to address the extent to which we should understand outcomes statistically. Chapters 2 (Bevan, Crema, and Palmisano, 2013) and 7 (Clark, 2013) also discuss ways to tackle uncertainty in the data and in any model.

Although motivated by the MBA S. Aegean, this chapter is more an overview of methods, and we do not attempt to address the archaeological record in detail—steps toward this are given elsewhere (Knappett, Evans, and Rivers, 2011).

Most of our network modeling is done using our own package, "ariadne" (Evans, Rivers, and Knappett, 2012c), written in Java and using open source libraries when necessary. We found that existing network packages were not ideal for networks embedded in space nor for the particular models we were interested in. Likewise, GIS packages were unnecessarily complicated and sophisticated for our spatial needs, while their network analysis tools were too limited.

## The Role of Geographic Space: First Thoughts

One attribute common to most network models is that they are construed dynamically. *Links* describe the flows between sites: goods, raw materials, people, and ideas, which are both the sinews and synapses of intersite relationships. In the most simple networks, links are *nondirectional* and switched either on or off—that is, all we are concerned with is whether two sites interact or not. At a more sophisticated level, links are both *directional* (that is, the relationship of A to B is not the same as B to A) and are *weighted* (that is, they vary from strong to weak, where the weight is a measure of the "flow").

In an archaeological context *nodes*, thought of as sites, have two different attributes. At one level we can interpret them as junctions for flows, without needing any of their local physical and geographical properties beyond, say, position, which determines their accessibility to other sites. In this sense they are rather like traffic intersections, or airports, acquiring their local significance only from the network as a whole. As such, they permit a variety of classifications, of which the simplest orders nodes by their "busyness"—the total flow through them. However, we should not ignore their local properties. In fact, our original interest in networks was triggered by the use of networks to explain the mismatch between local resources and exchange in the Early Bronze Age Cyclades (Broodbank, 2000). In some models (our first null models) sites are treated equally but, more realistically, are ranked by their *local* properties (for example, populations) as well as their properties derived from the *global* behavior of the network as a whole.

As we said earlier, our main interest is to see to what extent exchange networks reflect or transcend what we have termed geographical space. Such space has several facets, which we shall delineate later, but, by looking at the site map, could we first infer which sites are most significant, both from their intrinsic properties (for instance, resources) and their position in the network? We also need to identify the most important links, since they will reflect the distribution of artifacts. We begin this discussion by presenting two null models, each of which uses only physical geography, qualified by basic assumptions about travel, against which to test the more sophisticated hypotheses that we propose later.

### A Question of Distance: Fixed-Radius Model

The map of the S. Aegean whose networks we are considering (Evans, Rivers, and Knappett, 2012a, 2012b; Knappett, Evans, and Rivers, 2011) is given in Figure 5.1. On it we have labeled 39 of the most significant sites in the MBA; names and details of the sites are given in Table 5.1. In general we do not need them, since we are mainly using the resulting networks for comparative purposes and will not address the record in any detail. In the first instance what we

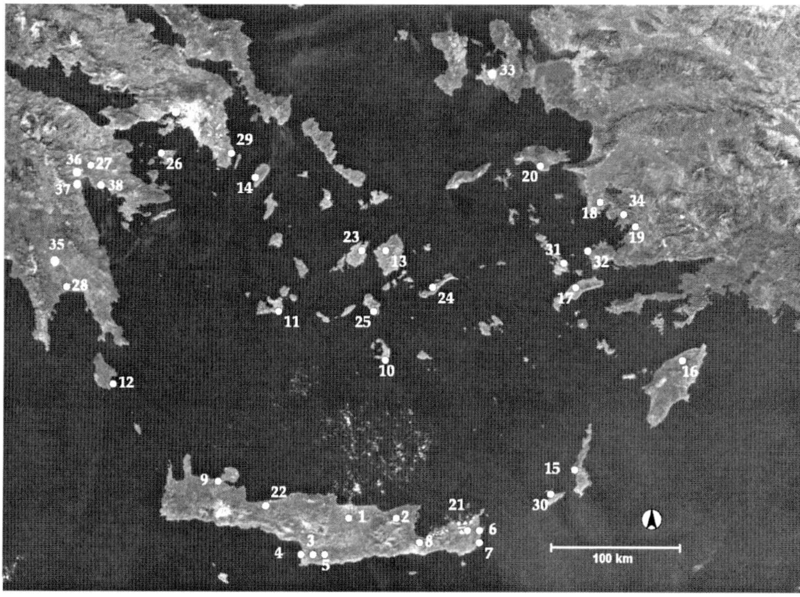


**Figure 5.1** Important sites for the MBA Aegean, including Knossos (1) and Thera (10). The sea journey from the northern Cretan coast to Thera is a little over 100 km.

**Table 5.1** Sites Enumerated in Figure 5.1 and the Size of Their Local Resource Base—with (S), (M), (L) denoting small, medium, or large in terms of their resource base (*inputs*), to be distinguished from their populations, which are *outputs*.

| | | |
|---|---|---|
| 1. Knossos (L) | 14. Kea (M) | 27. Mycenae (L) |
| 2. Malia (L) | 15. Karpathos (S) | 28. Ayios Stephanos (L) |
| 3. Phaistos (L) | 16. Rhodes (L) | 29. Lavrion (M) |
| 4. Kommos (M) | 17. Kos (M) | 30. Kasos (S) |
| 5. Ayia Triadha (L) | 18. Miletus (L) | 31. Kalymnos (S) |
| 6. Palaikastro (L) | 19. Iasos (M) | 32. Myndus (M) |
| 7. Zakros (M) | 20. Samos (M) | 33. Cesme (M) |
| 8. Gournia (L) | 21. Petras (L) | 34. Akbuk (M) |
| 9. Chania (L) | 22. Rethymnon (L) | 35. Menelaion (S) |
| 10. Thera (M) | 23. Paroikia (M) | 36. Argos (M) |
| 11. Phylakopi (M) | 24. Amorgos (S) | 37. Lerna (M) |
| 12. Kastri (M) | 25. Ios (S) | 38. Asine (S) |
| 13. Naxos (L) | 26. Aegina (M) | 39. Eleusis (M) |

see is four regional groupings: Crete, the Cyclades, the Peloponnese, and the Dodecanese. One of the tests for the importance of geography lies in the extent to which the exchange networks reflect these regional groupings.

To convert these qualitative regional groupings into something more quantitative we show a dendrogram of site separations in Figure 5.2. However, it

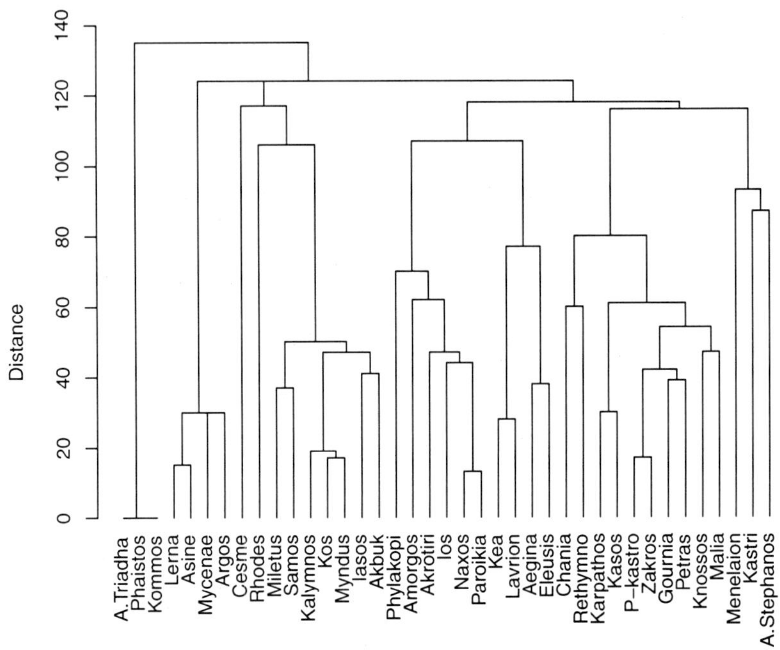


**Figure 5.2** Dendrogram for the sites of Figure 5.1, using an effective shortest distance between sites in which sea travel is counted in physical kilometers but land travel is penalized by a friction factor of 3.0. A horizontal line cutting the vertical axis at distance scale *D* cuts the dendrogram into a number of disconnected branches below the horizontal line. Each of these separate branches defines a cluster (or community) of sites within which any pair of sites is separated by an effective distance of *D* or less. It is not possible to move between sites in different branches without following a link of effective length greater than *D*. (Produced using the R statistics package: R Development Core Team, 2010.)

would have been too simplistic to have constructed the dendrogram with a compass, using just geographical distances. For exchange networks the relevant distances are those taken by the travelers who provide the exchange, which requires us to understand how they travel. For the period in question, the MBA, communication is by sailing ships that do not remain at sea at night. Thus our distances are not line-of-sight distances but correspond to the best sea routes, negotiating headlands where necessary. Even that knowledge is not sufficient, since land travel may produce routes that are shorter, and some sites are inland, notably Knossos and Mycenae. Given the slow and laborious nature of land travel in this era, we introduce a frictional coefficient that penalizes land travel over sea travel. This isolates many sites situated on large islands or on the mainland, so that they behave rather like islands. For instance, the direct routes between some sites on the northern and southern coast of Crete are not optimal for reasonable friction coefficients. Sea travel around the eastern or western tips, or at least to a narrower part of the island, is often easier in our terms. Given our choice of optimal routes, we have used a variety of frictional coefficients and find that our conclusions are largely insensitive to them. So, for the sake of argument, we have taken a frictional coefficient of three in all that follows. Given this relative insensitivity, we obtained our distance estimates by analyzing the physical geography by hand at a scale of around a few km, an appropriate level of error given the other uncertainties in our knowledge. One of these is how, or if, to take into account the prevailing winds and currents or the daily fluctuations in conditions. For the moment we assumed that these average out on a yearly cycle. Our effective distances thus reflect averaged time-of-travel along the least-time path, an example of the least-cost-path approach discussed in Chapter 8 (Herzog, 2013). We use the same estimates in all the models discussed in the following pages.

Our first model is preposterously simple. It asks what networks would occur if travel was restricted to journeys up to a specified distance, in the sense just described. This is the quantitative way of defining regional clusters. To identify the important travel distance scales for the sites of Figure 5.1, the simplest way is to connect sites only if they are less than a certain distance $D$ apart (for example, 100 km), as encoded in the dendrogram shown in Figure 5.2. A horizontal line at some fixed-distance scale $D$ cuts the tree into a number of smaller branches lying below the line, each defining a separate cluster of sites. Only sites within the same cluster are connected to one another by paths in which each section is less than the distance $D$ apart.

We should not think of $D$ too precisely, both because of the uncertainty in the calculation of effective distances and because of the simplistic nature of a discrete distance cutoff. However, in Figure 5.2 we see that a transition occurs if we go from below 110 km to above 130 km (the effective distance between Knossos and Akrotiri is about 130 km, once land travel is taken into account). At 110 km the sites are split into the four isolated zones we would expect: Crete, the Cyclades, the Dodecanese, and the Peloponnese (itself fractured into adjacent parts). However, almost all the sites are connected by routes involving steps of 130 km or more. This structure becomes very clear if, as a complementary approach, we show the network for different distance scales, as in Figure 5.3. The dark links correspond to distances up to the specified distance.

Simple as this model is, one thing we take from it is the very striking fact that the distance scale $D \approx 120$ km, at which we move from a fragmented network to one in which can be traversed easily, albeit in several steps, is more or less the maximum distance scale we would expect from MBA sailing vessels for a single journey, on the assumption that boats did just that in traveling from the northern Cretan coast to Thera.

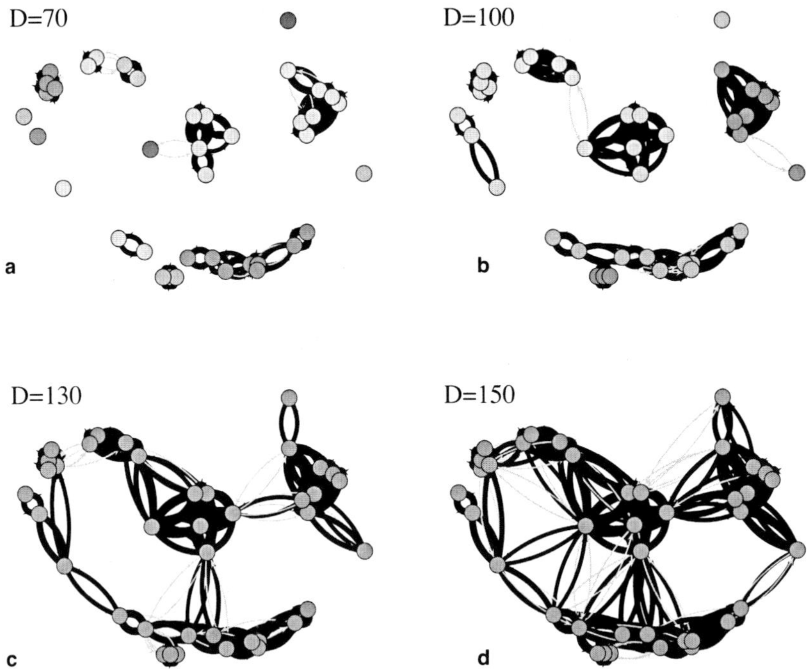


**Figure 5.3** Fixed-radius networks for the sites of Figure 5.1. Sites are placed at their geographical location; distances between sites are judged via shortest sensible routes with land travel penalized by a factor of 3.0 but without taking currents, winds, and so on into account. Two sites are linked (dark links) if the effective distance is less than *D*, where *D* = 70 km, 100 km, 130 km, and 150 km as indicated. Light grey links are edges that are longer than *D* but by no more than 20%. Sites of same shade are connected by routes via black links (that is, all hops are less than *D* km).

The arrival of sail in the MBA makes the Aegean a "Goldilocks" sea[1] in the sense that the technology is just right for the appearance of an active trading network with a handful of key sites. Too small a journey distance *D* and the network is overwhelmed by the prevailing geography, as is the case for the EBA, for which canoes have a distance scale more like 30 km, requiring the Cyclades and other regions to be relatively self-contained (Broodbank, 2000), as seen in the leftmost diagram in Figure 5.2. Too large a distance scale, as happens with modern ferries, and the whole region becomes over-connected in the sense that there are many more links than are necessary to make the network negotiable. Thus our MBA networks live in the transition between social organization being strongly influenced by geography and it being much less relevant. Moreover, this distance scale is more a function of interisland separation than a function of where sites arise, and, without it being a circular argument, for this reason we have taken MBA sites in Figure 5.1. Our choice of 39 sites is informed but debatable. However, we note that the inclusion of more sites would not affect the result that we move from an under-connected to an over-connected network at $D \approx 120$ km.

## Proximal Point Analysis (PPA)

A very different null model is provided by Proximal Point Analysis (PPA), again not quite simple enough to be done with ruler and compass but requiring nothing more than the effective distances and rudimentary ideas about social interaction. Rather than look for clustering in PPA we aim, primarily, to decide which sites have a more prominent role in regional interactions. This prominence is determined by connectivity, the number of links that a site has with its network neighbors. There are several variations on this theme, but, most simply, sites are taken to be similar (unweighted). If the premise of our previous model is that communities need to interact, but it is just too difficult to travel far (as controlled by $D$), the premise of PPA is that communities need to interact, but it is just too difficult to sustain more than a few important interactions—a Dunbar number for communities, rather than individuals (Dunbar, 1992).

However, the neighbors with which it does interact are defined by distance. To perform the analysis links we draw outward from each site to a specified number $k$ of nearest neighbors (typically three or four). By "nearest" is meant geographically closest, once headlands and land travel are taken into account as before. If, when this analysis is completed, directional arrows are removed from the links, some sites will emerge as being more connected than others, with five, six, or more links to other sites. These sites possess greater centrality in the network and are anticipated to be more dominant in regional interactions. In fact, our interest in archaeological networks was fired by the application of PPA to the Early Bronze Age Cyclades by Broodbank (Broodbank, 2000). Of course, as we have already implied, networks do not just form for the sake of it but arise because of a mixture of imperatives. Broodbank suggests that the EBA Cyclades were agriculturally marginal, requiring social storage networks.

Island archipelagos lend themselves to PPA network analysis because of the clear separation of sites. In particular, their early application was to anthropological networks in Oceania (Hage and Harary, 1991, 1996; Irwin, 1983; Terrell, 1977). Coastal sites are equally amenable to a PPA analysis (Terrell, 2010). However, some PPA models are for land-based networks (Collar, 2007) of sites that are discrete. Networks built on PPA have a strong emphasis on local geography because most neighbors are relatively close. However, there is also the contrary effect that even isolated sites will interact with the same number of neighbors, however far they are removed from them. The end result is that interactions are inclined to lie in strings of sites, including remote sites

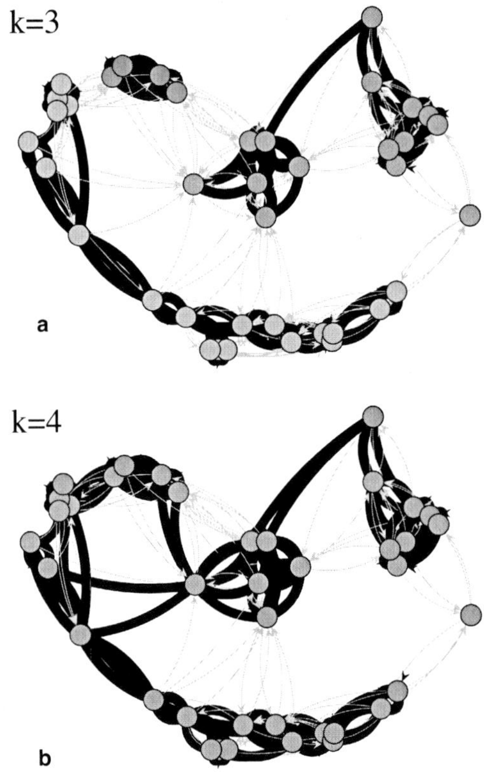


**Figure 5.4** PPA for the southern Aegean sites of Figure 5.1, joining each site to either three ($k = 3$) or four ($k = 4$) nearest neighbors (thick black links). Thin grey links indicate links to fifth ($k = 3$) or sixth ($k = 4$) nearest neighbors. Sites of the same shade are connected by routes via thick black links.

on the periphery, rather than in the clustering of the previous model. We have shown this arrangement for our sites in Figure 5.4, in which we join each site to its three or four nearest neighbors.

For $k = 3$ nearest neighbors we see a Western string encompassing Crete and the Peloponnese (with a separated northern part). The Cyclades and the Dodecanese, just simply connected to each other, have no connection to northern Crete. If we were to join each site to its four nearest neighbors, the Cyclades becomes connected to the East. This situation has a counterpart in the previous model on varying $D$. Unlike for that case, this connectivity is not stable to our initial choice of sites, our only other input—primarily because we have chosen so few (39) sites. If we were to double the number of sites, the Cyclades would become disconnected again for $k = 4$. However, we think of the number of connections that a site will make as something intrinsic to the nature of society, a Dunbar number, whereas the number of sites on the map is purely artificial, perhaps a reflection of our knowledge. We need to distinguish between conclusions that follow from the principles of the model and conclusions that are a consequence of our ignorance about the record, an issue that will recur again. However, this is not to damn PPA as a qualitative guide for those cases, in its applications such as contemporary and recent anthropology (see Terrell, 2010), for which the record is very good.

### Agency: Optimization

Our null models assumed that sites would like to interact directly with other sites for the purposes of exchange but are limited in doing so, because it is either difficult to travel too far or difficult to sustain too many exchanges. This need to establish a balance has an element of truth and suggests that social networks are optimal in some sense. We need to be careful that we do not create too simplistic a Panglossian "best of all possible worlds," but recent years have seen a revival of the idea that social networks do evolve to some form of best behavior, particularly in economic theory (Jackson, 2008). Although not always correct, the simplicity of the approach gives us a more sophisticated set of models than those we have considered so far.

There are two different ways to proceed. The first looks to identify the best network with the most likely network. The second of these is more social utilitarian and identifies best with most efficient. These further attributes of networks take us beyond the null models discussed previously and allow us to get away from the simple geographical/social determinism embodied in them.

## The Most Likely Networks

The first optimization approach looks to identify the best network with the most likely network, all other things being equal, within the constraints of our knowledge. There is a long literature to this approach, which maximizes "information," understood as Shannon entropy. To choose less likely networks would, in some sense, correspond to assuming information that we did not have (Batty, 2010). In particular, models of this type have been used extensively in modeling transport flows (Erlander and Stewart, 1990), and we pick out only those aspects that might be relevant to the flows of Aegean maritime networks. The discussion that follows is fairly technical—probably not enough for readers really wanting to understand the models and too much for readers uncomfortable with algebra. (The former can find more details about the approach elsewhere in our work [Evans, Knappett, and Rivers, 2009; Knappett, Evans, and Rivers, 2008].)

## Maximum Entropy: Basic Models

Define $F_{ij}$ as the flow from site $i$ to site $j$ where, for our Aegean network of Figure 5.1, $i$ and $j$ take values from 1 to 39. We can think of $F_{ij}$ as, say, the number of vessels per year traveling from $i$ to $j$, or some equivalent measure of exchange. The greater the flows through a site the more significant we expect it to be in the network, and an estimation of the flows is one of the main goals of network modeling. As a baseline, the null models in the simple form posed previously have identical (unweighted) flows when they exist, being switched on or off according to distance or neighborliness.

In the doubly constrained entropy model, the total outflow from a site $i$, $O_i$ (for example, the total number of vessels leaving $i$ in a year) and the total inflow to a site $j$, $I_j$ (for instance, number of vessels arriving) are inputs to be specified:

$$O_i = \sum_j F_{ij},\ I_j = \sum_i F_{ij} \tag{1}$$

Likewise the "cost" of sustaining each link has to be specified, say $c_{ij}$, which we assume to be a function of site separation, the effective distance $d_{ij}$ (once headlands and land travel are taken into account). Finally, the total cost of maintaining a given network is fixed at some value $C$:

$$C = \sum_{i,j} F_{ij} c_{ij}$$

The idea is that the most likely pattern of flows is that which maximizes the entropy $S$

$$S = -\sum_{i,j} F_{ij}(\ln(F_{ij}) - 1)$$

(for example, see Ball, 2004; Batty, 2010) subject to the constraints on flows and costs given previously. That is, one finds the most likely flows if trips between sites are themselves equally likely, provided the total pattern conforms to the specified constraints. This is the simplest assumption to make and has found wide applicability in a variety of transport models. Under these constraints one can prove that the optimal network of flows is of the form (Batty, 2010)

$$F_{ij} = A_i O_i B_j I_j \exp(-\beta c_{ij}) \tag{2}$$

Here the parameters $A_i$ and $B_j$ are self-consistently determined by the constraints on the input and output flows (1) as

$$\frac{1}{A_i} = \sum_j B_j I_j \exp(-\beta c_{ij}),\quad \frac{1}{B_j} = \sum_i A_i O_i \exp(-\beta c_{ij}), \tag{3}$$

while choosing the parameter $\beta$ is equivalent to setting the total cost $C$. This has the form of a generalized gravity model (for instance, see Jensen-Butler, 1972, as one of many summaries) in that the flow is related to a product of the attributes of the sites, falling off as some function of their separation. For this reason this model is usually termed the "doubly constrained" gravity model. This model is one step beyond more conventional gravity models, which specify the form of the flow to be that of equation (2) by fiat, where we define $A_i O_i$ and $B_j I_j$ to be the populations of $i$ and $j$, respectively, with no internal consistency—for example, see Alden (1979) for an application of simple gravity models to Toltec networks.

In archaeology we rarely have a direct handle on the flows in and out of a site. Given our lack of knowledge, we consider the simple default case, in which, on a site-by-site basis, inflow $I_j$ and outflow $O_j$ are equal to each other, dependent on site size. *Size* has many meanings—for example, the area occupied, size of key monuments (Renfrew, 1981; Renfrew and Level, 1979), or available local resources (Broodbank, 2000). We distinguish between population and resources (carrying capacities). Here we use the latter as our proxy for size and take the flows proportional to these resources. For our MBA Aegean example we have classified our sites as large, medium, or small (see Table 5.1 for details) and taken their resource bases $S_i$ to have relative values 1, 1/2, or 1/3, respectively. We find that outcomes are not very sensitive to different definitions of large, medium, and small.

In many problems, including archaeological ones, the costs are not known. Nonetheless, we expect that the longer the distance between two sites the higher the cost of making the journey. In our work we have chosen our costs such that $\exp(-\beta c_{ij}) = V(d_{ij}/D)$ where the $d_{ij}$ are the effective distances we introduced earlier, and $D$ is the distance scale for maritime travel introduced earlier (it plays a similar role to $\beta$). For the moment we can think of $V(x)$ as a single journey likelihood function or an ease-of-travel function that quantifies the fuzziness that we argued for in the fixed-radius model. We choose $V$ to have the form shown in Figure 5.5. This shape means that all short journeys, $d_{ij} << D$, are easy ($V$ close to 1) and have a cost close to the minimum possible. All long journeys, $d_{ij} >> D$, are difficult ($V$ close to 0) and are of high cost. For intermediate distances, $d_{ij} \sim D$, the function falls smoothly but rapidly.

By way of comparison, the networks of Figure 5.4 arising from our simple geographical model are obtained essentially by using a simple cutoff function in which $V(x) = 1$ for $0 < x < 1$ and $V(x) = 0$, $x > 1$ where $x = d_{ij}/D$. Our function in Figure 5.5 is a smoothed-out variant on this, which avoids making such a crude distinction between trips that are and are not possible and is better suited to our imprecise knowledge of what was possible—for example, the maximum distance traveled in a day and the errors in our estimates of ancient traveling times. We contrast this with the dashed curve in Figure 5.5, which corresponds to a travel cost for a single journey proportional to distance, for

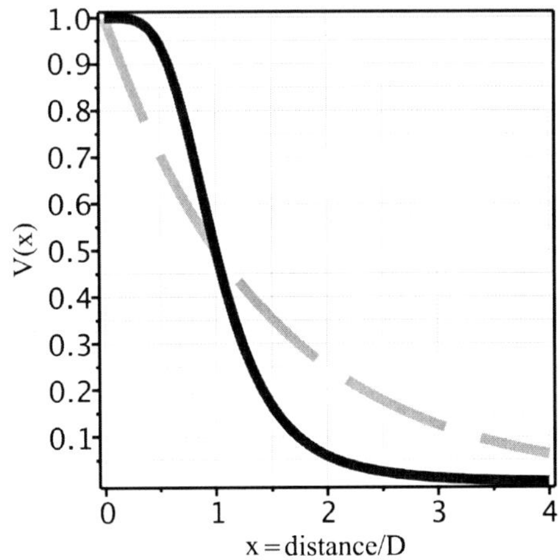


**Figure 5.5** Interaction potentials as a function of distance $d$, where $x = d/D$ and $D$ is a fixed-distance scale. By construction $V(0) = 1$ and $V(1) = 0.5$. The solid black line is $V(x) = 1/(1 + x^4)$ as used in this work. The dashed grey line is $V(x) = \exp(-\beta x) = 2^{-x}$ with $\beta = \ln(2)/D$ as used in many gravity models.

which this steep transition between manageable and unmanageable distances is lost.

In Figure 5.6 we show the maximum entropy gravity model for $D = 70$ km. We see that, despite the (intentionally) small 70 km distance scale, the model enforces connectivity across the southern Aegean, which we might not have expected. This connectivity arises because, with both site outflows and inflows determined by (nonzero) site size, they have to go somewhere, even for a site far removed from others. The outcome for $D = 70$ km is something rather like PPA in deemphasizing regions, with a strong western link and strong connections around the rim of the network, but with an infill of weak links. As $D$ increases to 150 km, so does the number of links, but, since the total flows remain fixed, links become weaker. An undesirable consequence of the model is that flows remain directly proportional to site size, and this is not what we would expect of exchange networks, which are characterized by a wide range of activity. This restricted behavior is not surprising, because, apart from replacing a cutoff function by a smoother interpolating function $V(x)$, there is no freedom in the model, once our simple assumptions about flows are taken into account.

## Maximum Entropy: An Enhanced Model

However, a variant of the gravity model does produce a much wider range of site activity. This model, originally designed for urban planning (Wilson,

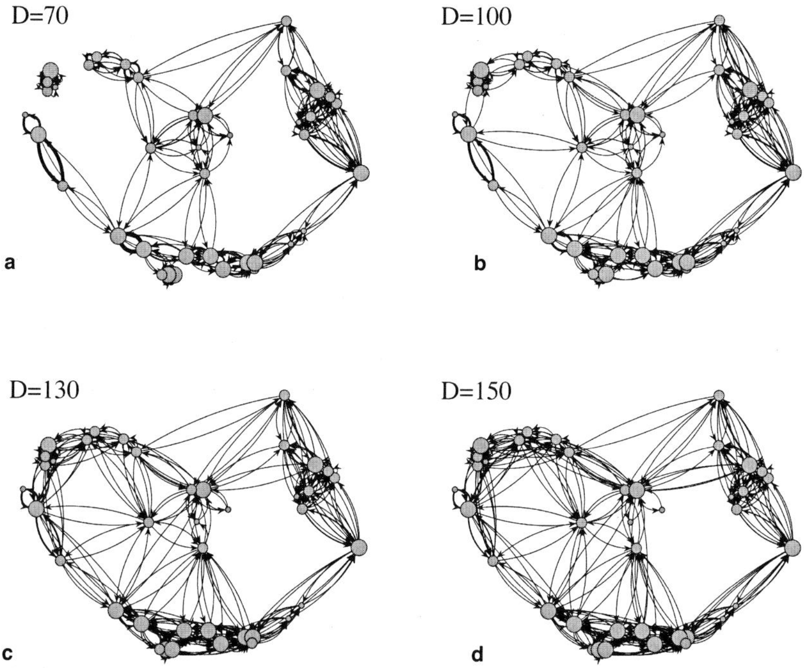


**Figure 5.6** The doubly constrained gravity model, a maximum entropy model, for $D$ = 70 km, 100 km, 130 km, and 150 km (as indicated). For each site $i$, outflow $O_i$ and inflow $I_i$ are both fixed equal to the fixed site size $S_i$ (taking values 1/3, 1/2, or 1.0; see Table 5.1). Only edges with flow over 0.0256 are included because this is approximately 1/39 and so indicates an above-average flow. Vertices of the same shade are connected by the edges shown. Note that the flows are symmetric, a necessary property of solutions if $d_{ij} = d_{ji}$ and $O_i = I_i$.

1967, 1970) was applied by Rihll and Wilson (1987, 1991) to an archaeological context, that of Iron Age Mainland Greek city-states. In this variant, entropy is maximized as before for fixed total cost, with outflows again proportional to site size. However, rather than impose inflows on a site-by-site basis *a priori* we choose a constraint that allows them to fluctuate. The result is that the flows are given by

$$F_{ij} = A_i O_i I_j^{\beta} V(d_{ij}/D), \tag{4}$$

where $\beta$ determines what is called site attractiveness. As before, the $A_i$ are determined self-consistently. The inflow $I_j$ is now identified as the attractiveness of site $j$, no longer a fixed input parameter. The new input parameter $\beta$ is usually chosen to be a little above 1.0. As we take larger and larger values for

$\beta$, we find that fewer and fewer sites attract larger and larger fractions of the total flow into their sites, which must therefore come from sites farther and farther away. Thus this parameter controls the nature of this "rich get richer" phenomenon. A smaller and smaller $D$ counteracts this effect, especially for the sharp exponential fall-off for $V(x)$ used by Rihll and Wilson, also displayed in Figure 5.5 but rejected by us for maritime networks.

We have applied the RW model to our context, using the $V(x)$ adopted before; the results are shown in Figure 5.7. The networks produced by this model tend to have starlike structures, as a small number of sites ("terminals" in the language of Rihll and Wilson, 1987, 1991) suck in most of the flow from a neighborhood and their output is directed between them. Conversely, most sites have no input, and all their output is focused on the nearest terminal site. The origins of this model in providing, among other things, a description of the replacement of corner shops by shopping malls in contemporary towns and cities is clear. We do not think this is a useful model for describing exchange in the MBA. Although we expect dominant trading centers to emerge, and exchange to be unbalanced at many sites, we would still expect to see some level of exchange to occur at all levels. Indeed, it does not matter how remote a site is, or how high its interaction costs are—in the RW model a site will always maintain a flow to the nearest terminal site. This model may have a role in archaeology if we adopt a sociopolitical rather than a socioeconomic framework, because a star is a possible representation of the domination of a local region by a single site if we think in terms of tribute rather than exchange. Thus it is perhaps better considered alongside models such as the XTent model (Bevan, 2010; Renfrew and Level, 1979).

## The Most Efficient Networks

As we mentioned earlier, an alternative approach to optimization is to adopt a social utilitarian principle and interpret best as most beneficial or most efficient (although these terms may not be synonymous). These models are common in socioeconomics and in any system of exchange in which there are identifiable costs and benefits, for which most beneficial or most efficient means achieving the greatest benefits relative to the costs. Even simple networks suggest a spectrum of benefits that encompass social storage, exogamy, acquisition of raw materials, distribution of prestige goods, cultural exchange, and trade, and it is not unreasonable to assume that networks evolve to accrue higher benefits.

### Utility Functions or Social Potentials

To each network we associate what in economics we would call a "cost/benefit" or "utility" function (Jackson, 2008) and in sociology a "social potential"

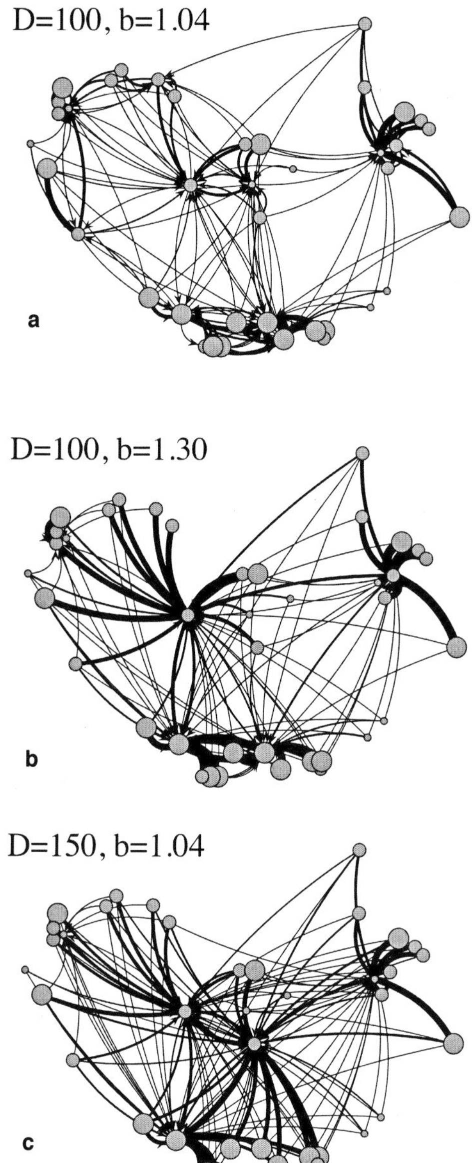


**Figure 5.7** Rihll and Wilson model for values $D = 100$ km on the top row, with $\beta = b = 1.04$ on the left and $\beta = b = 1.30$ on the right; $D = 150$ km on the second row, for $\beta = b = 1.04$. Note how the number of weak links decreases as $\beta$ increases. For each site $i$ only the outflow $O_i$ is fixed equal to the site size $S_i$ (taking values 1/3, 1/2 , or 1.0; see Table 5.1). Only edges with flow over $0.0256 \approx 1/39$ are shown, since these represent an above-average flow.

(for example, Bejan and Merkx, 2007). Because two of us are physicists, we call it a Hamiltonian $H$. Whatever it is called, it describes the costs minus the benefits of the network. If we assume that the network adjusts so as to increase the surplus of benefits over costs or increase utility, we are seeking to find the networks that minimize $H$. This notion is referred to as "strong efficiency" (see Jackson, 2008, for discussions of this and other definitions of efficiency).

At the very least, $H$ contains two types of term: that describing the benefits of exchange and local resources and that describing the cost of maintaining the resulting network. Since only relative magnitudes matter, the simplest choice is a two-term model, choosing just the direct benefits of exchange and a cost proportional to activity (for example, Jackson, 2008). This approach gives us a situation as for the RW model, with one parameter to vary. We have not been exhaustive, but, empirically, we find that such models lurch from over-connectedness (with the starlike structure of the RW model) to collapse ("boom to bust"), in which sites switch on or off without the variation that we would expect.

Neither these nor previous models pay due regard to ability of the communities to exploit the local resources, and so we have adopted a more sophisticated model in which the Hamiltonian utility function $H$, which characterizes each configuration of the system, separates into four terms:

$$H = -\kappa S - \lambda E + (jP + \mu F). \tag{5}$$

With all coefficients positive, $E$ represents the benefits of exchange and $F$ the cost of maintaining the network mentioned previously. In addition, the first term $S$ represents the benefits of local resources and $P$ the cost of maintaining the population. The parameters $\kappa$, $\lambda$, $j$, $\mu$, which control $H$, are measures of site self-sufficiency, constraints on population size, and so on. All other things being equal, increasing $\lambda$ enhances the importance of intersite interaction, whereas increasing $\kappa$ augments the importance of single-site behavior. However, increasing $j$ effectively corresponds to reducing population, and increasing $\mu$ reduces exchange.

As with the entropy models, beyond these generalized constraints, the model's inputs are the sites' fixed carrying capacities $S_i$ and the intersite "potentials" $V(d_{ij}/D)$ of Figure 5.5, again understood as a measure of the difficulty to travel from site $i$ to site $j$ in a single journey. The direct model outputs are again the flows $F_{ij}$ between sites $i$ and $j$. Unlike for gravity models the total output from a site is not fixed but is allowed to vary. First we allow the site weight (actual site size or population) $P_i$ to vary, though its behavior is linked to the fixed model input $S_i$, which now represents the fixed carrying capacity of each site. In another departure from the gravity model approach we also allow the total output flow from a site to be less than to its site weight, $0 \le \sum_j F_{ij} \le P_i$. So we may think of the combination $(F_{ij}/S_i)$ as the likelihood that an individual (or vessel) at site $i$ travels to site $j$. The advantage over the gravity model description is that if exchange between one isolated site and all others is very expensive, no interaction need occur, and that remote site can exist on its own resources and be of a reasonable size. As we have noted before, in a constrained gravity model every site will always have an output, regardless of how inefficient trade may be.

These outputs relate only to the *local* properties of the sites and links. We have also constructed outputs that reflect the effect of the network on individual sites. The most important of these is site rank, a variation of PageRank (Brin and Page, 1998), which is the basis of the algorithm used by Google™ to rank web pages. PageRank is a measure of the global flow of people/trade passing through a site, its "busyness," an attribute of how the network functions as a whole. Sites with high ranking in comparison to their site weights have high impact and are, literally, punching above their weight. (See our work elsewhere: Knappett, Evans, and Rivers, 2008, 2011.)

In detail, $S = \Sigma_i P_i(1 - P_i/S_i)$ describes the benefits of local resources—thus it is a sum of terms, one for each site, that describes the exploitation of the site as a function of its population. The detail is not crucial, as long as over-exploitation of resources incurs an increasingly nonlinear cost, whereas under-exploitation permits growth.

$E = \Sigma_{i,j} F_{ij} V(d_{ij}/D) P_j$ denotes the benefits from exchange. It is a sum of terms for every pair of sites. It takes into account the fact that direct long-distance single journeys are unlikely, if not impossible, to appear in our simulations. We expect exchange over long distances to be effected through a series of more manageable shorter steps. With the size of any flow out of site $i$, $F_{ij}$, limited by its population, it is likely to scale with $P_i$. Thus the form of $E$ is "gravitational," based on the premise that it is advantageous, in cultural exchange or trade, if both a site and its exchange partner, are large.

The final terms (in parentheses) enable us to control the total population size $P = \Sigma_i P_i$ and the total amount of exchange (and/or journeys made) $F = \Sigma_{i,j} F_{ij}$.

## Is Almost the Best Good Enough?

The act of finding the most efficient network looks highly deterministic, as in (2) and (3), but the reality should be more subtle. We are familiar in our personal lives with the experience of wanting to make the best choice, but finding very little to distinguish between several of the choices available and perhaps making a final choice with the toss of a coin. Our "efficient" models reflect this. We can think of $H$ as describing a landscape, both for our model and constrained gravity models (and interpret $H$ as minus the constrained entropy in these maximum entropy models). Each network that we can write down is a point on that landscape. What optimization does is to look for the lowest part of the landscape, its *global* minimum, the network describing that point being the "best" network. In practice, the landscape has many dimensions and is full of dips and bumps. As a result, there are several local minima offering significant improvements on our starting point but otherwise being comparably good.

As always, there are several ways to proceed. Consider the analogous optimal problem of putting a ball on our model landscape of hills, mountains, and plains and wanting to find the lowest point as the ball rolls under gravity. One possibility is to put the ball in some given initial position and then "shake" the landscape, giving the ball every incentive to roll downhill. After a while it gets trapped in some local minimum, which network we identify. We then repeat the process, either beginning from the same initial state or from a different one. Since we are trying to get as far downhill as possible, it shouldn't really matter where we begin. Final outcomes will vary even if we start from the same position each time, as we find networks that are comparably efficient, but, if the initial shaking has been good enough, they will be commensurate with the undiscovered best. However, locally they may differ significantly, and we have to interpret them statistically.[2] The implicit statistical fluctuations in the networks we construct reflect the normal variations in a real-world system. In principle this same approach could be used for gravity model solutions, which look to maximize entropy, but in practice a deterministic approach is typically used there so that the same initial conditions always lead to a unique solution.[3]

Two examples are given in Figure 5.8, in which we have taken $D = 120$ km and begin from the same initial conditions. We have chosen a set of parameter values that best seems to mimic what we see archaeologically, setting the costs of trade low with respect to its benefits, with a sufficient exploitation of resources. This combination stimulates connectivity and encourages a number of weak ties between clusters, this number increasing as the benefits of local resources increases. The outputs are the site weights (populations) $P_i$ and the link weights $F_{ij}$. In Figure 5.8, the sizes of the nodes are proportional to the former and the thicknesses of the lines to the latter. We will not display the site ranks. We merely observe that, for these networks, Thera and the states of northern Crete have among the highest exchange per capita, but do so slightly differently in the two cases. As for our original concerns regarding Knossos, statistically Knossos is capable of being among the most important sites, perhaps the most important in some histories.

If we were to set $D = 70$ km in our model for the same parameter values as in Figure 5.8 it becomes almost indistinguishable from the $D = 70$ km figure of our original geographic model of Figure 5.3 and totally unlike the contortions required by our constrained entropy models.

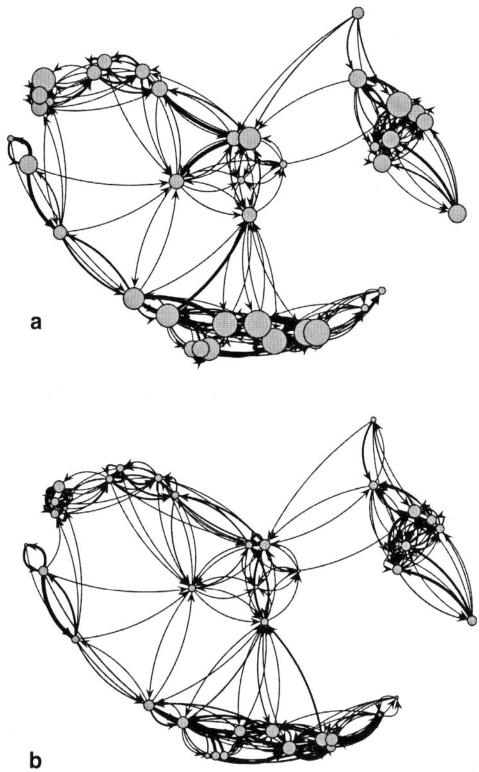


**Figure 5.8** Two runs of our model for values $D = 100$ km with identical input parameters ($\lambda = 3$, $\kappa = 1$, $\mu = 0.1$, $j = -2.0$). Akrotiri and Knossos are important in each. The size of vertex is proportional to the site weight $P_i$, with an edge shown if the flow along that edge, $F_{ij}$, is greater than 0.1.

## Does It Work?

As we have said, with the data as poor as they are, we might expect to find so many choices of parameters (that is, a large part of "parameter space") to be commensurate with the data that it would not be clear how to proceed.

This is where optimization comes to our rescue, for which the Hamiltonians are nonlinear. A nonlinear system is one in which, on prodding it, the response is not proportional to the strength of the prod. For our efficient networks the nonlinearities are the conventional ones that arise when the disproportionate benefits that large sites accrue from interacting with other large sites (the "gravitational" benefit) are swamped by the disproportionate costs of shortages due to high population. In the Rihll and Wilson models the nonlinearities lie in the constraints on the entropy, manifest in the nonlinear benefits of site "attraction."

Let us return to the efficient networks of Figure 5.8. Once the nature of the terms in (5) is given it looks as if, through the relative values of their coefficients, that we have added three new parameters to the geographic inputs. In practice this is not quite the case, given some trade-off between varying the total population, varying total exchange activity, and varying the relative strength of the benefits of exchange in comparison to the benefits of exploiting local resources. When we take these into account, our model turns out to be very sensitive to the parameter values (having fixed $D$), giving a picture of only healthy networks for very limited choices of them as we tread a fine line between boom and bust. This picture is commensurate with the observation by Broodbank, Kiriatzi, and Rutter:

> For the southern Aegean islands in the late Second and Third Palace periods, an age of intensifying trans-Mediterranean linkage and expanding political units, there may often have been precariously little middle ground to hold between the two poles of (i) high profile connectivity, wealth and population, or (ii) an obscurity and relative poverty in terms of population and access to wealth that did not carry with it even the compensation of safety from external groups. (2005: 95)

Empirically, this balance is a first sign that our model works—not primarily because we are looking for societal collapse, although that seems to happen all too easily, but because models incorporating instability tend to lead to settlements with a wide variety of population. If we change their form to make them structurally more stable it becomes increasingly difficult to generate a wide enough range of site sizes (populations) to match the record. More details are given elsewhere (Knappett, Evans, and Rivers, 2008, 2011).

This narrow path between boom and bust plays an important role in the evolution of exchange networks. As to how the networks evolve in time, systems evolve for a variety of reasons. The models we have considered here are not subtle enough to bootstrap themselves, and we have to enforce change exogenously, usually by varying the parameters smoothly, distorting the landscape. Thus, for example, as populations grow or total trade volume increases, the optimal network (lowest energy configuration) changes. It is not surprising that the model shows tipping points as valley bottoms rise and new valleys are formed. Boom and bust is a feature of nonlinear systems. Again, it can be useful to adopt the language of statistical ensembles, wherein the notion of a *phase transition* is familiar and these collapses constitute just such a change (Wilson, 2008; Wilson and Dearden, 2011).

In an empirical sense, what seems to keep the model networks together is its weak links. Such links, for which the exchange of goods and people is small, leave little or no archaeological trace, but it was proposed by Granovetter (1973, 1983) many years ago in a seminal paper that they play an important role in social networks for facilitating the exchange of information and for facilitating innovation. This proposition has been continually examined since and borne its weight. More generally, it has been argued (Csermely, 2004: 332) that "weak links stabilize complex systems," of which networks are but one type. There are different definitions of stability, but there are certainly many examples where the presence of many weak links does indeed aid stability (Csermely, 2004), including our models. If, for example, beginning from a typical network of Figure 5.9, we increase the costs of trade we find that the network puts its eggs in fewer baskets. The weak links are unrewarding to

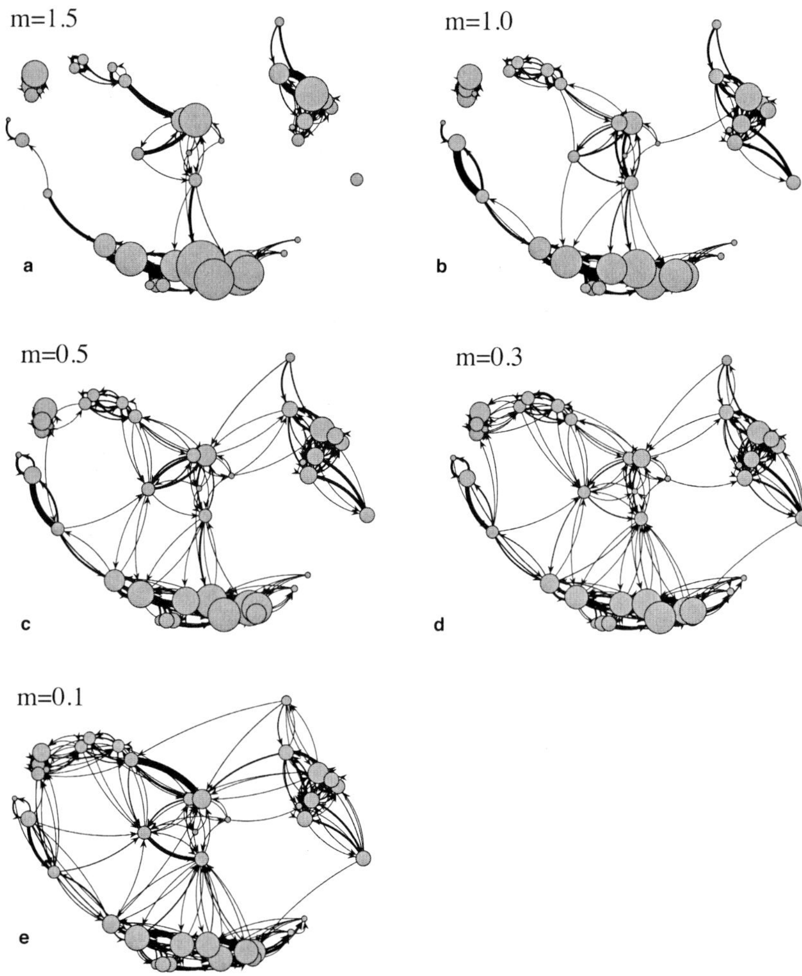


**Figure 5.9** The effect of increasing the cost of an edge in our model. From left to right, then top to bottom, the parameter $\mu$ is raised from 0.1, to 0.3, 0.5, 1.0, and finally 1.5. For $D = 120$ km ($\lambda = 3$, $\kappa = 1$, $j = -2.0$).

maintain, and trading becomes confined to fewer and fewer networks until it collapses. This is one way to understand the short-term stability of the southern Aegean network after the eruption of Thera before its ultimate collapse (Knappett, Evans, and Rivers, 2011).

We should not assume that instability is inevitable as the networks evolve. There is still enough room in the parameter space to stay in an area of stability and evolve with gentle growth or decline, but the opportunity both for collapse and rapid growth (equally unstable) is always there. It is in circumstances like this that social space and geographical space come together before going violently apart. The benign networks of Figure 5.8 in some sense go beyond the geography while being strongly conditioned by it, in giving details that we could not have predicted (albeit stochastically) with some outriders punching above their weight. However, regional geography reinstates itself in primitive form as we move toward instability, almost along the lines of our null models, with only a few strong links between collapsing regional clusters.

## Summary

We began this chapter with a general question: does the southern Aegean exchange network reflect the geographical space in which it is embedded? We have argued how conditional this answer is on MBA maritime technology, which makes the southern Aegean a Goldilocks environment for the appearance of trading networks, with single-journey distance correlated to the scale separating the main regional clusterings.

The bulk of our analysis has been devoted to a discussion of optimal networks, with optimal meaning most likely or most efficient. The former term, used in town planning and transport networks, adopts a microcanonical approach in arguing for a maximization of constrained network entropy, essentially information. It seems, in general, that the constraints in terms of inflows and outflows on a site-by-site basis that are conventionally imposed are too strong to permit networks that could match the record. Reasons why they will fail for some maritime networks include the highly implausible result that imposing constraints on a site-by-site basis requires sites to connect directly willy-nilly, even to sites that are inaccessible by simple sea travel. In this regard the models have something in common with PPA, which argues for a maximum number of connections, rather a minimum separation. (In addition, there is a tendency for sites to become important by becoming recipients of exchange, with little or no reciprocity.)

Most efficient or cost/benefit models are used widely in socioeconomics and avoid constraints on a site-by-site basis by adopting a canonical ensemble approach. As a result, we get networks that, in general, are strongly conditioned by geography. However, a characteristic of the optimal models we have discussed here is that they are prone to instabilities, and the links to geographic space make themselves starkly evident before they break down if we begin to walk off the delicate line between boom and bust that they embody.

We conclude with two points that we have not discussed so far. The first is that gravitational models help minimize the effects of our ignorance about site details because of a patchy record. These deserve a much greater discussion than space permits here, and we refer the reader to Evans, Knappett, and Rivers (2009), where it is analyzed for efficient networks in greater detail. Suffice it to say that Newtonian gravity is remarkable in that breaking up the masses into smaller parts does not change the result as long as the position of the centers of mass is unchanged. Our gravity models go some way in providing an island counterpart to this, in which the contribution of the communities on an island to the utility function largely depends only on its total population. With carrying capacity as given input and population as output, how this population is distributed is largely immaterial.

This insensitivity, coupled with the ensemble approach, leads to a way of considering the smallest scale of the social hierarchy, the household, or individual, which is complementary to that of Agent Based Modeling, as used in Chapters 6 (Premo, 2013) and 7 (Clark, 2013). This complementarity is familiar in describing such physical systems as gases, say, where we have the option to work with the individual gas atoms in a statistical way (statistical mechanics) or through the behavior of bulk properties of the system, such as pressure and free energy (thermodynamics). In this language our approach is thermodynamical and motivated the Boltzmann distribution quoted earlier.

## Notes

1. The notion of a Goldilocks economic environment, one that in terms of exchange is not too hot and not too cold, is ubiquitous in macroeconomics (for instance, see Gordon, 1998), where it was first coined to describe the U.S. economy in the late 1980s. It has a parallel usage in astrophysics, originally literally, for describing habitable planets but now more generally, in the context of anthropic approaches to the physical constants of the natural world.
2. Technically we adopt the standard procedure of using a Boltzmann distribution to assign to each network $G$ a probability $p(G).\exp(-\beta H(G))$ where $\beta$ is a large constant, the inverse volatility.
3. For those readers familiar with statistical mechanics, $j$ and $\mu$ are chemical potentials, and, in our cost/benefit analysis we are working with a Grand Canonical ensemble (with volatility playing the role of temperature). However, maximum entropy models correspond to working with microcanonical ensembles. We have already discussed the advantage of canonical ensembles over microcanonical ensembles in that, for the former, only the network-wide total of exchange/trade is fixed, and we do not have to impose inflows or outflows on a site-by-site basis, for which we don't have the knowledge anyway (and for which we chose as proportional to the carrying capacity by default). As a result, we do not find ourselves in

the peculiar position of having to enforce long single journeys over unreasonable distances to balance flows, as happens in Figure 5.7. The other global constraint, on population, has no counterpart in entropy models.

## References

Alden, J. R. 1979. Reconstruction of Toltec period political units in the Valley of Mexico, in Renfrew, C., and Cooke, K. (Eds.), *Transformations: Mathematical Approaches to Culture Change*, pp. 169–200. New York: Academic Press.

Ball, P. 2004. *Critical Mass: How One Thing Leads to Another.* London: Heinemann.

Barton, C. 2013. Stories of the past or science of the future? Archaeology and computational social science, in Bevan, A., and Lake, M. (Eds.), *Computational Approaches to Archaeological Spaces*. Walnut Creek, CA: Left Coast Press.

Batty, M. 2005. Network geography: Relations, interactions, scaling, and spatial processes in GIS, in Fisher, P., and Unwin, D., *Re-presenting GIS*, pp. 149–70. Chichester: John Wiley.

———. 2010. Space, scale, and scaling in entropy-maximising, *CASA Working Papers 154*. London: Centre for Advanced Spatial Analysis (UCL).

Bejan, A., and Merkx, G. W. 2007. *Constructal Theory of Social Dynamics*. New York: Springer.

Bevan, A. 2010. Political geography and palatial Crete, *Journal of Mediterranean Archaeology* 23(1): 27–54.

Bevan, A., Crema, E., Li, X., and Palmisano, A. 2013. Intensities, interaction and uncertainties: Some new approaches to archaeological distributions, in Bevan, A., and Lake, M. (Eds.), *Computational Approaches to Archaeological Spaces*. Walnut Creek, CA: Left Coast Press.

Brin, S., and Page, L. 1998. The anatomy of a large-scale hypertextual web search engine, *Proceedings of the Seventh International Conference on World Wide Web*, 7: 107–17.

Broodbank, C. 2000. *An Island Archaeology of the Early Cyclades*. Cambridge: Cambridge University Press.

Broodbank, C., Kiriatzi, E., and Rutter, J. B. 2005. From Pharaoh's feet to the slave-women of Pylos? The history and cultural dynamics of Kythera in the Third Palace Period, in Dakouri-Hild, A., and Sherratt, E. S. (Eds.), *Autochthon: Studies Presented to Oliver Dickinson on the Occasion of His Retirement*, pp. 70–96. British Archaeological Reports International Series 1432 series. Oxford: Archaeopress.

Clark, T. 2013. An introduction to integrative distance analysis (IDA), in Bevan, A., and Lake, M. (Eds.), *Computational Approaches to Archaeological Spaces*. Walnut Creek, CA: Left Coast Press.

Collar, A. 2007. Network theory and religious innovation, *Mediterranean Historical Review* 22: 149–62.

Csermely, P. 2004. Strong links are important, but weak links stabilize them, *Trends in Biochemical Science* 29: 331–34.

Dunbar, R. I. M. 1992. Neocortex size as a constraint on group size in primates, *Journal of Human Evolution* 22: 469–93.

Erlander, S., and Stewart, N. F. 1990. *The Gravity Model in Transportation Analysis: Theory and Extensions*. Utrecht: Brill.

Evans, T., Knappett, C., and Rivers, R. 2009. Using statistical physics to understand relational space: A case study from Mediterranean prehistory, in Lane, D., Pumain, D., van der Leeuw, S., and West, G. (Eds.), *Complexity Perspectives on Innovation and Social Change*, pp. 451–79. Berlin: Springer Methods Series, Vol. 7.

Evans, T., Rivers, R., and Knappett, C. 2012a. Interactions in space for archaeological models, *Advances in Complex Systems* 15: 1150009.

———. 2012b. Thirty-nine Minoan sites, data files, doi: 10.6084/m9.figshare.97395

———. 2012c. ariadne, program files, doi: 10.6084/m9.figshare.97746

Gordon, R. 1998. Foundations of the Goldilocks economy: Supply shocks and the time-varying NAIRU, *Brookings Papers on Economic Activity* 2: 297.

Granovetter, M. S. 1973. The strength of weak ties, *American Journal of Sociology* 78(6): 1360–80.

———. 1983. The strength of weak ties: A network theory revisited, *Sociological Theory* 1: 201–33.

Hage, P., and Harary, F. 1991. *Exchange in Oceania: A GraphTheoretic Analysis*. Oxford: Clarendon Press.

———. 1996. *Island Networks: Communication, Kinship and Classification Structures in Oceania*. Cambridge: Cambridge University Press.

Herzog, I. 2013. The potential and limits of optimal path analysis, in Bevan, A., and Lake, M. (Eds.), *Computational Approaches to Archaeological Spaces*. Walnut Creek, CA: Left Coast Press.

Irwin, G. 1983. Chieftainship, Kula, and trade in Massim prehistory, in Leach, J. W., and Leach, E. (Eds.), *The Kula: New Perspectives on Massim Exchange*, pp. 29–72. Cambridge: Cambridge University Press.

Jackson, M. O. 2008. *Social and Economic Networks*. Princeton, NJ: Princeton University Press.

Jensen-Butler, C. 1972. Gravity models as planning tools: A review of theoretical and operational problems, *Geografiska Annaler. Series B, Human Geography* 54: 68–78.

Knappett, C. 2011. *An Archaeology of Interaction: Network Perspectives on Material Culture and Society*. Oxford: Oxford University Press.

Knappett, C., Evans, T., and Rivers, R. 2008. Modelling maritime interaction in the Aegean Bronze Age, *Antiquity* 82: 1009–24.

———. 2011. Modelling maritime interaction in the Aegean Bronze Age, II: The eruption of Thera and the burning of the palaces, *Antiquity* 85: 1008–23.

Premo, L. S. 2013. Multilevel selection and the evolution of food sharing in fragmented environments: A spatially explicit model and its implications for Early Stone Age archaeology, in Bevan, A., and Lake, M. (Eds.), *Computational Approaches to Archaeological Spaces*. Walnut Creek, CA: Left Coast Press.

R Development Core Team. 2010. R: A language and environment for statistical computing. R Foundation for Statistical Computing, Vienna, Austria. ISBN 3-900051-07-0. URL: www.R-project.org

Renfrew, C. 1981. Space, time, and man, *Transactions of the Institute of British Geographers, New Series* 6(3): 257–78.

Renfrew, C., and Cooke, K. L. (Eds.). 1979. *Transformations: Mathematical Approaches to Culture Change*. London: Academic Press.

Renfrew, C., and Level, E. V. 1979. Exploring dominance: Predicting polities from centres, in Renfrew, C., and Cooke, K. L. (Eds.), *Transformations: Mathematical Approaches to Culture Change*, pp. 145–67. London: Academic Press.

Rihll, T. E., and Wilson, A. G. 1987. Spatial interaction and structural models in historical analysis: Some possibilities and an example, *Histoire & Mesure* 2: 5–32.

________. 1991. Modelling settlement structures in Ancient Greece: New approaches to the Polis: I, in Rich, J., and Wallace-Hadrill, A. (Eds.), *City and Country in the Ancient World*, pp. 59–95. London: Routledge.

Smith, A. T. 2003. *The Political Landscape: Constellations of Authority in Early Complex Polities*. Berkeley and Los Angeles: University of California Press.

Terrell, J. 1977. *Human Biogeography in the Solomon Islands*. Chicago: Field Museum of Natural History.

________. 2010. Language and material culture on the Sepik Coast of Papua New Guinea: Using social network analysis to simulate, graph, identify, and analyze social and cultural boundaries between communities, *The Journal of Island and Coastal Archaeology* 5: 3–32.

Wilson, A. G. 1967. A statistical theory of spatial distribution models, *Transportation Research* 1: 253–69.

________. 1970. *Entropy in Urban and Regional Modelling*. London: Pion.

________. 2008. Boltzmann, Lotka, and Volterra and spatial structural evolution: An integrated methodology for some dynamical systems, *Journal of the Royal Society Interface* 5: 865–71.

________. 2008. Phase transitions in urban evolution, *CASA Working Papers 141*. London: Centre for Advanced Spatial Analysis (UCL).

Wilson, A. G., and Dearden, J. 2011. Phase transitions and path dependence in urban evolution, *Journal of Geographical Systems* 13: 1–16.

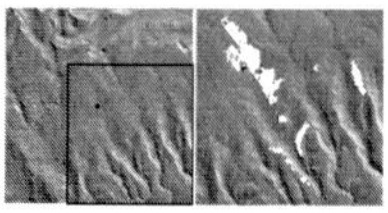

# 6 Multilevel Selection and the Evolution of Food Sharing in Fragmented Environments: A Spatially Explicit Model and Its Implications for Early Stone Age Behavior

*L. S. Premo*

Anthropological research on socioeconomic cooperation (food sharing, division of labor, and so on) frequently employs the assumption that humans act in ways that may appear altruistic but ultimately benefit individual and/or inclusive fitness. Models of mechanism that emphasize individual-level selection have strongly influenced the human behavioral ecological models of circumstance that seek to explain the presence of food sharing in small-scale hunter-gatherer societies. Questions of when and why hominins began sharing food have long been of interest to researchers studying the Late Pliocene and Early Pleistocene fossil and archaeological records in East Africa. Paleolithic archaeologists have routinely drawn on the models of circumstance influenced by individual-level selection, such as tolerated theft, risk reduction reciprocity, showing-off, and costly signaling, when inferring hominin behavior from the archaeological record of the Early Stone Age (for example, Isaac, 1978; Leakey, 1971; O'Connell et al., 2002; Potts, 1982, 1984, 1988, 1991; Rose and Marshall, 1996; Schick, 1987). However, the notion that altruists benefit from their actions does not square with the biological definition of altruism, which states that "a behavior is altruistic when it increases the fitness of others and decreases the fitness of the actor" (Sober and Wilson, 1998: 17). And yet to date relatively few researchers have employed multilevel selection theory to

*Computational Approaches to Archaeological Spaces* by Andrew Bevan and Mark Lake, Eds., 127–150 

investigate the evolution of altruism in hominins (but see Boehm, 1996, 1999; Bowles, 2006, 2009; Choi and Bowles, 2007; Premo, 2005, 2006; Richerson and Boyd, 2005; Wilson, 1998) even though it seems particularly well suited for investigating the evolution of altruism in the very kinds of geographically structured populations that likely characterized much of hominin prehistory (Bowles, 2009; Bowles and Gintis, 2011; Premo, 2006; Wilson, 1998).

Here I consider the evolution of food sharing in hominins from a multilevel selection perspective. I begin with a brief review of the models of mechanism that have been proposed to explain the evolution of altruism, including kin selection, reciprocal altruism, and multilevel selection. Next I introduce some of the models of circumstance that have made use of these mechanisms. Finally I present and discuss data collected from a spatially explicit evolutionary agent-based model designed to investigate the effect of geographically fragmented woodlands resources on the evolution of altruistic food sharing during the Pliocene and Pleistocene in East Africa.

## Models of Mechanism for the Evolution of Altruism

Charles Darwin stressed that natural selection operates at the level of the individual: "Natural selection will never produce in a being anything injurious to itself, for natural selection acts solely by and for the good of each" (1859: 201). Yet he also noted that many animals—from insects to humans—are puzzlingly altruistic; that is, individuals often act in ways that benefit another's fitness at some cost to their own. By speculating that self-sacrificial traits, which are deleterious at the individual level, could be beneficial at the level of an insect "community" (1859: 202, 236) or a human "tribe" (1871: 166) Darwin recognized the possibility that *groups* could play an important evolutionary role. He also noted that humans act altruistically toward those within their group more regularly than toward those outside their group (Darwin, 1871: 87), hinting that relatedness or maybe even culturally mediated population structure might play an important role in the evolution of altruism in humans.

Ronald A. Fisher, J. B. S. Haldane, and Sewall Wright recognized the significance of Darwin's speculations concerning the possibility that altruism might evolve owing to group-level benefits. Armed with an understanding of Mendelian inheritance and a more sophisticated population genetics toolkit, they were able to explore Darwin's musings with analytical models. Ultimately each separately concluded that altruistic traits could not be explained by selection within a single population. They argued that the evolution of altruism requires some sort of intergroup selection (Fisher, 1930; Haldane, 1932; Wright, 1945). Of the three, Sewall Wright (1945) provided the more complete mathematical model of the evolution of altruism. Over a decade later, George C. Williams and Doris Williams (1957) modified Wright's fitness equations to address the evolution of altruistic traits in populations structured by sibgroups. In their model, each individual spends a portion of its life in a small group with its siblings, isolated from individuals in other sibgroups, before melding back into one large gene pool for reproduction. Williams and Williams's model was the first complete mathematical model of the evolution of altruism via differential group fitness. It shows that sibgroups with a larger proportion of altruists contribute a proportionally greater number of healthy young adults to the general gene pool. G. C. Williams's (1966) famous rebuttal of Wynne-Edwards's (1962) brand of "naive group selection" made the case that selection does not favor individual altruists simply because they provide group-level benefits. Williams correctly points out that within-group selection does not favor an individual who sacrifices something of his or her fitness for the good of the group, because the altruist will always have a lower individual fitness than those in the same group who are more selfish. Could it be that seemingly "altruistic" traits such as alarm calling, feeding restraint, and food sharing evolved via individual selection and that individuals who appear to be acting altruistically actually enjoy hidden fitness benefits?

William Hamilton (1963) was among the first to develop an analytical model of the evolution of altruism that did not include intergroup selection. He argued that regardless of whether a trait is beneficial or deleterious to the *individual*, it would spread as long as it ultimately benefits copies of the *gene(s)* responsible for its display. To maximize one's inclusive fitness (formally defined in Hamilton, 1964a, 1964b) one should act altruistically only toward those who are likely to share copies of one's genes. Because one's close relatives are more likely than distantly related individuals to possess the same genes, a good heuristic is to act altruistically toward close kin. John Maynard Smith (1964) christened Hamilton's inclusive fitness model as "kin selection."

Hamilton's model was extremely influential and useful for explaining cooperation in many animal species (for a recent critique, see Nowak, Tarnita, and Wilson, 2010). But the ethnographic record, let alone everyday experience, is replete with data suggesting that humans regularly act altruistically toward nonrelated individuals. In 1971 Robert Trivers tackled the issue of the evolution of altruism among non-kin. He proposed that seemingly altruistic behaviors could evolve in a population of unrelated individuals when two conditions are met: (1) an altruistic act's benefit to the recipient exceeds its cost to the donor, and (2) any recipient of an altruistic act habitually reciprocates the favor to the donor at some time in the future. Hence individuals should act altruistically not just toward close kin but also toward those from whom they will receive the benefits of an equivalent behavior in the future as well as toward those from whom they had received benefits in the past. Trivers introduced *direct* reciprocity, whereby recipients eventually pay back the benefits to the donor directly. Richard Alexander (1979, 1987) later introduced *indirect* reciprocity. For "altruism" to evolve via indirect reciprocity, individuals should act altruistically toward those that are likely to act altruistically toward others, though not necessarily the original donor. It should be clear that Hamilton's inclusive fitness model (kin selection) and (direct and indirect) reciprocal altruism were quite intentionally constructed to, as Trivers (1971: 35) put it, "take the altruism out of altruism."

But Hamilton, Trivers, and Alexander were not the only ones doing innovative work on the evolution of altruism in the wake of Williams's (1966) influential book. George Price's (1970, 1972, 1995) revolutionary model harkens back to the models of differential group fitness by Wright, and Williams and Williams. In Price's model, henceforth referred to as the Price Equation, change in the allele frequency in a structured population (that is, a metapopulation) is calculated as the sum of two terms: the average evolutionary change *within* groups and the average evolutionary change caused by differential fitness *between* groups. The Price Equation provides an elegant tool for quantifying the relative strengths of within-group and between-group selective pressures in any structured population. Thanks to Price, Darwin's seemingly contradictory speculation that natural selection might favor the spread of altruistic traits in a structured population despite selecting against them at the level of the individual finally had an intuitive mathematical expression. Hamilton (1975) applied Price's work to the issue of altruism in humans. He was quick to recognize that, because common descent is but one mechanism by which phenotypes can be positively assorted within groups in a metapopulation, Price had provided a model for the evolution of altruism that was more general than kin selection. Hamilton had previously written that group selection should be "treated with reserve so long as it remains unsupported by mathematical models" (1963: 354). Hamilton apparently found the support he needed in the Price Equation, because he redefined kin selection as a subclass of group selection (Hamilton, 1975: 141).

Williams's (1966) case against Wynne-Edwards's application of naive group selection to population self-regulation had such a strong and long-lasting effect on evolutionary biologists that relatively few paid serious attention to Price's mathematics and Hamilton's revisions at the time. The few who did maintained that some utility remained in the original notion of group selection as articulated by Darwin and others long before Wynne-Edwards. David Sloan Wilson (1975, 1977, 1983, 1987) led a small but persistent cadre of researchers who continued to work on what is now known as multilevel selection theory. Multilevel selection theory is built on the idea that natural selection concurrently operates at two levels of the biological hierarchy: *within* groups and *between* groups (Sober and Wilson, 1998; Wilson and Sober, 1994). Within-group selection promotes phenotypic traits that allow one to maximize one's individual fitness, regardless of how those traits affect the fitness of others. Between-group selection promotes phenotypes that are beneficial to the fitness of others while costly to that of the donor. The relative strengths of these opposing selective forces are what ultimately determine which allele becomes more common at the metapopulation level. The Price Equation is useful because it measures the relative strengths of these two components of selection separately. Among other things, the Price Equation makes it easier to see when between-group selective forces outweigh their within-group counterparts, which is generally thought to occur but rarely in nature.

Multilevel selection theory employs the notion of *trait group* (Wilson, 1975). A trait group is defined as "a set of individuals that influence each other's fitness with respect to a certain trait but not the fitness of those outside the group" (see Sober and Wilson, 1998: 92). When between-group selection is stronger than within-group selection, trait groups with greater proportions of altruists have a higher fitness than trait groups with lesser proportions of altruists. As a result, trait groups characterized by a greater proportion of altruists yield a greater number of offspring than trait groups with fewer altruists, thereby slightly increasing the proportion of altruists in the metapopulation even though within-group selection reduces the proportion of altruists within all groups (see Sober and Wilson, 1998: Figure 1.1 and Box 1.2). Multilevel selection theory applies to trait groups the same three components of natural selection—genetic variation, heritability, and differential reproductive success—that Darwin (1859) applied to individuals (Wilson, 1998). Multilevel selection models demonstrate how altruistic traits can increase in frequency at the metapopulation level despite decreasing in frequency within each trait group (for more, see Wilson, D. S., 2002; Wilson, E. O., 2012; Wilson and Wilson, 2007, 2008).

## Food-Sharing Models of Circumstance

Human behavioral ecologists have made use of various *models of circumstance* to explain why altruistic traits such as food sharing are prevalent in recent hunter-gatherer societies. Such models of circumstance describe the social and ecological conditions that may facilitate the spread and/or maintenance of an altruistic trait such as food sharing (Winterhalder, 1996, 1997). Human behavioral ecology began during a time when invoking selection at any level above the individual was considered anathema by most. As a result, many models of circumstance concerning food sharing invoke within-group selection only. In short, they are built around the assumption that for a seemingly "altruistic" trait to be present in human societies, it must provide some benefit to the donor.

Nicholas Blurton Jones argued that when explaining social behavior one must "rely only on individual interests and not on the long-range benefits to the group" (1987: 32). He proposed that the earliest forms of food sharing might be better conceptualized as "tolerated theft" (Blurton Jones, 1984, 1987). From the successful hunter's point of view, units of meat beyond the point of diminishing returns are not worth risking possible injury at the hands of those who have much to gain by obtaining what are—from their perspective—highly valuable units of food. Blurton Jones's (1987) influential model proposes that "altruistic" individuals tolerate the theft of diminishing returns units of food in order to selfishly avoid unnecessary physical confrontations with those for whom the very same units of food hold greater value. Though it may appear altruistic to an observer, tolerated theft is favored by individual-level (that is, within-group) selection.

In many hunter-gatherer societies, individual hunters fail to bring home game far more often than they succeed. Hillard Kaplan and Kim Hill (1985) proposed that reciprocal food sharing might help reduce the risk associated with failing to procure sufficient resources for days or even weeks at a time. According to the risk reduction reciprocity model, successful hunters should decide to act altruistically only toward those they believe will reciprocate the act when the roles are reversed as well as toward those that previously shared food with them. In a group of "altruists," hunters reduce the risk they each face by reciprocally sharing unpredictable resources. Kaplan and Hill (1985) described food sharing (among the Aché, at least) as an adaptation that increases individual fitness and is favored by within-group selection.

To address the types of intragroup food sharing that she believed tolerated theft and risk-reducing reciprocity could not explain, Kristin Hawkes invoked the evolutionary mechanism of sexual selection in the show-off hypothesis (Hawkes, 1990, 1991). Intersexual selection favors traits that increase an individual's likelihood of being chosen as a mate by members of the opposite sex. According to the show-off hypothesis, the male strategy of sharing highly variable, widely distributable resources with females can be favored by sexual selection. Females enhance their individual fitness by providing "show-off" males with chances at paternity in order to secure resources that are otherwise unattainable. Here, food sharing is not used to avoid fights or to reduce risk, but to mediate conflicting reproductive strategies in such a way that individuals from both sexes are able to increase their inclusive fitness (Hawkes, 1991).

Costly signaling theory holds that a trait that is costly to an individual can transmit a signal of the individual's fitness (Grafen, 1990; Zahavi, 1975). When communicated successfully to a potential mate or ally, costly signals can result in fitness benefits to both the sender and the receiver. Such a signaling system can be evolutionary stable if variation in the signal is strongly correlated with variation in the underlying quality it represents. Signals remain "honest" if they are costly and, thus, difficult to fake. Like the show-off hypothesis, costly signaling involves intersexual selection. But, unlike the show-off hypothesis, females do not mate with males in direct exchange for portions of meat, per se. Rather, females use a male's ability to obtain meat as a phenotypic indicator of his genetic quality. Assuming food sharing qualifies as a costly signal, females can make use of the information encoded in a food-sharing event to acquire greater fitness benefits than what is provided by the food itself (Bird, 1999; Boone, 1998; Smith and Bliege Bird, 2000; Sosis, 2000). From the perspective of the donor (sender), one generously shares food only in the self-interest of building an attractive status and accumulating social capital, both of which in theory can be used to increase one's individual fitness via enhanced mating opportunities and beneficial coalitions with other group members (Bliege Bird, Smith, and Bird, 2001; Gurven et al., 2000).

Interesting insights have been gained by applying these models of circumstance to hunter-gatherer data. However, the individual-level selection perspective to which they all subscribe is restrictive in light of our current understanding of multilevel selection theory. We now know that the benefits bestowed on levels above the individual—specifically, at the level of trait groups—can play a crucial role in explaining the presence of altruistic traits in structured populations. What is more, explanatory models that make use of multilevel selection allow altruistic traits to retain the characteristic that defines them: self-sacrifice. New models of circumstance based on multilevel selection hold great potential for reworking, refocusing, and reenergizing behavioral ecological research on the evolution of biologically altruistic traits in hominin societies.

## Modeling Multilevel Selection and Hominin Food Sharing

*The human adaptation on which all else depended was sharing, cooperation.*
—Lancaster (1978: 89)

Previous theoretical work with an agent-based model called SHARE (Simulated Hominin Altruism Research Environment) showed that, when starting from equally mixed starting populations of altruists and egoists, relatively simple altruistic food-sharing strategies were more likely to evolve to fixation within the so-called *transitional zone* of ecological patchiness (see Premo, 2005, 2006, 2007, 2008). Geographically fragmented (or "clumped") distributions of reliable food resources can structure an otherwise freely mixing population of foragers. Under so-called transitional levels of ecological patchiness, this population structure facilitates positively assorted interactions (that is, altruists interact with other altruists at a higher rate than one would expect in a randomly mixing population) while allowing for successful groups to export some of their offspring to nearby resource patches. As has been shown elsewhere, positive assortment enhances the strength of between-group selection, depresses the strength of within-group selection, and provides an environment in which altruistic traits can be favored via the differential contributions of trait groups to the metapopulation (Wilson, 1977; Wilson and Dugatkin, 1997). Previous work with SHARE also showed that the selective influence of ecological patchiness was inversely related to the sophistication of the food-sharing strategy modeled. When altruists were given perfect knowledge of others' food-sharing tendencies (that is, whether the individual asking for food was altruistic or selfish), food sharing spread to fixation more frequently regardless of ecological patchiness (Premo, 2006).

However, because these previous simulations initialized the starting populations of foragers with equal proportions of altruists and egoists, the results summarized previously cannot be used to address the so-called problem of origination: how can altruistic food sharing establish a toehold in a population composed entirely of selfish nonsharers? In the remainder of this chapter I address the problem of origination in food sharing from a multilevel selection perspective. I model altruistic food sharing first as a discrete trait and then as a continuous (polygenic) trait. The results of these simulation experiments may have interesting implications for behavioral interpretations of Early Stone Age archaeological data in East Africa.

## The Model

The model reported in this chapter was programmed in NetLogo (Wilensky, 1999), a free, user-friendly, and increasingly powerful platform for programming and executing spatially explicit agent-based models. (The commented source code is freely available on request from the author.) Table 6.1 provides the values used for the parameters discussed in the following sections.

**Table 6.1** Standard and Experimental Variable Values

| *Parameter* | *Value(s)* |
|---|---|
| *Standard Variables* | |
| Woodlands patch size | $6 \times 6$ |
| Total cells with woodlands resources | 5,184 |
| Forager metabolic rate (energy units per step) | 2 |
| Forager fertility threshold (energy units) | 100 |
| Forager cost of reproduction (energy units) | 50 |
| Forager maximum (energy units) | 110 |
| Forager food share threshold (energy units) | 50 |
| Forager death probability (per time-step) | 0.0001[a] |
| Probability of sharing food ($s$) | |
| (when modeled as a discrete trait) | $s = 0$ (selfish), or $s = 1$ (altruistic) |
| (when modeled as a continuous trait) | $0 \leq s \leq 1$ |
| Mutation rate ($\mu$) | 0.01 |
| Mutation size (continuous trait only) | 0.2[b] |
| *Experimental Variables* | |
| Gap size ($g$) | 2, 4, 6, 8, 10 |
| Food-sharing trait type | Discrete, Continuous |
| Food-sharing strategies | Simple, Indirect reciprocity |
| Offspring dispersal | Local, Random |

[a]This is the probability that a forager will die when in a patch of woodlands; this value is 3 times greater when the forager is between patches.
[b]This value provides the standard deviation of the normal distribution (mean = 0) from which each mutation size is drawn randomly.

In the current model, an artificial population of hominin forager agents populates a grid that is wrapped into a torus to avoid edge effects. This grid represents a rather simple environment. The environment contains two types of resources that foragers can procure for food: closed woodlands resources and carcasses. Woodlands resources are systematically distributed into regularly shaped and regularly spaced "patches" that represent local concentrations of closed habitat. Each patch of closed habitat contains 36 cells of woodlands resources. Woodlands resources grow logistically at a rate of 0.2 per time-step and asymptote at a maximum value of 10 energy units. Patches of closed habitat are separated by "gaps" of open habitat. Open habitat does not contain spatially and temporally reliable woodlands resources. The size of the gaps ($g$) between woodlands patches is positively correlated with level of ecological patchiness in dependable resources—holding patch size constant, ecological patchiness increases with $g$. The total number of cells containing woodlands resources is held constant over all runs. Carcasses provide a less reliable resource because they appear randomly over space and through time with a constant probability of 0.0001 per cell per time-step. Although carcasses can provide as many as 100 units of energy when they first appear, their energy level decays logistically at a rate of 0.02 per time-step. Carcasses are removed from the environment when their energy level drops below 10.

In the model, foragers use local search rules to find food. Forager agents can perceive and communicate only with agents located on cells within their Moore neighborhood (the forager's current cell plus the 8 cells surrounding it). If at least one of the surrounding 8 cells has an energy value (woodlands resource energy + carcass energy) that is greater than or equal to the metabolic cost of a single time-step, then the forager will move to the surrounding cell that contains the highest value of energy, ties broken randomly. Once in that cell, the forager procures 90% of the energy present, adding that amount to its personal energy budget. If sufficient energy is not available in any of the 8 surrounding cells, then the forager simply moves to one of those cells chosen at random. Regardless of the amount of food found, the metabolic cost of searching for food and moving to a new cell is 2 energy units per time-step. If a forager metabolizes the last of its energy while searching for food, then it is removed from the simulation. To avoid dying from starvation, a forager begins asking for help from others as soon as its energy level falls below 50 units. A forager in need can ask for help only once per time-step and only from a randomly chosen forager from within its Moore neighborhood. Forager agents are also culled stochastically. To model a scenario in which the closed habitat offers small-bodied hominins some protection from predators, the probability of dying while in open habitat (0.0003) is three times greater than the probability of dying while in a woodlands patch (0.0001). Each forager faces this stochastic risk of mortality during each time-step. All simulations are initialized with 400 foragers. During the course of each simulation run, the size of the forager population fluctuates slightly but rarely dips below the initial size.

Each forager agent is capable of carrying up to 110 units of energy that has been procured but not yet consumed. This makes possible food transfers between forager agents. Foragers reproduce asexually as soon as they have accumulated at least 100 units of energy. With local offspring dispersal, offspring appear on a cell within the parent's Moore neighborhood. By contrast, with random dispersal, offspring are equally likely to appear in any unoccupied woodlands cell in the environment. Each reproductive event costs the parent 50 energy units, which are immediately subtracted from the parent's energy budget. Each offspring begins with 50 units of energy.

The forager trait of interest in this study is the probability of sharing ($s$) with a conspecific in need. I model $s$ as a haploid locus. Each offspring inherits its parent's $s$ with a probability of $1 - \mu$, where $\mu$ represents a mutation rate of 0.01. When modeling food sharing as a discrete trait (where $s$ can take one of two binary values: 0 or 1), each mutation results in the offspring displaying the opposite value of its parent. Thus, a mutation in the offspring of a selfish parent ($s = 0$) will result in a new altruistic forager ($s = 1$). By contrast, when a mutation occurs when food sharing is modeled as a continuous trait (where $s$ can be any real number bound by 0 and 1), the value of $s$ that will be displayed by the offspring is calculated as the sum of its parent's $s$ and a real number drawn from a normal distribution with a mean of 0 and a standard deviation of 0.2.

When a forager's energy level dips below the food share threshold (50 units), the forager is considered to be in need of assistance. A "starving" forager will ask a forager it has selected randomly from the subset of foragers that are located on patches within its Moore neighborhood to share "excess" food. When modeling food sharing as a discrete trait, I investigate two different sharing strategies. With the first, sharing is a simple function of $s$: when a random number drawn from a uniform distribution bound by 0 and 1 is less than or equal to $s$, then the donor will share if it has any energy in excess of the food share threshold. With the second, altruistic foragers[1] are given the ability to recognize whether the prospective recipient is selfish ($s = 0$) or altruistic ($s = 1$), and they use this information in making their decision about whether to share with the prospective recipient, provided there is excess energy to share. According to Alexander's formulation of indirect reciprocity, individuals should share only with those who are likely to share with others. Nowak and Sigmund (1998, 2005) explored the notion of image scoring, whereby altruists gain a good reputation for helping, and cheaters gain a bad reputation for being selfish. If individuals are preceded by their reputation, then indirect reciprocity can evolve even if individuals rarely observe the pairwise interactions of others. Thus, the second strategy employed here is an especially strong form of indirect reciprocity, whereby each altruist has perfect knowledge of the bad reputations of selfish foragers as well as the good reputations of fellow altruists.

As a result of how indirect reciprocity is represented in this model, altruistic foragers who have energy in excess of the food share threshold will share only with other altruists who are in need, and selfish foragers never share food when asked. The social environment created by the second strategy is favorable to the spread of altruistic traits. Just one sharing strategy is used when $s$ represents a continuous trait. Here, sharing is again a simple function of $s$: when a random number drawn from a uniform distribution bound by 0 and 1 is less than or equal to $s$, then the forager will share some of its food with the prospective recipient.

Regardless of whether $s$ is modeled as a discrete trait or a continuous trait or, in the case of the former, whether the simple sharing strategy or indirect reciprocity is used, the number of energy units that a forager (the donor) shares with a conspecific in need (the recipient) is equal to the donor's current energy level minus the food share threshold. (Note that foragers cannot share food if their energy levels do not exceed the food share threshold.) The difference between the food share threshold and the donor's current energy level is added to the energy budget of the recipient, and the energy budget of the donor is reduced to the food share threshold. Thus the cost to the donor and the benefit to the recipient, as measured by number of energy units, are symmetrical. Helping is no small gesture, for the donor who provides food to a needy recipient will find itself in that position in the following time-step. In all cases, the donor gains nothing by helping others.

## Results

For each simulation, mean $s$ is calculated for the population of foragers during each time-step and compared to the previous maximum mean $s$ observed during the course of that run. The maximum value of mean $s$ is collected from each simulation, as is the "final" value of mean $s$ after the 50,000th time-step of each run. I use Kruskal-Wallis $H$ tests (and report the $\chi^2$ and P value for each) to assess whether $g$ has an effect on mean $s$. For the purposes of this exercise, visual inspection of the data is sufficient for a coarse-grained assessment of the direction (positive or negative) and magnitude of the effect of $g$ on mean $s$.

Before investigating how ecological patchiness ($g$) affects the evolution of food sharing (mean $s$) when the value of $s$ differentially affects fitness, it is important to investigate the evolutionary dynamics of the model in the *absence* of selection. This not only provides an independent check of the model, but it also provides a baseline to which we can later compare the results of simulations that include selection. To investigate the evolution of $s$ when modeled as a selectively neutral trait, the value of $s$ must be decoupled from the probability of sharing food. To do so, the model described previously is run with one significant modification: foragers never share food regardless of the value of $s$ (forager share threshold is set to 0). As a result, there is no fitness difference between a forager with $s = 0$ and $s = 1$, and fluctuation in mean $s$ is due to drift alone. Whether $s$ is represented as a discrete or continuous trait, the results show that the model meets two fundamental expectations. First, the central tendency of mean $s$ for any set of simulations is 0.5 (Figure 6.1). Second, ecological patchiness ($g$) has no effect on mean $s$ (discrete trait: $\chi^2 = 1.05$, $P = 0.90$; continuous trait: $\chi^2 = 2.39$, $P = 0.66$).

*Food Sharing as a Discrete Trait* Here I investigate whether ecological patchiness affects the evolution of food sharing in the presence of selection (that is, food share threshold = 50) when $s$ is represented as a simple discrete trait. Holding constant the size of closed habitat patches and the total number of cells with woodlands resources, 20 unique simulations were run for 50,000 time-steps for each of 5 values of $g$ ($g = 2, 4, 6, 8$, and 10). This procedure was repeated for both of the food-sharing strategies described previously—simple sharing and indirect reciprocity—resulting in a total of 200 runs. Each run was initialized with a pure population of 400 selfish foragers—$s = 0$ in all foragers at time 0 ($t_0$). Population-level variation in $s$ is introduced by mutation. Mutations occur at birth with probability $\mu = 0.01$. The relative frequency of altruists (mean $s$) can be thought of as representing a measure of the altruistic strategy's success in at least partially invading a pure population of selfish nonsharers.

In the presence of selection, simple altruistic foragers were in the minority throughout the duration of these simulations. The altruistic case did not evolve to fixation in any simulated population. In fact, final mean $s$ values are consistently lower here than when $s$ represents a selectively neutral discrete trait (compare Figure 6.1 [left panel] with Figure 6.2). This is not surprising. Given that according to the simple sharing strategy altruistic foragers always share "excess" food when asked and selfish foragers never do, altruists are generally at a disadvantage relative to their selfish counterparts, who enjoy the spoils of food sharing without ever paying the associated costs. Nevertheless, these results show that ecological patchiness ($g$) has a significant effect on the success of the simple food-sharing strategy when success is measured as the relative frequency of altruists in the total population (final mean $s$: $\chi^2 = 70.54$, $P < 0.001$; max mean $s$: $\chi^2 = 92.91$, $P < 0.001$). Figure 6.2 shows that in patchier environments simple altruistic foragers exist at higher frequencies than in less patchy environments. Although the effect of $g$ on mean $s$ as measured at the end of each simulation is significant, it is not a very large. The effect of ecological patchiness on maximum mean $s$ is larger.

As one might expect, when altruistic foragers are able to base sharing decisions on perfect knowledge of the sharing probability displayed by prospective recipients, as in the case of the indirect reciprocity strategy used here, altruists are more successful at displacing the selfish individuals regardless

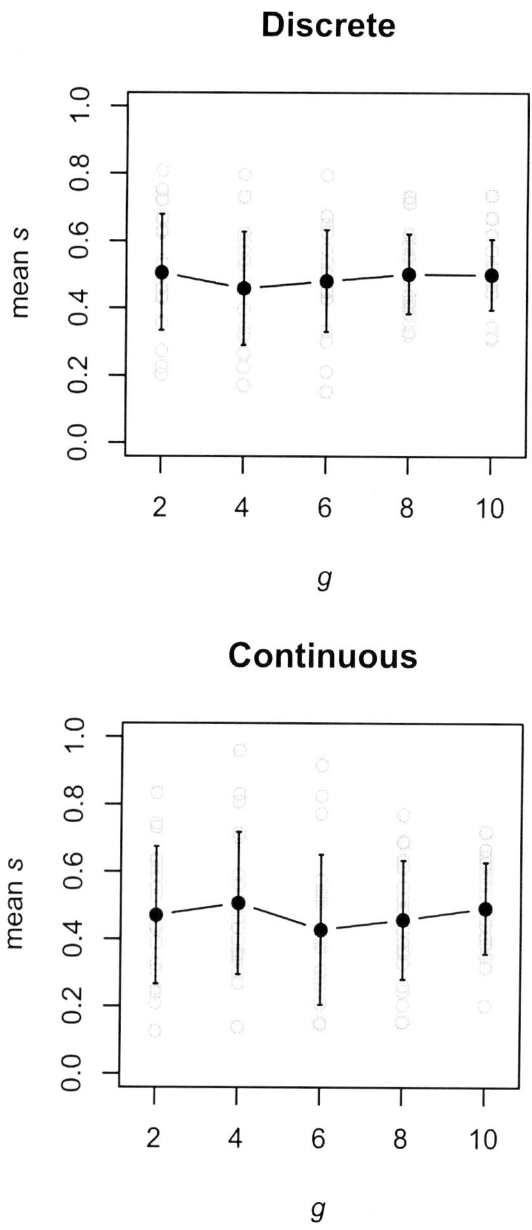


**Figure 6.1** Ecological patchiness ($g$) does not affect the evolution of food sharing ("final" mean $s$) under neutral conditions, regardless of whether $s$ is modeled as a discrete trait (top panel) or as a continuous trait (bottom panel). Solid black symbols represent the mean ($\pm 1$ sd) of 20 unique simulations, each of which is plotted in grey.

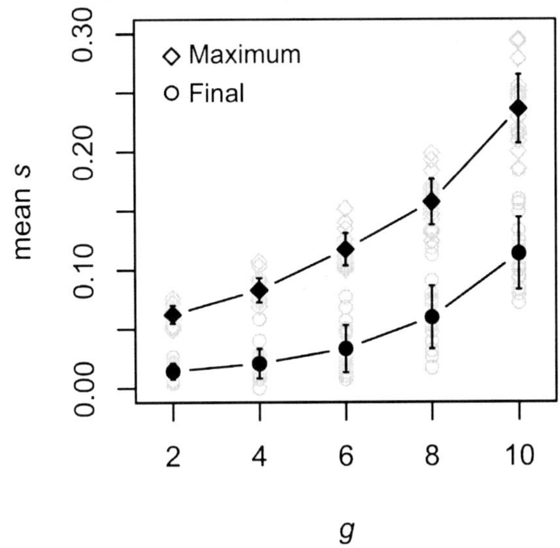


**Figure 6.2** The effect of ecological patchiness ($g$) on the evolution of the simple strategy of food sharing when modeled as a discrete trait. Solid black symbols represent the mean ($\pm 1$ sd) of 20 unique simulations, each of which is plotted in grey. Circles: mean $s$ collected after the 50,000th time-step of each simulation. Diamonds: maximum value of mean $s$ during each simulation.

of patchiness level (compare Figures 6.2 and 6.3). But note that ecological patchiness does not have an effect on the evolution of food sharing by indirect reciprocity (final mean $s$: $\chi^2 = 4.92$, $P = 0.30$; maximum mean $s$: $\chi^2 = 4.06$, $P = 0.40$). Because with indirect reciprocity altruistic foragers are able to direct their sharing toward other altruists and refuse to share with selfish foragers, increased ecological patchiness does not increase the level of positive assortment beyond that already provided by the sophisticated sharing strategy.

*Food Sharing as a Continuous Trait* Here I investigate whether ecological patchiness affects the evolution of food sharing in the presence of selection (that is, food share threshold = 50) when $s$ is modeled as a continuous (polygenic) trait. Holding constant the size of closed habitat patches and the total number of cells with woodlands resources, 20 unique simulations were run for 50,000 time-steps for each of 5 values of $g$ ($g = 2, 4, 6, 8$, and 10), resulting in a total of 100 runs. Each run was initialized with a pure population of 400 selfish foragers ($s = 0$ in all foragers at $t_0$). Population-level variation in $s$ ($s$ is a real number bound by 0 and 1 in this case) is introduced by mutation. Mutations occur at birth with probability $\mu = 0.01$. When modeling food sharing as a continuous trait, mean $s$ serves as a proxy for the evolution of altruism from pure selfishness.

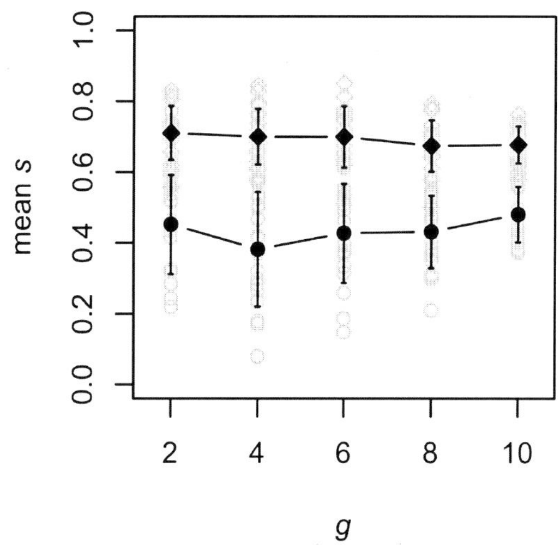


**Figure 6.3** There is no effect of ecological patchiness on the evolution of food sharing by indirect reciprocity when *s* is modeled as a discrete trait. Solid black symbols represent the mean (±1 sd) of 20 unique simulations, each of which is plotted in grey. Circles: mean *s* collected after the 50,000th time-step of each simulation. Diamonds: maximum value of mean *s* during each simulation.

No one would confuse these foragers with Mother Theresa—mean *s* remains well below 0.5 in all these simulations (Figure 6.4). Nevertheless, ecological patchiness (*g*) has a significant effect on mean *s* even in the presence of the very simple, probabilistic sharing strategy (final mean *s*: $\chi^2 = 70.88$, $P < 0.001$; maximum mean *s*: $\chi^2 = 86.70$, $P < 0\ .001$). As Figure 6.4 illustrates, selection favors a slightly higher level of altruism in patchier environments. Although the magnitude is small, it may seem rather surprising that *g* has any effect at all on mean *s* given that the simple sharing strategy provides no means for foragers with a higher value of *s* to protect their interests from lower *s* foragers or to punish less altruistic foragers for rarely (or never) sharing their "excess" food. As before, *g* has a larger effect on maximum mean *s* than on final mean *s*, although here the difference is less marked than when *s* is represented as a discrete trait (compare Figures 6.2 and 6.4).

*The Role of Spatial Association among Kin* Assuming relatively high fidelity in vertical transmission of traits from parent to offspring, local offspring dispersal increases the level of positive assortment in spatially explicit interactions, such as food sharing, relative to what one would expect to see under random mixing. Thus, spatial association among kin may be key to understanding the positive effect of ecological patchiness on the evolution of food sharing observed under some conditions (see also Pepper and Smuts, 2000, 2001,

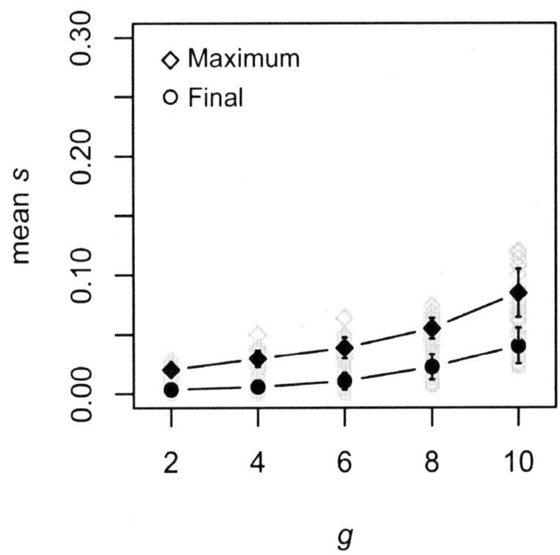


**Figure 6.4** The effect of ecological patchiness on the evolution of food sharing when *s* is modeled as a continuous trait. Solid black symbols represent the mean (±1 sd) of 20 unique simulations, each of which is plotted in grey. Circles: mean *s* collected after the 50,000th time-step of each simulation. Diamonds: maximum value of mean *s* during each simulation.

2002). It is clear that local offspring dispersal cannot facilitate the evolution of food sharing in relatively homogeneous environments, where parents and their offspring can easily drift apart while following their simple foraging strategies (see $g = 2$ in Figures 6.2 and 6.4). However, by constraining the movements of foragers to within just a few stands of woodlands resources, such that parents and their offspring remain in close spatial proximity for longer periods, larger gap sizes may allow individuals who are similar owing to common descent to interact with one another more frequently than they would in a panmictic (that is, randomly mixing) population.

The degree to which the spatial association among relatives explains the effect of patchiness on food sharing in this model can be assessed by collecting data from additional simulations in which each offspring is born into a randomly chosen unoccupied woodlands patch cell rather than into a cell adjacent to its parent. Figure 6.5 shows that, although there is a significant effect for both the discrete trait treatment (final mean *s*: $\chi^2 = 21.81$, $P < 0.001$; maximum mean *s*: $\chi^2 = 65.75$, $P < 0.001$) and the continuous trait treatment (final mean *s*: $\chi^2 = 18.68$, $P = 0.001$; maximum mean *s*: $\chi^2 = 53.84$, $P < 0.001$), the magnitudes of these effects are diminished relative to those observed in the case of local offspring dispersal. Even though foragers do not have the ability to recognize closely related kin or act differently toward them in this model, foragers are more likely to interact with those who are more closely

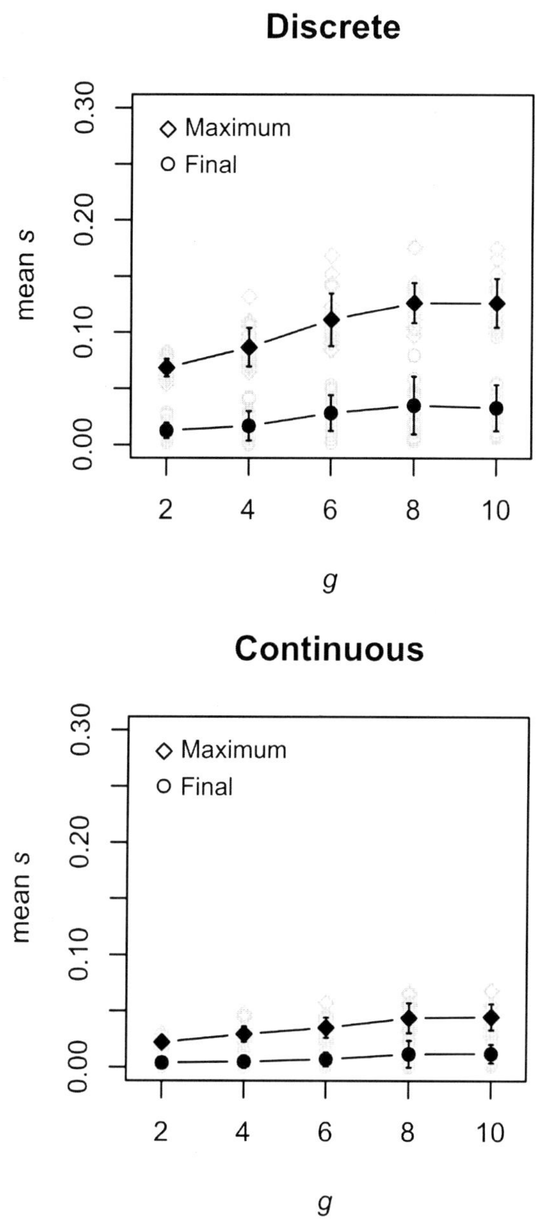


**Figure 6.5** Random offspring dispersal diminishes the positive effect of ecological patchiness ($g$) on the evolution of food sharing (mean $s$), regardless of whether $s$ is modeled as a discrete trait (top panel) or as a continuous trait (bottom panel). Solid black symbols represent the mean ($\pm 1$ sd) of 20 unique simulations, each of which is plotted in grey. Circles: mean $s$ collected after the 50,000th time step of each simulation. Diamonds: maximum value of mean $s$ during each simulation.

related—and, consequently, more likely to possess a similar *s* value—when *g* is large and offspring dispersal is local than when *g* is small and/or offspring dispersal is random. Clearly, the kin-based positive assortment facilitated by local offspring dispersal increases the magnitude of the positive effect of *g* on mean *s*.

## Discussion

What do these findings mean for interpretations of Early Stone Age archaeological material concerning food sharing and social organization in Plio-Pleistocene hominins? First, is it worth reiterating that the model discussed here does not attempt to emulate the past *in silico*. The model's experimental design is meant to elucidate the effect of between-group selection in the face of environmental fragmentation while assuming very little in terms of hominin behavioral complexity, with the exception of the case of indirect reciprocity. For example, the model assumes that an individual cannot take another's food by force (tolerated theft), it assumes that foragers cannot preferentially share with those that shared with them in the past (direct reciprocity), that they cannot recognize kin (inclusive fitness) or trade food for sex (show-off), and that they cannot preferentially mate with or attribute status to generous donors (costly signaling). Because of these assumptions, one should not mistake the results presented here as a refutation of any of the individual selection-based models of circumstance for the evolution of food sharing in hominins. A model can falsify only those hypotheses that it has been built to address. The present model was designed to address the extent to which ecological patchiness might serve as a catalyst for the evolution of food sharing, and the results speak only to that issue.

Although the exact timing and rate of grassland expansion in East Africa remain topics of debate (Cerling, 1992; Cerling et al., 2010, 2011; Levin et al., 2004; Quade et al., 2004; Vrba, 1985), most researchers agree that Plio-Pleistocene hominin species in this region were affected by the fragmentation of ancestral woodlands. One possible consequence is that increasingly secluded patches of ancestral closed habitat helped to subdivide hominin populations into evolutionarily meaningful trait groups. Assuming these hominins possessed only simple sharing strategies, the results presented above suggest that the geographic population structure provided by woodlands patches alone could have facilitated modest increases in the relative frequency of altruistic food sharing. If a simple form of food sharing were already present in hominin populations, the degradation of closed habitat could have accelerated its spread through the differential reproduction of groups that contained a greater proportion of food sharers. The model results also show that ecological fragmentation would not have affected the spread of more complex sharing strategies, such as indirect reciprocity.

The finding that even a simple altruistic food-sharing strategy—one that does not include punishment, cultural norms, or memory—can establish itself, even if only at a relatively low frequency, in a population of selfish foragers is what makes these results concerning the effect of ecological patchiness interesting. By purposefully omitting *a priori* assumptions about cognitive and behavioral complexity in Plio-Pleistocene hominins, the model allows us to gain a better understanding of just how influential ecological fragmentation may have been for the evolution of altruistic food sharing in the absence of complex foraging and redistribution strategies. How complex hominin foraging and redistribution strategies were during this period is an empirical question that is not dealt with in the current model. But the model results suggest that modern human-like behaviors are not necessary for altruistic food sharing to gain a toehold in species that are likely to have behaved quite differently from extant human foragers. Fragmented ancestral environments may have been sufficient for even very simple sharing strategies to begin to invade selfish populations.

The unwieldy size of the parameter space that often accompanies even relatively simple, heuristic models is a problem inherent to the kind of basic, theory-building research carried out here. I have investigated four experimental parameters with this model (Table 6.1). But it is possible that the dynamics I report may be sensitive to some of the parameters that were not varied in the present sensitivity analysis. Those of particular interest for future research include the food share threshold, the rate at which woodlands resources regenerate, the carcass probability, and the forager metabolic rate. In general, the parameter values that increase the cost of sharing or the frequency at which altruistic foragers share with selfish foragers will enhance the effect of ecological patchiness on mean $s$. As probabilistic sharing is made less costly and/or less frequent—by increasing the rate at which woodlands resources regenerate or increasing the availability of carcasses on the landscape—the positive effect of ecological patchiness will decrease and eventually disappear. It is not as easy to predict how changes to other parameters would influence the effect of ecological patchiness on the evolution of food sharing. For instance, although an increased forager metabolic rate would likely increase the benefits of sharing it would also likely increase the overall number of requests for help and decrease the time that foragers spend on average in any single woodlands patch, which would in turn reduce the degree to which increased ecological patchiness enhances positive assortment among foragers in the presence of local offspring dispersal.

Finally, the positive relationship between $g$ and mean $s$ shown in Figures 6.2, 6.4, and 6.5 probably does not increase monotonically over a larger range of gap size values. In fact, previous studies have shown that, although altruists benefit from the population structure provided by resource distributions that fit within the so-called transitional zone of ecological patchiness, they do not do as well in extremely fragmented environments marked by gap sizes that are large relative to the size of the resource patches (Pepper and Smuts, 2000; Premo, 2006, 2007). If the spatial structure of food resources did play a role in the evolution of food sharing among hominins during the Pliocene and Pleistocene, then the positive effects would have been greatest within such transitional zones of ecological patchiness. This finding leaves us with an interesting new question: did the ecological shift from closed woodland to savanna in East Africa, which took place during the Pliocene and Pleistocene, provide such a resource structure for thousands of generations of hominin foragers? My hope is that future paleoenvironmental reconstructions in the region can attain the level of temporal and spatial resolution required to address that question empirically.

## Conclusion

Darwin's speculation on how altruism evolved in species as different as bees and humans did not close the door on the possibility that altruistic traits spread by virtue of group-level benefits. However, following the dismissal of "naive group selection" in the 1960s, evolutionary biologists began to look for means by which seemingly "altruistic" traits benefit the donor's individual and/or inclusive fitness. The emphasis on individual-level selection is apparent in the models of circumstance that human behavioral ecologists have used to explain the presence of food sharing in small-scale human societies. Many of these models of circumstance have informed archaeological reconstructions of Plio-Pleistocene hominin behavior.

Although generally underutilized by paleoanthropologists, multilevel selection theory provides a useful framework for investigating social evolution in species of hominins for whom we should avoid assuming modern-like behaviors *a priori*. Simulation results show that increased environmental fragmentation increases the extent to which an altruistic food-sharing strategy can gain a toehold in a pure population of selfish nonsharers. Given that ecological patchiness may provide a parsimonious explanation for the evolution of food sharing among foragers who exhibit simple procurement and redistribution strategies, it may be time to take a fresh look at Isaac's (1978) hypothesis that hominins could have been regularly sharing food as early as 2 million years ago, at a time when woodlands were dissected by the expansion of open grasslands in East Africa. In reevaluating Isaac's provocative hypothesis, we would be wise to employ null models that incorporate our current understanding of paleoenvironmental trends and multilevel selection theory to generate testable expectations of the Early Stone Age archaeological record rather than to continue to apply referential models based on the behaviors of extant primates or individual-level selection models built to take self-sacrifice out of altruism.

## Acknowledgments

I thank Susan Anton and Cara Wall-Scheffler for helpful comments following a presentation of an earlier version of this chapter at a conference in Burgos, Spain. Amelie Schmolke provided many useful comments on an earlier draft. Finally, I thank Steve Kuhn, Mary Stiner, and Steve Lansing for their guidance and encouragement with this project.

## Note

1. I need not give this ability to selfish foragers ($s = 0$) because they never share, regardless of whether the prospective recipient is selfish or altruistic.

## References

Alexander, R. D. 1979. *Darwinism and Human Affairs.* Seattle: University of Washington Press.

———. 1987. *The Biology of Moral Systems.* New York: Aldine de Gruyter.

Bird, R. L. 1999. Cooperation and conflict: The behavioral ecology of the sexual division of labor, *Evolutionary Anthropology* 8: 65–75.

Bliege Bird, R. L., Smith, E. A., and Bird, D. W. 2001. The hunting handicap: Costly-signaling in human foraging societies, *Behavioral Ecology and Sociobiology* 50: 9–19.

Blurton Jones, N. G. 1984. A selfish origin for human food sharing: Tolerated theft, *Ethology and Sociobiology* 5: 1–3.

———. 1987. Tolerated theft: Suggestions about the ecology and evolution of sharing, hoarding, and scrounging, *Social Science Information* 26(1): 31–54.

Boehm, C. 1996. Emergency decisions, cultural-selection mechanics, and group selection, *Current Anthropology* 37(5): 763–93.

———. 1999. *Hierarchy in the Forest: The Evolution of Egalitarian Behavior.* Cambridge, MA: Harvard University Press.

Boone, J. 1998. The evolution of magnanimity: When is it better to give than to receive? *Human Nature* 9(1): 1–21.

Bowles, S. 2006. Group competition, reproductive leveling, and the evolution of human altruism, *Science* 314: 1569–72.

———. 2009. Did warfare among ancestral hunter-gatherers affect the evolution of human social behaviors? *Science* 324: 1293–98.

Bowles, S., and Gintis, H. 2011. *A Cooperative Species: Human Reciprocity and Its Evolution.* Princeton, NJ: Princeton University Press.

Cerling, T. E. 1992. Development of grasslands and savannas in East Africa during the Neogene, *Global and Planetary Change* 5: 241–47.

Cerling, T. E., Levin, N. E., Quade, J., Wynn, J. G., Fox, D. L., Kingston, J. D., Klein, R. G., and Brown, F. H. 2010. Comment on the paleoenvironment of *Ardipithecus ramidus, Science* 328: 1105.

Cerling, T. E., Mbua, E., Kirera, F. M., Manthi, F. K., Grine, F. E., . Leakey, M. G., Sponheimer, M., and Uno, K. T. 2011. Diet of Paranthropus boisei in the Early Pleistocene of East Africa, *Proceedings of the National Academy of Sciences of the United States of America* 108: 9337–41.

Choi, J.-K., and Bowles, S. 2007. The coevolution of parochial altruism and war, *Science* 318: 636–40.

Darwin, C. 1859. *On the Origin of Species, A Facsimile of the First Edition*. London: John Murray.

________. 1871. *The Descent of Man and Selection in Relation to Sex*. London: John Murray.

Domínguez-Rodrigo, M., Egeland, C. P., and Barba, R. 2007. The home-base debate, in Dominguez-Rodrigo, M., Barba, R., and Egeland, C. (Eds.), *Deconstructing Olduvai: A Taphonimic Study of Bed I Sites*, pp. 1–10. Dordrecht: Springer Press.

Fisher, R. A. 1930. *The Genetical Theory of Natural Selection*. Oxford: Clarendon Press.

Grafen, A. 1990. Biological signals as handicaps, *Journal of Theoretical Biology* 144: 517–46.

Gurven, M., Allen-Arave, W., Hill, K., Hurtado, M. 2000. "It's a Wonderful Life": Signaling generosity among the Ache of Paraguay, *Evolution and Human Behavior* 21: 263–82.

Haldane, J. B. S. 1932. *The Causes of Evolution*. London: Longmans, Green and Co.

Hamilton, W. D. 1963. The evolution of altruistic behavior, *The American Naturalist* 97(896): 354–56.

________. 1964a. The genetical evidence of social behavior I, *Journal of Theoretical Biology* 7: 1–16.

________. 1964b. The genetical evidence of social behavior II, *Journal of Theoretical Biology* 7: 17–52.

________. 1975. Innate social aptitudes of man: An approach from evolutionary genetics, in Fox, R. (Ed.), *Biosocial Anthropology*, pp. 133–55. London: Malabry Press.

Hawkes, K. 1990. Why do men hunt? Some benefits for risky strategies, in Cashdan, E. (Ed.), *Risk and Uncertainty*, pp. 145–66. Boulder, CO: Westview Press.

________. 1991. Showing off: Tests of an hypothesis about men's foraging goals, *Ethology and Sociobiology* 12: 29–54.

Isaac, G. L. 1978. Food sharing and human evolution: Archaeological evidence from the Plio-Pleistocene of East Africa, *Journal of Anthropological Research* 34: 311–25.

Kaplan, H., and Hill, K. 1985. Food sharing among Ache foragers: Tests of explanatory hypotheses, *Current Anthropology* 26(2): 223–46.

Lancaster, J. B. 1978. Carrying and sharing in human evolution, *Human Nature* 4: 82–89.

Leakey, M. 1971. *Olduvai Gorge: Excavations in Beds I and II, 1960–1963, Vol. 3*. Cambridge: Cambridge University Press.

Levin, N. E., Quade, J., Simpson, S. W., Semaw, S., and Rogers, M. 2004. Isotopic evidence for Plio-Pleistocene environmental change at Gona, Ethiopia, *Earth and Planetary Science Letters* 219: 93–110.

Maynard Smith, J. 1964. Group selection and kin selection, *Nature* 201: 1145–47.

Nowak, M. A., and Sigmund, K. 1998. Evolution of indirect reciprocity by image scoring, *Nature* 393: 573–77.

________. 2005. Evolution of indirect reciprocity, *Nature* 437: 1291–98.

Nowak, M. A., Tarnita, C. E., and Wilson, E. O. 2010. The evolution of eusociality, *Nature* 466: 1057–62.

O'Connell, J., Hawkes, K., Lupo, K., and Blurton Jones, N. 2002. Male strategies and Plio-Pleistocene archaeology, *Journal of Human Evolution* 43: 831–72.

Pepper, J. W., and Smuts, B. B. 2000. The evolution of cooperation in an ecological context: An agent-based model, in Kohler, T., and Gumerman, G. (Eds.), *Dynamics in Human and Primate Societies*, pp. 45–76. Oxford: Oxford University Press.

———. 2001. Agent-based modeling of multilevel selection: The evolution of feeding restraint as a case study, in Pitt, W. C. (Ed.), *Swarmfest 2000, Proceedings of the 4th Annual Swarm User Group Conference: Natural Resources and Environmental Issues*, Vol. VIII, pp. 57–68. Logan, UT: S. J. and Jessie E. Quinney Natural Resources Research Library.

———. 2002. A mechanism for the evolution of altruism among nonkin: Positive assortment through environmental feedback, *American Naturalist* 160(2): 205–13.

Potts, R. 1982. *Lower Pleistocene Site Formation and Hominid Activities at Olduvai Gorge, Tanzania*. Ph.D. dissertation, Department of Anthropology, Harvard University.

———. 1984. Home bases and early hominids, *American Scientist* 72: 338–47.

———. 1988. *Early Hominid Activities at Olduvai*. New York: Aldine de Gruyter.

———. 1991. Why the Oldowan? Plio-Pleistocene toolmaking and the transport of resources, *Journal of Anthropological Research* 47: 153–76.

Premo, L. S. 2005. Patchiness and prosociality: An agent-based model of Plio/Pleistocene hominid food sharing, in Davidsson, P., Takadama, K., and Logan, B. (Eds.), *Multi-Agent and Multi-Agent-Based Simulation*, pp. 210–24. Berlin: Lecture Notes in Artificial Intelligence 3415, Springer-Verlag.

———. 2006. *Patchiness and Prosociality: Modeling the Evolution and Archaeology of Plio-Pleistocene Hominin Food Sharing*. Unpublished Ph.D. dissertation. Department of Anthropology, University of Arizona, Tucson. University Microfilms, Ann Arbor.

———. 2007. Exploratory agent-based models: Towards an experimental ethnoarchaeology, in Clark, J. T., and Hagemeister, E. M. (Eds.), *Digital Discovery: Exploring New Frontiers in Human Heritage*, pp. 29–36. Budapest: CAA 2006. Computer Applications and Quantitative Methods in Archaeology. Archeolingua Press.

———. 2008. Exploring behavioral terra incognita with archaeological agent-based models, in Frischer, B., and Dakouri-Hild, A. (Eds.), *Beyond Illustration: 2D and 3D Technologies as Tools of Discovery in Archaeology*, pp. 46–56. Oxford: Archaeopress, British Archaeological Reports International Series 1805.

Price, G. R. 1970. Selection and covariance, *Nature* 227: 520–21.

———. 1972. Extension of covariance selection mathematics, *Annals of Human Genetics* 35: 485–90.

———. 1995. The nature of selection, *Journal of Theoretical Biology* 175: 389–96 [written circa 1971, published posthumously].

Quade, J., Levin, N., Semaw, S., Stout, D., Renne, P., Rogers, M., and Simpson, S. 2004. Paleoenvironments of the earliest stone toolmakers, Gona, Ethiopia, *Geological Society of America Bulletin* 116(11/12): 1529–44.

Richerson, P. J., and Boyd, R. 2005. *Not by Genes Alone*. Chicago: University of Chicago Press.

Rose, L., and Marshall, F. 1996. Meat eating, hominid sociality, and home bases revisited, *Current Anthropology* 37(2): 307–38.

Schick, K. 1987. Modeling the formation of Early Stone Age artifact concentrations, *Journal of Human Evolution* 16: 789–807.

Smith, E. A., and Bliege Bird, R. L. 2000. Turtle hunting and tombstone opening: Public generosity as costly-signaling, *Evolution and Human Behavior* 21: 245–61.

Sober, E., and Wilson, D. S. 1998. *Unto Others: The Evolution and Psychology of Unselfish Behavior*. Cambridge, MA: Harvard University Press.

Sosis, R. 2000. Costly-signaling and torch fishing on Ifaluk Atoll, *Evolution and Human Behavior* 21: 223–44.

Trivers, R. L. 1971. The evolution of reciprocal altruism, *Quarterly Review of Biology* 46: 35–57.

Vrba, E. S. 1985. Ecological and adaptive changes associated with early hominid evolution, in Delson, E. (Ed.), *Ancestors: The Hard Evidence*, pp. 63–71. New York: Alan R. Liss.

Wilensky, U. 1999. NetLogo, http://ccl.northwestern.edu/netlogo/, Center for Connected Learning and Computer-Based Modeling, Northwestern University. Evanston, IL.

Williams, G. C. 1966. *Adaptation and Natural Selection*. Princeton, NJ: Princeton University Press.

Williams, G. C., and Williams, D. C. 1957. Natural selection of individually harmful social adaptations among sibs with special reference to social insects, *Evolution* 11: 32–39.

Wilson, D. S. 1975. A theory of group selection, *Proceedings of the National Academy of Sciences of the United States of America* 72(1): 143–46.

________. 1977. Structured demes and the evolution of group-advantageous traits, *American Naturalist* 111: 157–85.

________. 1983. The group selection controversy: History and current status, *Annual Review of Ecology and Systematics* 14: 159–87.

________. 1987. Altruism in Mendelian populations derived from kin groups: The Haystack Model revisited, *Evolution* 41: 1059–70.

________. 1998. Hunting, sharing, and multilevel selection: The Tolerated-Theft Model revisited, *Current Anthropology* 39(1): 73–97.

________. 2002. *Darwin's Cathedral: Evolution, Religion, and the Nature of Society*. Chicago: University of Chicago Press.

Wilson, D. S., and Dugatkin, L. A. 1997. Group selection and assortative interactions, *American Naturalist* 149(2): 336–51.

Wilson, D. S., and Sober, E. 1994. Reintroducing group selection to the human behavioral sciences, *Behavioral and Brain Sciences* 17: 585–654.

Wilson, D. S., and Wilson, E. O. 2007. Rethinking the theoretical foundation of sociobiology, *The Quarterly Review of Biology* 82: 327–48.

________. 2008. Evolution "for the Good of the Group," *American Scientist* 96: 380–89.

Wilson, E. O. 2012. *The Social Conquest of Earth*. New York: Liveright.

Winterhalder, B. 1996. Social foraging and the behavioral ecology of intragroup resource transfers, *Evolutionary Anthropology* 5: 46–57.

________. 1997. Gifts given, gifts taken: The behavioral ecology of nonmarket, intragroup exchange, *Journal of Archaeological Research* 5(2): 121–68.

Wright, S. 1945. Tempo and mode in evolution: A critical review, *Ecology* 26: 415–19.

Wynne-Edwards, V. C. 1962. *Animal Dispersion in Relation to Social Behavior*. Edinburgh: Oliver and Boyd.

Zahavi, A. 1975. Mate selection: A selection for handicap, *Journal of Theoretical Biology* 53: 205–14.

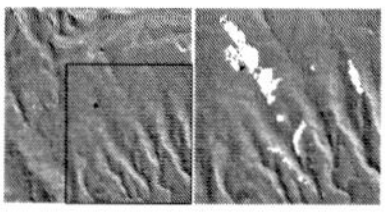

# 7 Stories of the Past or Science of the Future? Archaeology and Computational Social Science

*Michael Barton*

Since it emerged as a formal field of study from 18th- and 19th-century antiquarianism, archaeology has focused on the past, a fact explicitly recognized in the Greek neologism adopted for its name. From its inception, archaeology has had as its primary goal the systematic (and, to many, scientific) reconstruction of past societies; that is, the most common outcome of archaeological practice has long been the construction of narratives that recount in history-like fashion some aspect of past human lives and societies. Within the past half-century, this basic aim has been expanded to encompass explanations why people and societies in these archaeological reconstructions acted and changed in the way they did—although debate continues within archaeology as to the relative importance of reconstruction (or history) and explanation and about what constitutes adequate explanation (Barton and Clark, 1997; Dunnell, 1982; Hegmon, 2003; Killick, 2004; Pauketat and Alt, 2005; Wylie, 1992, 2000). With ongoing advances in methods for data collection and analysis, archaeologists endeavor to create and explain narratives of the past with increasing detail, accuracy, and insight. But regardless of the methods used or the relative emphasis on historical accounts or explanation, the reconstruction of the past still underlies all modern archaeology.

Taking an approach that utilizes multiple lines of convergent evidence from increasingly sophisticated analytical protocols (Killick, 2004; Wylie, 2000)

*Computational Approaches to Archaeological Spaces* by Andrew Bevan and Mark Lake, Eds., 151–178 

has been successful in reconstructing the manufacturing processes and uses of a variety of ancient material culture, as well as physical aspects of resource processing and consumption. However, conceptual issues still can make such reconstructions problematic, especially for nearly extinct technologies such as chipped stone that fall beyond the everyday experience of archaeologists (Barton, 1991, 1997; Bleed, 2001). Moreover, when we attempt to extend reconstructions beyond resource use, and the production and physical use of material culture to those issues of individual and social dynamics across space and time—the issues that most interest most archaeologists and the larger public—we encounter increasingly insurmountable problems with the archaeological record.

A major tenet of the "new archaeology" of the mid-20th century (and one that was more fundamental to its original call for a revolution in the field than the "processual" approach that involved the attempt to make the field more scientific through the discovery of general laws of human behavior) was its exhortation to make people rather than artifacts the primary subject of archaeological inquiry (Binford, 1962). That is, archaeology should be a sort of ethnography of the past, a perspective that is largely implicit in much of the Anglo-American archaeological literature today. Yet much of the dynamics of society occur as interactions between individuals, as webs of meaning (Geertz, 1973) that leave no material traces. Even with participant observation that is a hallmark of anthropological methods, anthropologists have long wrestled with the difficulties of interpreting the meanings of behaviors or even the sense of identity in societies with very different cultural knowledge and histories (Abu-Lugbod, 1991; Geertz, 1973; Keesing, 1974; Taylor, 1971). For prehistoric societies, a goal of interpreting such interactions and their meanings "place archaeologists in the role of ethnographers of a lost 'ethnographic present,' struggling hopelessly to overcome the problems posed by the fact that the people they would like to talk to are long dead and most of the residues of their lives long decayed" (Shennan, 2002: 9).

The standard archaeological response to this dilemma has been to refer to the rich world of material culture. Indeed, the new archaeology introduced the idea that artifacts not only provide information about technological behaviors but can also serve as a proxy for ethnographic observation (Longacre, 1964). Certainly, material culture is deeply entwined with human social life and can even serve as an active agent of social reproduction (Hegmon, 1998; Newton, 1981; Pauketat and Alt, 2005). Although some of the social significance of material culture cannot be known without conversation with the users (Killick, 2004), the spatial arrangements of material objects can sometimes provide clues to the social contexts in which they were embedded. Unfortunately, the spatial organization of material culture when it was an active participant in human society is not usually preserved when it enters the archaeological record, most commonly as trash (Barton et al., 2002; Schiffer, 1983). Moreover, most of the material record originally produced by human social practices has long since been lost to natural and cultural formation processes. Of that small part that remains, the great majority is undiscovered or inaccessible, or cannot be recovered owing to time and money constraints. Excavations represent a very tiny window on a mostly missing record. Archaeologists do a truly amazing job at recovering and analyzing the data available from the archaeological record. However, they cannot analyze a record that they cannot recover or that no longer exits.

So how then can we reconstruct past human lives and societies in reasonably complete detail across space and time, so as to trace and explain the dynamics of human society when our informants cannot speak to us, when most of the objects they made and used are gone, while those that remain are jumbled into palimpsests of trash, and we have resources to recover a only tiny fraction of those? I fear the answer is that we cannot. Certainly, we can say that each and every artifact that we find was produced by a once living individual or a group of individuals working together, but this is an intellectually trivial insight. And when multiple lines of evidence converge, we can even infer with reasonable confidence the nature of a particular event that took place at a particular place and time. But beyond this, the archaeological record is absent, and our reconstructions are simply speculations—careful, statistically informed in some cases and imaginatively subjective in others, but reconstructions remain speculative narratives in all cases.

Perhaps others may be more confident in their ability to accurately imagine the past or perhaps they don't mind that their stories of the past are a sort of historical fiction. But these very real constraints of the archaeological record on our ability to reconstruct the past have concerned me increasingly over the course of my professional career. At the same time, I have become increasingly convinced that the human past is essential to understanding human society today and planning for our long-term future (Barton et al., 2004; Barton, Ullah, and Mitasova, 2010; Fisher, Hill, and Feinman, 2009; Redman and Kinzig, 2003; van der Leeuw, 2000; van der Leeuw and Redman, 2002). Fictional narratives will not be helpful in such an endeavor, no matter how compelling. The importance of the past to present and future makes it all the more imperative that we find new ways to make more effective use of the incomplete, discontinuous, and highly altered archaeological record in order to provide reliable insights into the long-term dynamics of human society and the interactions between the social and natural worlds. This has led me to explore alternative ways of collecting and analyzing archaeological data and ultimately to an evolving perspective on the practice of archaeology.

Over the last two decades a variety of new digital technologies, which I will gloss here under the phrase *computational modeling*, have become available that offer a new way to allow archaeology to move forward in producing valuable insights from the human past and that can be applied broadly to the understanding of human social dynamics in spite of the limitations of the archaeological record. I illustrate some of the potentials and limitations of computational modeling with examples from my research over the past decade. But although these new technologies may offer a way to create more systematic and transparent reconstructions of the past, these reconstructions are not necessarily any more accurate than traditional narratives and still cannot replace the lack of data that characterizes the archaeological record. What computational modeling does hold a promise to do, however, is provide archaeologists with the tools to better understand and explain the spatial-temporal dynamics of human societies, with important benefits to archaeology and to social science more broadly. Before reviewing how I have used models in archaeological research, I briefly review the ways in which models are widely used in archaeology today.

## Models in Archaeology

A model is an abstract representation of real-world phenomena. Models are inherent in the way we make sense of our world in our everyday lives and are equally important in scientific practice. Models in science are commonly created to simplify very complex reality so that we can better identify and comprehend critical relationships among entities and key processes that drive the operation of real-world systems. Models are generally evaluated in science by their ability to account for a constrained set of relevant empirical observations. Especially since changing their focus from describing artifacts to describing and explaining human societies, models have become pervasive in archaeology. In the great majority of cases, these models take the form of narrative prose that may or may not be supported by statistics or graphs. Narratives are a format that we find intuitively easy to understand; they can help us to mentally envision humans of the past and the world these people inhabited. However, narratives can be difficult to evaluate scientifically in a systematic and convincing manner, because they can employ inherently ambiguous natural language.

It is significant that archaeological models (and particularly narrative models) are usually constructed inductively from empirical data, rather than deduced from theory. In fact, the more empirical data that is used to create these inductive models, the more convincing they seem to be to most archaeologists and to many others who read their accounts. Such archaeological inference is found throughout the literature and by far dominates the process of model-building. As noted previously, when based on the convergence of multiple and distinct lines of evidence, such inference can provide robust knowledge about past physical events and the manufacture or use of artifacts (Wylie, 2000).

However, the inferential approach to model-building also creates models that are difficult to test in an effective and convincing way. An obvious issue is that the data used to create a model cannot also be used as an independent test of the model. This leaves much explanation in archaeology epistemologically equivalent to the interpretive approaches, or hermeneutics, of the humanities. As Taylor (1971) explains in his classic paper: "A successful interpretation is one which makes clear the meaning originally present in a confused, fragmentary, cloudy form. But how does one know that this interpretation is correct? Presumably because it makes sense of the original text: what is strange, mystifying, puzzling, and contradictory is no longer so, is accounted for." This is equivalent to saying that the measure of a good model is that it can sensibly account for the empirical data. Although this statement may seem perfectly reasonable to many archaeologists—and certainly to the general public—a difficulty of this approach is how we define "sensibly" and how we can compare two models that both account for available data in a way that is sensible to different adherents.

Such epistemological issues are exacerbated by problems of "equifinality" that arise when inferential models are generated from the limited archaeological record. In spite of the seeming vast quantity of artifacts (mostly broken pottery and chipped stone) that fill museums and repositories, the archaeological record is so sparse temporally and spatially, relative to the kinds of processes described in many narratives, that multiple, quite different models can be (and often are) constructed to account for the same data—as I illustrate below. An interpretive approach, inherent in inferential model-building, has no mechanism to evaluate which of alternative competing models is better—or even a way to systematically define what "better" means. This situation can lock archaeological explanation into what Taylor calls the "hermeneutical circle," in which the ultimate recourse for convincing others of the superiority of one inductive model over another is to assert the scholarly "authority" of the model author (see Clifford, 1983).

I do not mean to imply that induction or inference has no role in science. Careful analysis of large, rich datasets commonly offers important insights into dynamic processes and relationships. Moreover, induction can be important in generating provisional models—that is, hypotheses—although in most sciences, theory plays an equally or even more important role. It is significant that models are constructed in other scientific fields so as to predict data that have not yet been collected rather than to account for observations already made. And models are then evaluated on their ability to predict empirical phenomena (Dunnell, 1982; van der Leeuw, 2004). However, archaeology is seriously under-theorized, and, consequently, this kind of interplay between induction and testing is little practiced, in spite of it being part of the proscriptive literature for decades. It is also worth noting that the kind of robust archaeological inference based on triangulating multiple lines of evidence derived from different scientific disciplines is most successful when the evidence derives from fields with a robust tradition of hypothesis testing (for example, Wylie, 2000: 232–33).

These pragmatic issues inherent in inferential model-building may not be problematic to those who are content with an archaeology whose primary goal is the creation of plausible reconstructions about an unknowable past. However, if insights about the human past are to play a meaningful role in understanding human society today and inform policy decisions for the future, narrative fictions are an insufficient for such contributions, no matter how plausible. This suggests that archaeology needs alternatives to inferential narratives for creating and evaluating models about human social practice. Moreover, as discussed shortly, it also suggests that although reconstruction of past societies will remain an important activity of archaeologists, we can realistically aspire to more ambitious goals.

In the next sections I draw on examples from my research to illustrate the transition from an archaeology focused on inferring stories of the past to one that seeks to test transparent models of the dynamics of human societies.

## Example 1: Long-Term Land-Use in Eastern Spain

The landscapes of Mediterranean Europe have been occupied and reoccupied by hunter-gatherers at least since the Middle Pleistocene, by subsistence farmers since the Early Holocene, and by urban empire-builders for over three millennia. These inhabitants have strewn the surface with lithic debris, sherds of broken vessels, and fragments of building materials. This artifactual palimpsest has been plowed, eroded and redeposited, piled into terraces and other structures, and modified physically and chemically. Consequently, most Mediterranean archaeologists focus their research efforts on carefully excavating in those rare prehistoric towns, farmsteads, and caves that have somehow escaped the normal alterations of subsequent human activities.

Beginning over two decades ago, colleagues from the University of Valencia (Spain) and I began a project to systematically collect and analyze the palimpsest of archaeological data scattered across the landscapes of eastern Spain (Figure 7.1).[1] Using a patch-based survey methodology, we systematically recorded information across broad swaths of a series of upland valleys, rather than seeking to locate "sites." We augmented the intensive survey data with information from coring, geophysical prospection, and hand excavation (Barton et al., 1999; Barton et al., 2002; Barton et al., 2004; Bernabeu Aubán et al., 1999; Bernabeu Aubán et al., 2000; Bernabeu Aubán et al., 2006; Bernabeu Aubán, Guitart, and Pascual, 1989; García Puchol, Barton, and Bernabeu Aubán, 2008). This work has involved innovative statistical and GIS-based analyses to "unmix" the altered palimpsests that constitute much of the archaeological record of the Mediterranean region. The result has been a more comprehensive, empirically grounded view of human land-use and its changes through time than is afforded by more normal site-based studies in this region.

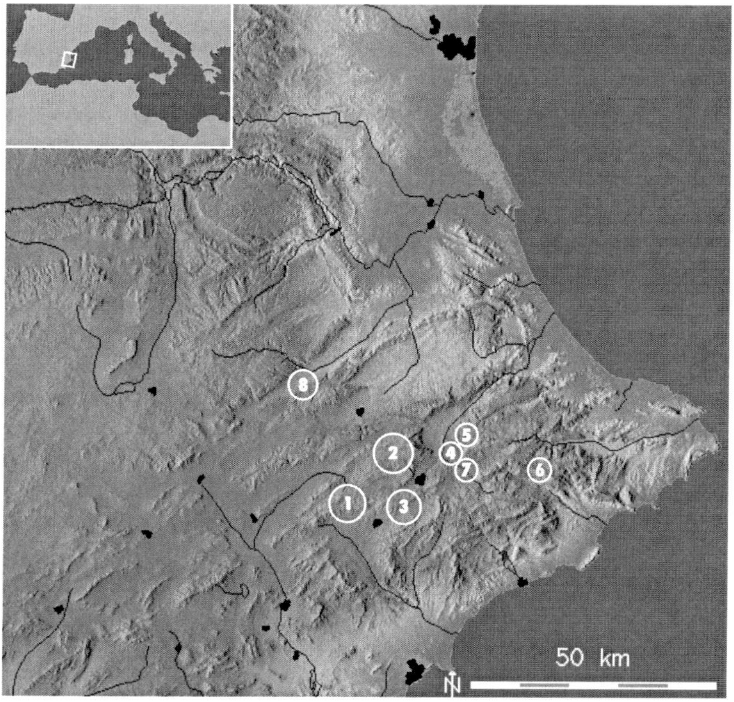


**Figure 7.1** Locations of Alicante Prehistoric Ecology and Land-Use project in northern Alicante Province, Spain: (1) Polop Alto Valley, (2) Middle Serpis Valley, (3) Penaguila Valley, (4) Alcalá Valley, (5) Gallinera Valley, (6) Gorgos Valley, (7) upper Ceta Valley, (8) Canyoles Valley.

We recognized at the outset of this research that millennia of artifactual accumulation, interspersed with human and natural formation processes of varying severity, made it impossible to accurately reconstruct the settlement system of any particular social group at any particular time. Rather, we interpreted our results as models of cumulative socioecological landscapes produced by varying patterns of land-use and social organization (Barton et al., 1999; Barton et al., 2002). This approach allowed us to create a narrative of past human occupation of this region in terms of land-use intensity, clustering and dispersion, and persistence and to compare human action at regional scales in different valleys across time (Barton et al., 2004a; Barton et al., 2004b).

Models of changing spatial configurations of land-use intensity for the Polop Alto Valley for the past 100,000 years are shown in color Figure I.1. Our approach involves a systematic and replicable way to combine spatially and temporally dispersed archaeological data into comprehensive and comparable geospatial models of human land-use. This is more transparent methodologically and more empirically grounded than the more common subjective interpretations of land-use based on excavations of a few square meters from a few sites in a region. Nonetheless, our models of the social processes that generated the land-use patterns we observe are completely narrative and inferential. We do examine alternative lines of evidence for mobility strategies to bolster our inferences. However, even with a patch-based approach to collecting archaeological data across valleys, the available data remain very sparse, and using surface collections necessitates a coarse temporal resolution many orders of magnitude greater than the scale of real human generations (although the mismatch between archaeological chronological frameworks and the temporal scale of human action is not limited to surface collections). Hence, alternative narratives could be developed to fit the data we present, with no way to assess whether they constitute better reconstructions than ours. Our models are in no way predictive, in that they cannot account for human land-use in as yet unstudied locales; in fact, we generate different land-use models for each valley we study (Barton et al., 2004a, 2004b). This situation makes it impossible to test the accuracy of our inferential models and makes them of limited use for making decisions about future land-use beyond offering general insights about socioecological path dependence and the interaction of environment, society, and history.

More sophisticated, scientific data collection and computational analysis methods can lead to new insights about past societies. Moreover, the methodological transparency and replicability that come with applying such methods systematically are desirable. However, when they are used for inferential reconstructions, computational models still cannot create narratives of the past that can be demonstrated to be any more accurate than ones created subjectively. And, because they are created to account for known data, these inductively created models are unable to reliably forecast the outcomes of human social practices. These shortcomings are a significant reason that society has difficulty in making use of the "lessons of history," even though we have been admonished to do so since antiquity. If the lessons of history are simply interpretations of often debatable (and regularly debated) plausibility, how can we feel confident that we are even getting accurate lessons to apply? And if these lessons are simply reconstructions of particular past circumstances with no predictive power, how can we use them for decision making even if they are reasonably accurate?

## Example 2: Technology and Ecology in Upper Pleistocene Europe

Studies of change over long time spans are a hallmark of archaeology and other historical sciences. In addition to narratives, change through time is often represented in a variety of graphical formats that can range from images of artifacts arranged in a proposed chronological order to graphs of radiocarbon samples with error ranges. Estimating the age of material culture is a task that an inferential approach often performs well. Careful age inference often involves assessing the degree of agreement between several very different kinds of information, including stratigraphy, seriation, and radiometric methods. Such archaeological dating is cited by Wylie (2000) as an example of using of multiples lines of evidence to make strong inferential arguments.

Change—or more often simple variation—across space is also of considerable importance in archaeological research. Since most archaeological practice takes place in the contexts of sites, variation in artifact densities, artifact types, features, or other material residues of human activities are very commonly represented across the space of sites. In addition to narratives, maps are the most common way in which variation across sites is represented. Variation is similarly represented at regional scales in the form of maps that usually show the distribution of sites occupied "contemporaneously" (for example, Hill et al., 2004), but occasionally other kinds of data also are displayed in map form (for instance, Ebert, Camilli, and Berman, 1996; Kvamme, 2007; Weatley, 1995; Williams, Limp, and Briuer, 1990).

Because of the nature of the archaeological record and archaeological practice, there is a common tendency to model change through time as a sequence of discrete packets of time-temporal "phases"— for example—during which nothing (or nothing of importance) changes. Change takes place in the time between these packets (see Dunnell, 1982). Moreover, artifacts, features, and sites inferred to be included within the same time packet are considered to be contemporaneous and treated as if individual humans living within the same temporal phase potentially could have interacted. Sometimes such contemporaneity is reasonable, but often temporal phases can be over a century or even millennia in length—far beyond human generation spans. Spatial variation commonly has been treated as discrete, especially at a regional level represented by maps showing site distributions. However, the increasing use of GIS has provided archaeologists with tools to create models of the more continuous use of space that is probably closer to the way in which people used landscapes (for example, Bell and Lock, 2000; Carey et al., 2006; Hill, 1998; Indruszewski and Barton, 2008; Kvamme, 2007; Llobera, 2001). An example of such continuous spatial models of prehistoric human land-use is discussed in the previous section and is shown in color Figure I.1.

In reality, human social practice changes simultaneously and more or less continuously across both temporal and spatial dimensions. Archaeologists have been rather less successful in representing such combined spatial-temporal change in forms other than narrative reconstructions that take the form of stories about past societies. Very often, such narratives present a story of past life during a time phase, then a story of life during a subsequent time phase, and yet another reconstructed past during the next phase. Sometimes processes internal or external to a society are invoked to account for differences between the reconstructed societies portrayed in each phase. When an attempt is made to model spatial-temporal change in a more formal or quantitative way, it usually takes the form of a series of graphs of change through time in material culture or a more inferential measure, such as political structure—with each graph representing a discrete geographic place—or a series of maps that show the spatial distribution of sites or other archaeological data—with each map representing a distinct temporal packet or phase. Although both of these kinds of representations can be seen in my own work (for example, Barton et al., 2004a, 2004b), neither does a particularly good job of representing continuous change in both space and time dimensions or interactions between spatial and temporal change.

Recently, several colleagues and I sought to characterize the spread of a particular hunting technology across Pleistocene Europe in the millennia leading up to the last glacial maximum (LGM). Careful reanalysis and calibration of several hundred radiocarbon dates for the first occurrence of shouldered points (*punta de muesca* or *pointe à cran*) suggested that these might indicate a radiation of this technology from eastern Europe as the climate became increasingly glacial. Using GIS methods, I created a computational model that combined spreading rates derived from a spatial analysis of the radiocarbon data with information about the rates at which humans on foot likely traversed terrain of varying ruggedness. An initial interpolation of the spatial-temporal distribution of shouldered points (color Figure I.2a) indicated that this technology spread from a single region in central Europe. I then used the location of the earliest date for the first appearance of shouldered points as a starting point and calculated an anisotropic cost-surface map that estimated costs for a human to walk across variable terrain from that point to any other point in Europe as a way of weighting the spread with real-world values about the ease and likelihood of foot travel. Next I calculated a regression equation that used the cost-surface values to estimate date of first occurrence for each known site with shouldered points. The regression gave a fairly good fit with $R^2 = 0.6$ and $p < 0.0001$. Finally, I used the regression equation and the cost-surface map in a map algebra expression to estimate the first appearance of shouldered points in every $30' \times 30'$ grid cell of Europe (Tiffagom et al., 2007). The resulting map is a model of the combined spatial-temporal spread of shouldered points across glacial Europe (Figure I.2b).

This computational model can be transformed back into a narrative. These points probably represent the tips of compound weapons used for hunting large steppe ungulates. This technology, and presumably the hunting practices that utilized it, first appeared in central Europe around 27,000 years ago. From that time and place it spread north and then both east and west across the "mammoth steppe" in front of the advancing continental ice sheets. As the ice sheets reached their farthest southern extents and the environment of Europe became the most glacial of the last 100,000 years, this technology then spread south across the rest of Europe to the Mediterranean.

This quantitative, computational model of ecological change among Pleistocene hunters of Europe is considerably less complex than the land-use models of eastern Spain discussed in the previous section. However, it is a more robust model in a couple of important respects. It is predictive in that it predicts the time of arrival for shouldered point hunting technology in nearly 22 million grid cells across all of Europe, of which only 69 have radiocarbon estimates. This predictive ability also makes it testable in a way that the land-models were not. To test this model, it is necessary only to locate more Upper Pleistocene sites with shouldered points and secure radiocarbon dates. Those dates could then be compared with the dates predicted by the model, and its accuracy can be systematically evaluated. Finally, this model is sufficiently general that it potentially could be applied to other technologies or items of material culture in other regions of the world as a way to represent the spatial-temporal changes in their distributions. It should easily differentiate between the dispersal of a form of material culture from a single source or from multiple sources.

However, this model is largely a descriptive one. Although it incorporates real-world human movement parameters, it tells us nothing about social or ecological process that drove the spread of shouldered point technology. It was necessary to return to an inferential narrative story to imbue this computational model with such information. Moreover, the model is very closely tied to data from the very sparse archaeological record (that is, only 0.00032% of the $30' \times 30'$ grid cells have data). Not only could new sites and dates produce a new regression equation for predicting the remaining grid cells, but a single new site discovered with an earlier first-appearance date for shouldered points than that of Jarosov II in the Czech Republic (the starting point for the original cost surface) could completely alter the spatial-temporal pattern of the model by becoming a new starting point for the walking costs map.

## Example 3: Resilience and Agriculture in the Southwestern United States

In the late 1990s I began a comparative project in the American Southwest to look at the socioecology of the shift from hunting and gathering to agricultural dependence in a region that bears considerable resemblance ecologically to the study area in eastern Spain discussed previously (Example 1) but that had experienced a different social history. My team carried out patch-based survey and limited subsurface testing in the vicinity of the middle Chevelon Creek drainage in north-central Arizona (Figure 7.2). Analyses of the survey data suggested important changes in the distribution of small farming settlements through time that we interpreted in a model of the reorganization agricultural ecology (Peeples, Barton, and Schmich, 2006).

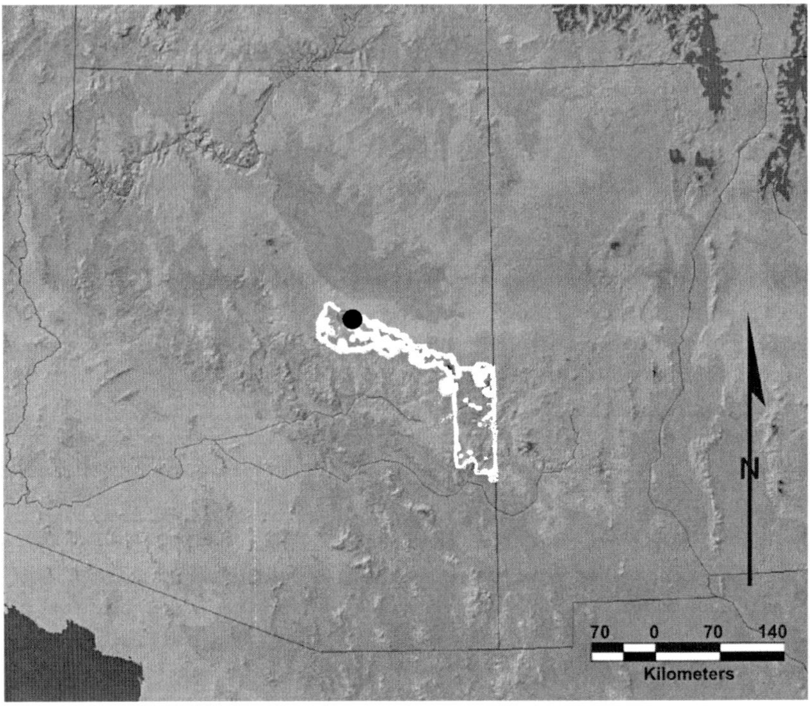


**Figure 7.2** Location of Mogollon Region Small Sites projects (survey and testing) in the state of Arizona. Black dot is location of fieldwork, white outline marks boundary of Apache-Sitgreaves National Forests.

In this model we suggested that the earliest part-time agriculturalists in this region focused their subsistence activities on the dynamic environments of deeply incised canyons—like that of Chevelon Creek. Small gardens of maize, squash, and beans—planted on terraces along streams—became one of a diverse suite of resources utilized by ancient inhabitants. Although the canyons provided a rich array of food resources, no particular resource or particular locale was continuously dependable, owing to regular flooding that continually altered the narrow canyon bottoms. This discouraged permanent settlement or labor investment in any particular resource, including domestic plants, and kept population densities low. Such Late Archaic sites are relatively uncommon in this area and tend to be highly clustered near the canyons (color Figure I.3).

However, sometime before 1000 C.E., a new strain of maize was introduced into the region that could be grown on the uplands between the canyons. Dominated by pinyon pine and juniper woodlands, these uplands were a much more stable landscape, in which swidden cultivation of maize could produce predictable yields from year to year. These uplands also had a much lower biodiversity than the canyon lands, and woodland clearance for cultivation would have reduced the availability of the only abundant wild crop in this environment, pinyon nuts. Together, this predictability and low biodiversity encouraged increasing dependence on cultivated crops. However, upland landscapes also suffered from thin soils that developed very slowly in the rocky terrain. A few years of cultivation would have exhausted the productive capacity of a field, and the agricultural potential of the lands around a small farmstead could have been exhausted in a generation or less, forcing a family to abandon its farm and establish a new residence in uncleared woodland. The thin soils recovered too slowly to permit the abandoned land to be cultivated by others, and abandoned farms were never reoccupied. After a few generations of successful upland agriculture, all available land had lost its ability to be farmed successfully, and the region was permanently abandoned. An important empirical implication of this model is that the repeated process of swidden cultivation, abandoning an unproductive farmstead, and establishing a new farm would have created a cumulative archaeological landscape of regularly spaced prehistoric sites—a pattern that we observed in this region (color Figure I.4).

In our original publication, this model was largely narrative and supported by GIS-based maps and statistical analyses. We presented a traditional archaeological reconstruction of ancient land-use in the middle Chevelon Creek drainage, but we also framed our model within the concepts of resilience theory (Gunderson and Holling, 2001; Redman, 2005; Redman and Kinzig, 2003) and suggested that it could offer more general insight into the socioecology of small-scale farming and the long-term consequences of human-environmental interactions. Although aspects of this model are potentially testable, the model's narrative format makes it difficult to test in any robust way. Untested, the model may serve as a useful cautionary tale, but its ability to contribute to land-use decisions for small-scale farming remains limited.

A few years ago, I carried out a test of one aspect of the model we developed for early agriculture in the middle Chevelon Creek region as an exercise in building computational models. I created a computational model in NetLogo (Wilensky, 1999)—a platform for the creation of agent-based models (ABM)—to test whether land-use dynamics in which small-holders exhausted the fields around a farmstead and moved to another could produce a cumulative archaeological landscape of regularly spaced sites. I created a group of agents, representing small-holder farming households, and gave each agent a set of simple land-use rules. In brief, each agent seeks to farm the most productive land in the vicinity of its farmstead. The harvest from cultivation a plot of land returns an abstract "energy currency" to the household. If a household accumulates sufficient excess energy, it will produce a daughter household that will establish a new farm at some distance from its original home. If a plot becomes unproductive, a new plot of land is cleared and cultivated. If a household loses enough energy because of unproductive farming, it will abandon its farmstead and establish a new one in a more productive locale. Besides seeking to maximize farming returns, each household seeks to minimize farming costs (including clearing new land or traveling to a farm plot) and to minimize the cost of moving and establishing a new farm. The environment in which the household agents live and farm is a simple one, with soil that loses fertility (and consequently the amount of energy produced by a harvest) each time it is farmed and vegetation that starts as woodland (that must be cleared to be farmed) that regrows slowly when a plot is not cultivated.

When this portion of our narrative is translated into a computational model in this way, it does indeed produce a cumulative archaeological landscape of regularly spaced sites. The computational model also offers additional, useful insights into this aspect of small-scale agriculture. A few farming household agents will be successful at swidden farming in the ABM and even produce daughter household agents that will also be successful initially. Declining fertility and agricultural returns will force agents to move to new locations, however. Over time, with fewer and fewer areas of potentially productive land available, agents will have to move more frequently and will not accumulate enough surplus to produce daughter household agents. Ultimately, all farmsteads will be abandoned, leaving a regularly spaced pattern of sites in the virtual "archaeological record." Each site has a short occupation span, with more recent sites occupied for shorter spans than older sites, and there is little overlap in the occupation spans of many of the sites. The vegetation returns to woodland as more and more farmsteads are abandoned, and soils become permanently depleted across most of the region.

Other socioecological dynamics also can produce an evenly spaced spatial distribution of farmsteads. If soils can regain fertility when they are left fallow, household agents can continue to occupy a farmstead permanently if they control enough land to rotate their fields over the time needed for a plot of land to regain sufficient fertility before it must be cultivated again. Because daughter household agents must seek unoccupied land on which to establish a new farm, over time, the landscape will become populated by evenly spaced farmsteads around which agent households rotate fields. The vegetation becomes highly diverse with patches of woodland and fallowed plots in various stages of forest succession. While farmsteads are evenly dispersed on the landscape, they are not often abandoned, and the occupations of most farmsteads overlap each other temporally. If something else (social or environmental changes) leads to regional abandonment, the "archaeological record" will be one of regularly spaced sites, most of which would have a long occupation span and with considerable temporal overlap among the occupations of different sites.

Translating the narrative that we developed to account for the archaeological record along Chevelon Creek into a computational algorithm offers several important benefits. First, it allows us to test or validate the model—or at least a portion of the model in this case. I was able to show that the kinds of land-use practices we proposed indeed can produce the archaeological record that we observed empirically. As van der Leeuw (2004) observes, ability to potentially falsify a computational or other formal model is one of the most powerful advantages of such an approach over inferential models, which cannot be evaluated in this way. Testing an independently derived model against empirical data is a rigorous way to differentiate better from worse models (that is, those that are better able to forecast real-world outcomes), rather than debating the "sense-making" merits of alternative inferential models. Second, it offers a way to investigate invisible past socioecological processes (van der Leeuw, 2004). Once a formal model like this one has been validated, it can be treated as a hypothesis about prehistoric land-use that can be tested in other locales. This provides a way to transparently evaluate the likelihood that particular socioecological processes did in fact take place in the past.

Finally, the computational model I developed is about small-scale farming, not about past societies. If we can show through validation against the archaeological record that such a model—or perhaps a more sophisticated version of such a model—can reliably forecast long-term outcomes of different land-use practices under different environmental conditions, it may be useful for land-use planning today. In fact, even the simple, abstract model I constructed in NetLogo offers an important insight in this regard. Under the environmental conditions that prevailed along Chevelon Creek, households had to regularly shift residences to make a living; and eventually they had to abandon the region entirely. But the computational model also shows that the very same kinds of land-use practices led to the prehistoric abandonment of middle Chevelon Creek also could produce long-term, sustainable agriculture under slightly different conditions, where thicker soils could regain their fertility through fallowing. The computational model is not about what happened in the past but about the dynamics of human socioecological systems. The Chevlon case helps to validate the general model and illustrates one possible outcome of a suite of land-use practices in one particular place and time. However, the model makes clear that this is not the only possible outcome, and in so doing it becomes a potentially valuable planning tool.

## Example 4: Socioecological Dynamics in the Mediterranean

The last example is from the Mediterranean Landscape Dynamics (MedLand) project. Although this large-scale (six years) National Science Foundation supported project[2] involved numerous archaeologists and paleoecologists, and worked extensively with archaeological and paleoenvironmental data, it was not designed to learn about the past. Rather, from the outset, the goal of this project was to build a sophisticated modeling laboratory to carry out virtual experiments (*sensu* Bankes, Lempert, and Popper, 2002; van der Leeuw, 2004) on the long-term, recursive interactions between society, land-use, and environmental change. Rather than reconstruct past societies from archaeological data in this virtual environment, we use the rich archaeological record of the Mediterranean region to validate and parameterize the components of the MedLand Modeling Laboratory (MML).

Because it aims to simulate high-resolution, real-world landscape dynamics and land-use practices, the MML is a hybrid modeling environment with components that utilize different modeling approaches (Mayer and Sarjoughian, 2007, 2008, 2009; Mayer et al., 2006). These include GIS-based modeling of landscape surface processes such as hydrology, erosion/deposition, and vegetation succession; an ABM of farming households and their land-use practices; and regression-based paleoclimate and paleovegetation models (Figure 7.3). The details of the MML are published elsewhere (Barton, Ullah, and Bergin, 2010; Barton, Ullah, and Mitasova, 2010; Ullah, 2011; Ullah and Bergin, 2012), so I will offer only a brief synopsis here.

Agropastoral land-use can be modeled stochastically or in an agent-based modeling (ABM) platform. When modeled stochastically, farming and grazing patches are randomly distributed within catchments—calculated using GIS routines to account for terrain and suitability for farming and/or herding —around communities (Barton, Ullah, and Mitasova, 2010; Ullah, 2011). Alternatively, individual households can be simulated as virtual agents, organized into villages. Agents choose land to farm or graze on the basis of their need for farming/herding returns (calories that affect birth and death rate), the suitability of land for particular agricultural activities, and costs to use the land (including access on foot and vegetation clearance) (Barton, Ullah, and Bergin, 2010; Mayer and Sarjoughian, 2009; Mayer et al., 2006; Ullah and Bergin, 2012). Agents also can collect fuel wood, the need for which varies according to household size and activities. Whether modeled stochastically using GIS routines or by using ABM technology, household land-use can alter the vegetation cover and soil characteristics.

These anthropogenic changes to landscapes, in turn, affect the results of other surface processes on landscapes in terms of location and intensity of erosion and deposition. These are simulated in a surface process model that uses 3-dimensional algorithmic implementations of the unit stream power erosion/deposition (USPED) model—based on the well-known universal soil loss equation (USLE/RUSLE)—and the reach-average shear stress equation (Degani, Lewis, and Downing, 1979; Flannagan, Laflen, and Meyer, 2003; Haan, Barfield, and Hayes, 1994; Mitasova, Mitas, and Brown, 2001; Mitasova et al., 2004; Moore and Burch, 1986; Singh and Phadke, 2006; Tucker and Hancock, 2010; Warren et al., 2005; Wischmeier, Johnson, and Cross, 1971; Wischmeier and Smith, 1978). Climate and natural vegetation values also are input to the landscape module and can come from modern values or, for

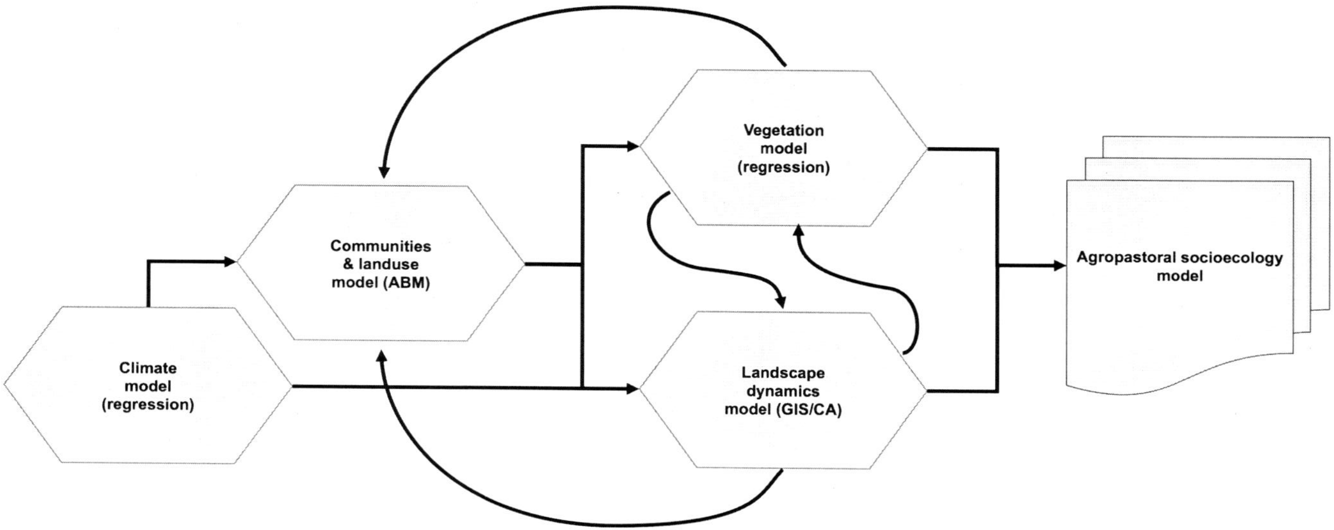


**Figure 7.3** Schematic representation of components in MedLand Modeling Laboratory

ancient settings, from the results of paleovegetation and paleoclimatic modeling. The landscape simulation can change the terrain and soil depth, as well as allow vegetation to regrow and soils to regain fertility if a patch of land is not used. This provides feedback to land-use in the ABM when households make new decisions about which land to farm and graze.

Initial results of using the MML to study the dynamics of ancient land-use have been promising. In one set of modeling experiments, we parametrized the MML with values derived from Neolithic farming communities of northern Jordan and examined the consequences of varying land-use (intensive vs. shifting cereal cultivation, and ovicaprine grazing vs. no grazing) and community size on soil loss and vegetation over the course of two and ten generations, 40 and 200 years, respectively (Barton, Ullah, and Bergin, 2010; Barton, Ullah, and Mitasova, 2010). Some of the modeling results confirm widely held expectations about such land-use: shifting cultivation can cause greater soil loss than repeated cultivation of a few plots (for example, with manuring); extensive ovicaprine grazing will cause more erosion than farming without associated grazing; larger communities with more people farming and grazing more land will have a greater impact on the landscape than smaller communities (color Figure I.5). Other results were less intuitively obvious because of the complex interactions between land-use and landscape dynamics. Notably, when community size is below some threshold, whose value is determined by local environmental conditions, the amount of soil loss can be substantially offset by soil accumulation (that is, by the redeposition of sediments eroded from other parts of a catchment), so that the economic effects of mixed agropastoral land-use can be negligible or even beneficial. However, if communities pass the threshold size, the consequences change qualitatively such that soil loss greatly exceeds soil accumulation within a land-use catchment. This imbalance continues over the long-term, with the potential for leaving a catchment unsuitable for farming. One mitigating strategy is, not surprisingly, to reduce community size through emigration or fissioning. Another less obvious solution discovered in these experiments is to increase the area devoted to grazing relative to cultivation, moving zones of soil loss into uncultivated uplands, and providing more sediment for redeposition in the areas around farmed fields. Conservation measures, like terracing, also could be instituted but could require some degree of social reorganization to ensure the availability of sufficient labor for terrace construction and long-term maintenance. This kind of investment in landesque capital and intensification of land-use has often been accompanied by the growth of inequalities in social power and prestige.

As inferred independently from the archaeological record, Neolithic settlement and land-use generally follow patterns suggested by our modeling experiments (Kuijt and Goring-Morris, 2002; Legge and Harris, 1996; Martin, 1999; Quintero, Wilke, and Rollefson, 2004; Rollefson and Kohler-Rollefson,

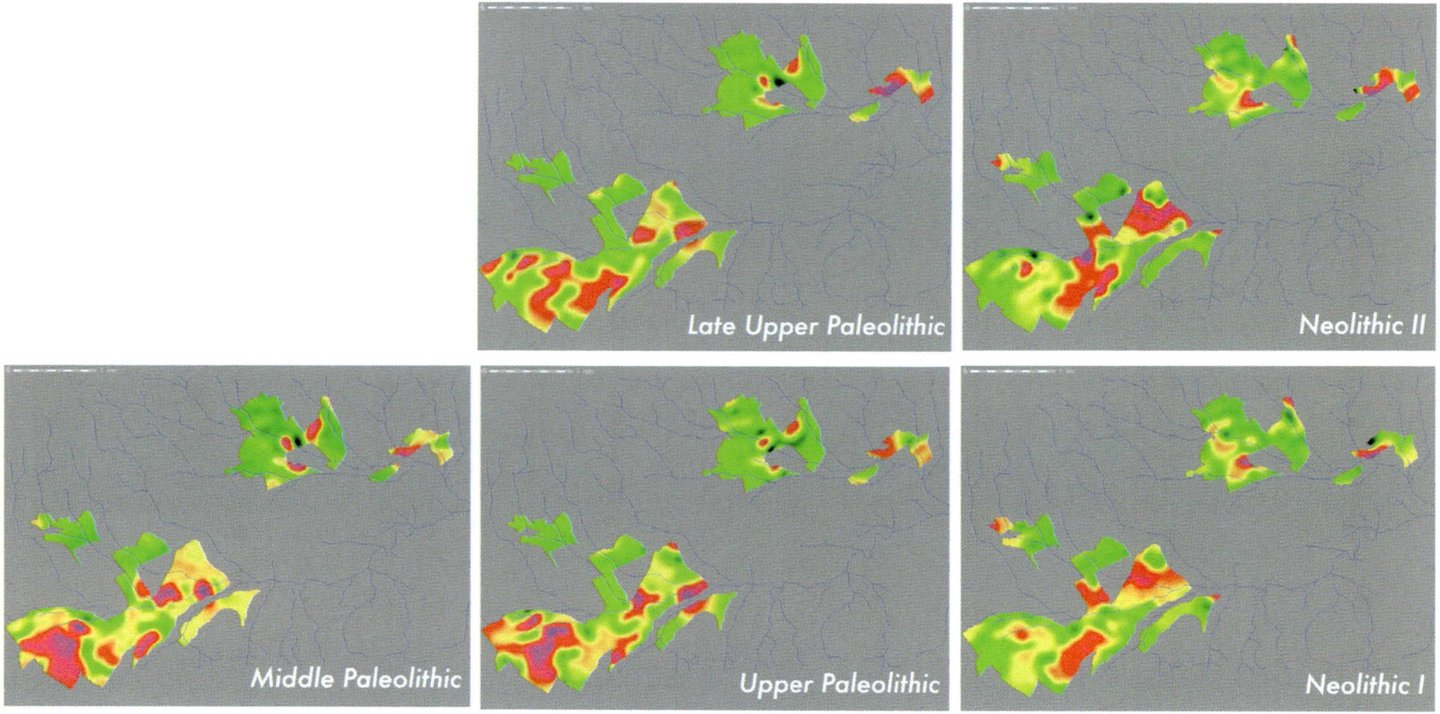


*Figure I.1* *Settlement Intensity Index for time periods from the Middle Paleolithic through Neolithic II-B in the Polop Alto Valley. High values represent more intensive settlement (see Barton et al., 1999; Barton et al., 2002; Barton et al., 2004b for calculation of settlement intensity index).*

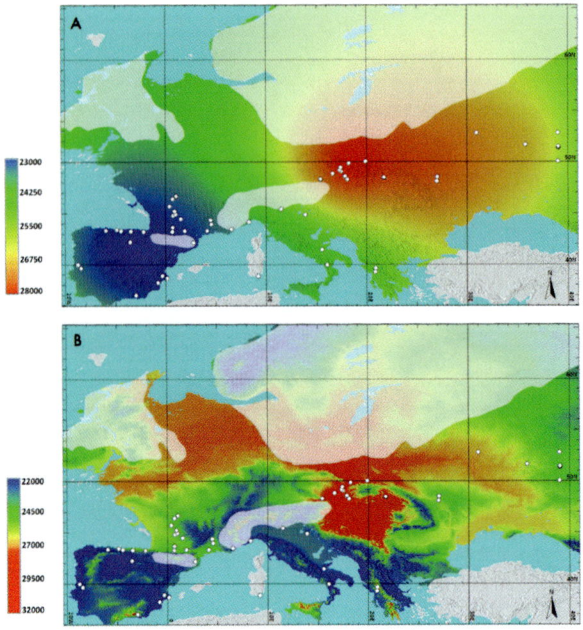


*Figure I.2* *Spatial-temporal distribution for the dissemination of shouldered point hunting technology in western Eurasia during the Late Pleistocene: (A) Interpolation of the spatial-temporal distribution of the first occurrence of shouldered at a series of Upper Paleolithic archaeological sites (indicated by circles); (B) Map of spatial-temporal spread of shouldered points combining interpolation and anisotropic cost surface of walking costs.*

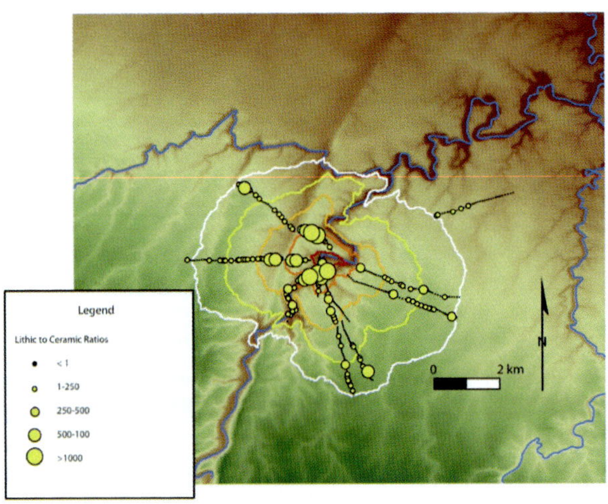


*Figure I.3* *Density of lithics in survey patches along transects extending outward from Chevelon Crossing, Middle Chevelon Creek, Apache-Sitgreaves National Forests, Arizona (from Peeples, Barton, and Schmich, 2006).*

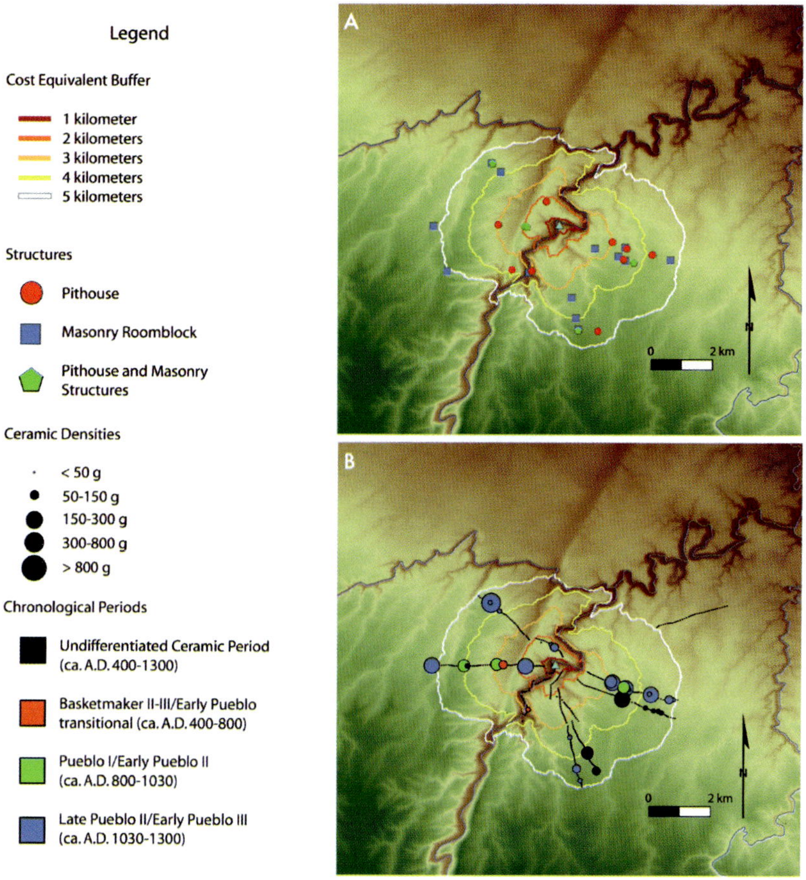


*Figure I.4 Locations of structures (A) and densities of ceramics (B) and in survey patches along transects extending outward from Chevelon Crossing, Middle Chevelon Creek, Apache-Sitgreaves National Forests, Arizona (from Peeples, Barton, and Schmich, 2006).*

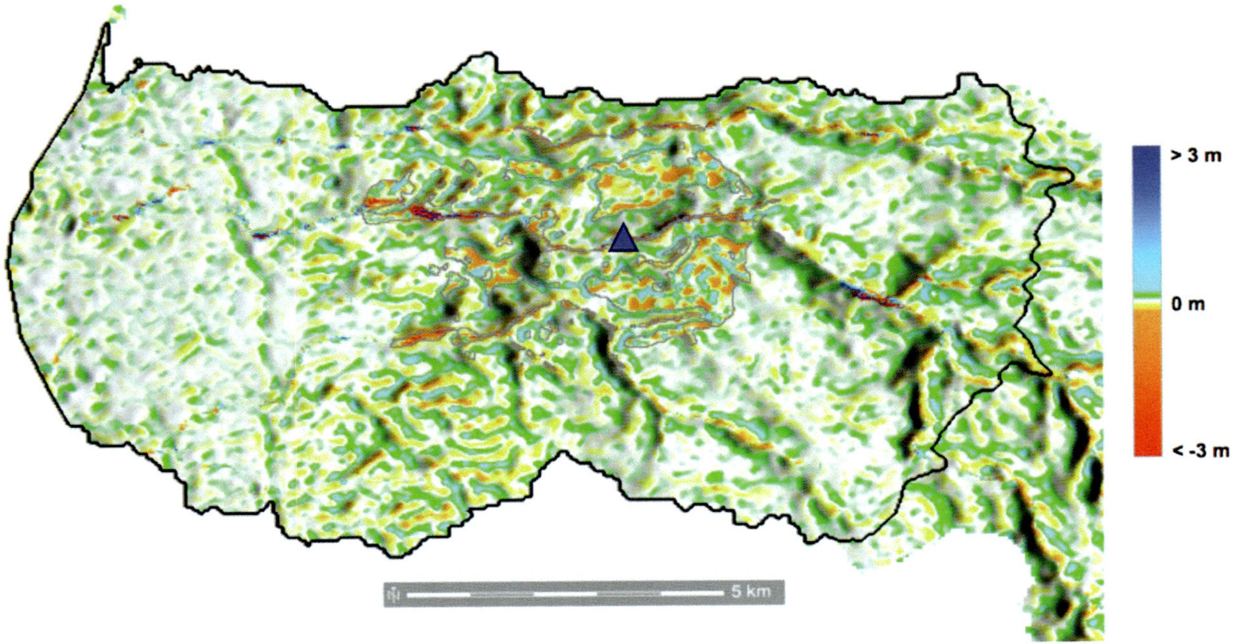


*Figure I.5* Cumulative erosion and deposition around a Neolithic village after two generations (40 years) of shifting cultivation and ovicaprine grazing. Black line indicates outer limits of grazing catchment, gray line indicates outer limit of shifting cultivation catchment. Village indicated by blue triangle (from Barton, Ullah, and Bergin, 2010).

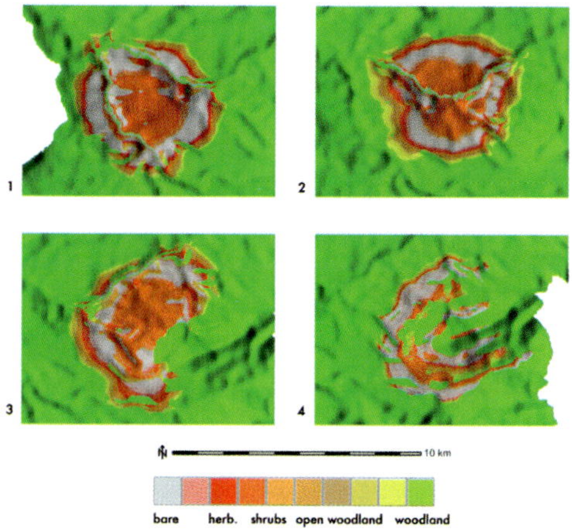


*Figure I.6 Maps of anthropogenically altered land-cover around each of the four village locations after 100 years of shifting cultivation and ovicaprine grazing. Numbers refer to locations of village in Figure 7.4 (from Ullah and Bergin, 2012).*

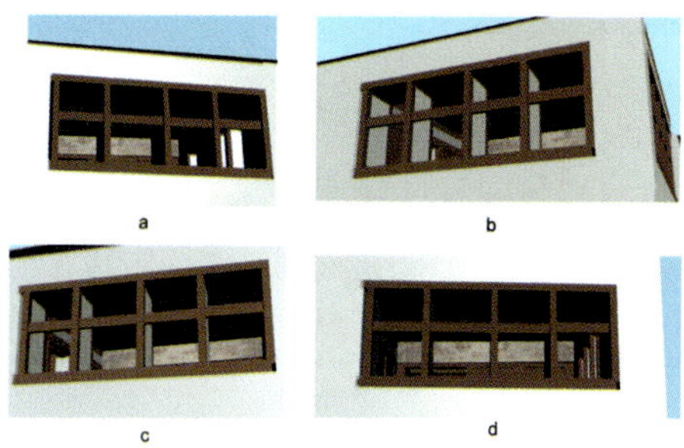


*Figure I.7 3D model of Room 5 (first floor) of the West House, Akrotiri, Thera; (a–d): views from the public space outside the building (Square of the Ship Procession) (reconstruction by E. Paliou, 2011, Figure 3).*

*Figure I.8 Views of the wall paintings of the adorants from different viewpoints in Room 3 (Xeste3, ground floor) (reconstruction by E. Paliou, Paliou, Wheatley, and Earl, 2011, Figure 5).*

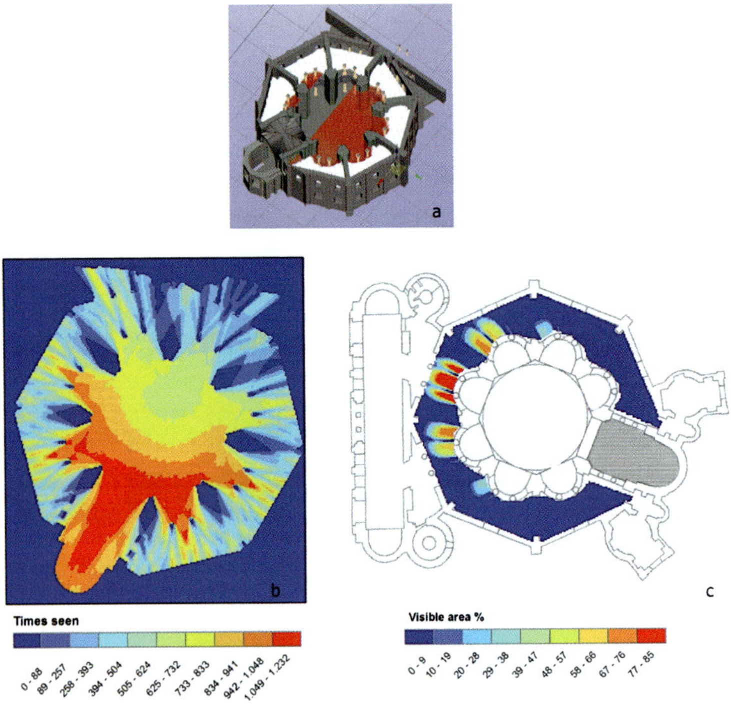


*Figure I.9* (a) 3D model of San Vitale (created by I. Giannakopoulou, modified by E. Paliou, based on Deichmann, 1989); the sampled area (10,270 viewpoints, sampling every 20 cm) at the matroneum is indicated in white; (b) times-seen of the ground floor from the matroneum; (c) visible area (%) of the chancel (hatched) as seen from the matroneum.

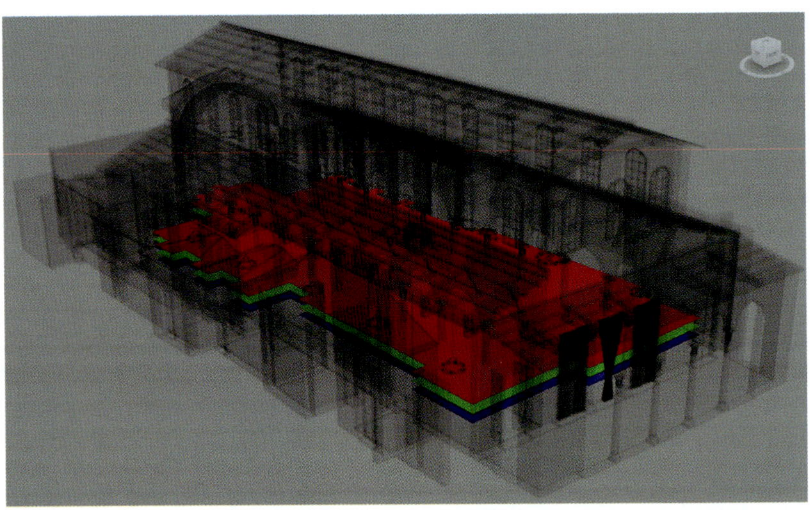

*Figure I.10* Planar geometry defining observers (top 2 m; middle 1.5 m; bottom 1 m)

*Figure I.11 Light meter indicating that the amount of light entering from the clerestory windows and the other openings linearly decreases as it reaches the floor of the Basilica*

*Figure I.12 Pseudocolor exposure of the nave of the Basilica at 1 P.M.; the lighting values show high levels of illumination at the upper walls of the nave gradually decreasing to the floor. The Ambon is well lit during the day, whereas the sanctuary is the area of the nave that is less lit.*

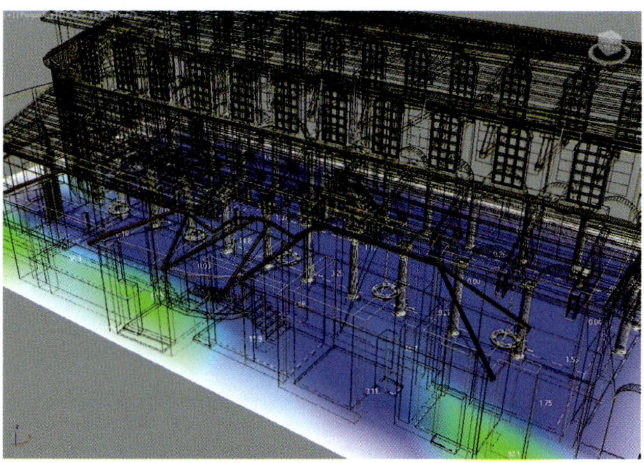

*Figure I.13 Light meter indicating the amount of light entering through the side openings mainly affecting the aisles*

*Figure I.14 Pseudocolor exposure indicating the reflections caused by the shiny surfaces*

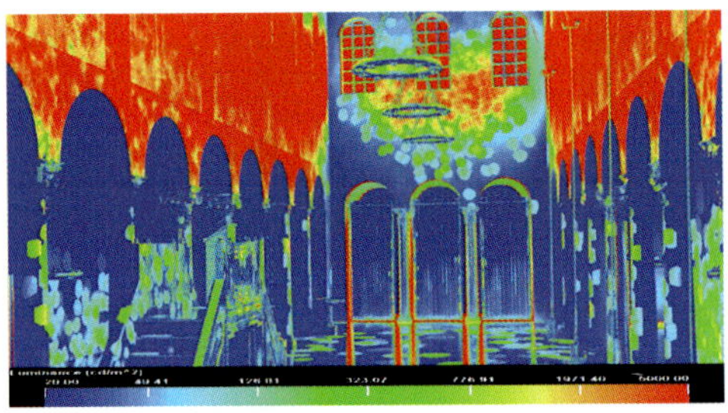


*Figure I.15 Pseudocolor exposure of an alternative structural model with the curtains of the triforium closed*

*Figure I.16 Pseudocolor exposure of an alternative structural model with the curtains of the triforium and the side doors open*

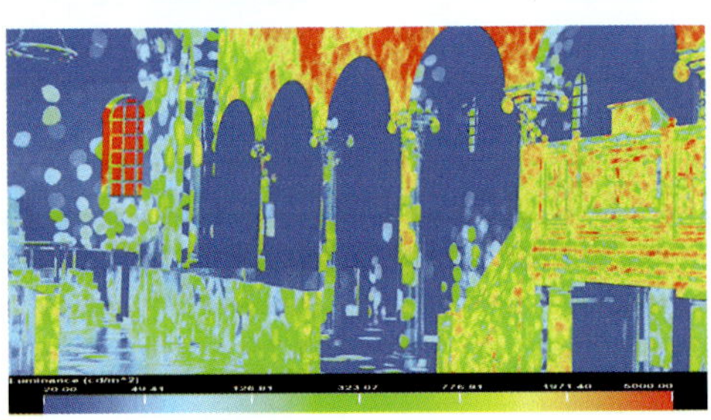


*Figure I.17 Pseudocolor exposure of an alternative structural model with windows at the east aisle*

1992; Rosen, 2008; Simmons, 2007; Twiss, 2007). In the earliest Prepottery Neolithic (PPN-A), most communities were very small and practiced mixed agropastoral subsistence strategies. These mixed farming/herding strategies continued into the PPN-B, as many communities grew in size, with some in apparently favorable localities having thousands of inhabitants. In the Late Neolithic, however, the large towns were abandoned or greatly reduced in size, with most of the regional population again living in very small communities. Additionally, some groups may have begun to rely more on animal herding, marking the beginning of mobile pastoral economies. Finally, the Late Neolithic record of Mesopotamia, at least, is interpreted to suggest increased social inequality along with investments in landesque capital. Note, however, that currently, we can only make comparisons between our model results and the *inferred* prehistoric record of southwestern Asia; we are not yet able to evaluate model results directly against empirical archaeological data.

A second experiment carried out in the MML involved studying the results of situating a small farming village in different topographic contexts within the Rio Penaguila and upper Rio Serpis Valleys of eastern Spain, the location of one of the earliest known farming communities in the Iberian Peninsula (Bernabeu Aubán et al., 2003; Bernabeu Aubán and Orozco Köhler, 2005). In four different experimental runs (Figure 7.4) a simulated village, populated by household agents, was set alternatively in an alluvial plain (for easy access to land for farming and grazing), in a canyon bottom (for seclusion), at the base

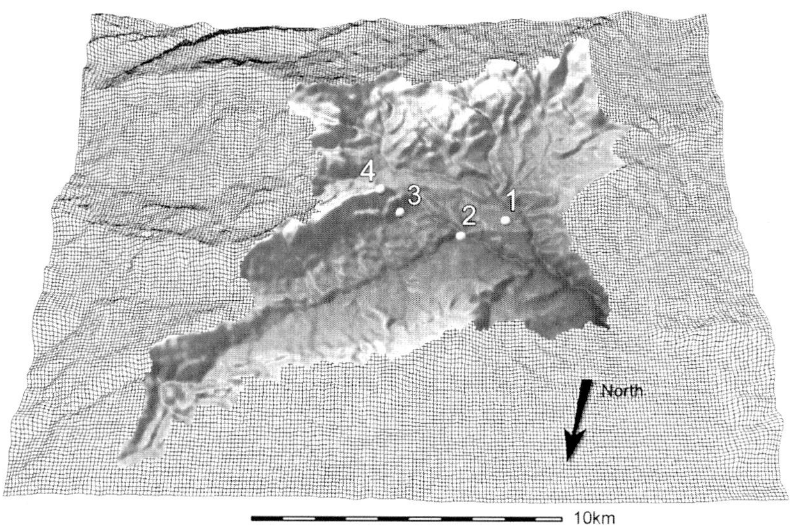


**Figure 7.4** Four different locations for modeled village and its agropastoral catchment. Setting is in the Rio Penaguila and upper Rio Serpis valleys of northern Alicante Province, Spain (from Ullah and Bergin, 2012).

of a cliff (for defensibility), and on a topographic prominence (for maximum visibility). In each locale, all other initializing parameters besides geographic setting were kept the same for the village. The agents farmed and grazed the land around each site for 100 years, and resulting data were collected on population size, economy, vegetation cover, and erosion/deposition (color Figure I.6). When the village was located in the alluvial plain it was more successful initially than when placed in the other settings, as measured in terms of population growth and agricultural returns. However, this success also led to a recursive, self-amplifying growth cycle of increasing population, expanding cultivation and grazing, soil degradation and loss, and even more expansion of cultivation and grazing. When situated in the other locales, the same village grew more slowly and experienced more variable economic returns. However, the smaller and more stable population also had much less detrimental impact on the surrounding landscape (Ullah and Bergin, 2012). Archaeological sites are commonly found in a variety of topographic settings—including ones analogous to those in the MML experiment—and archaeologists regularly speculate on the reasons for settlement placement. But as far as I know, there has been no study to assess the ecological impact of prehistoric settlements in such different settings. This set of experiments, then, provides a set of hypotheses about the long-term ecological consequences of socially mediated site placement that could be tested against the archaeological record.

Unlike the other models presented, the MML is not designed to tell us about the past but is rather a much more generic modeling environment or laboratory where we can ask a wide variety of questions about human socioecological systems. As such, it can also be used to simulate land-use/landscape dynamics in prehistoric contexts. Also, unlike the other models presented, the MML is not designed to match empirical data. The other models were inherently valid in that they accounted for a set of empirical data. But as already noted, such inherent validation of inferential models—even computational ones—does not necessarily make them more accurate. The MML is very different conceptually in this regard. Because it is designed on the basis of our knowledge of human land-use practices and surface processes as we observe them today, the MML can create realistic scenarios that do not match the prehistoric record. For example, we regularly run what we call control models for our experiments in which landscape simulations are run without a human presence—because the environment changes whether people are involved or not. This approach allows us to calibrate the results of experiments in which we *do* simulate human land-use to better measure net anthropogenic consequences of alternate suites of activities (Barton, Ullah, and Bergin, 2010; Barton, Ullah, and Mitasova, 2010). Results generated by the MML can be treated as hypotheses about human-environmental interaction. When these complex hypotheses are tested repeatedly and validated against situations for which we know the results (that is, from the archaeological record), it can give us increasing confidence that the models are accurately simulating relevant processes and can forecast the likely real-world results.

## Discussion

The past dozen years of exploring ways to embed computational model-building into archaeology has led me to a changing perspective on archaeological practice. While the work we are now doing with the MML involves a variety of activities that traditionally would be considered archaeology, our goal is not to collect empirical data of the archaeological record in order to reconstruct past societies. Rather, we are using the archaeological record to verify and to improve our models of human socioecological dynamics. The goal of this research is to gain a better understanding of the operation of human society and to use that understanding to create formal or computational models that can more reliably forecast outcomes of human decisions and social practice—in the past, present, or future.

As noted, the archaeological record is one of shreds and patches (to paraphrase Hamlet) and in many cases is inadequate to reconstruct the past at the level of detail needed to create reliably accurate models of ancient social process. However, it can be very useful as a dataset against which we can test the long-term and large-scale results of computational models created from ethnographic knowledge of the small and intimate practices of human actors. It is this potential to forecast large-scale, long-term consequences of day-to-day decision making that can make an archaeology refocused in this way increasingly valuable to social science and to society more generally.

For the most part, social science has largely remained an observational science since its inception. Much like natural philosophy and natural history of the 18th and 19th centuries, social scientists collect data on human social behavior "in the wild" and inductively create narrative, statistical, or formal models to account for observed behaviors. The practical, and in recent decades ethical, restrictions on carrying out controlled social experiments has precluded the testing of most social science models, retarded the development of comprehensive social theory, and severely limited the ability of social sciences to produce reliable forecasts about the consequences of social practice—even when models are highly formal and quantified (Buchanan, 2009; *The Economist*, 2009; Farmer and Foley, 2009). Hence, social science has been saddled with the unwarranted label of being a soft science (Diamond, 1987). Very recent microscale social experiments, scaled up through formal and quantitative modeling, are beginning to revolutionize our perception of social dynamics (Bowles and Gintis, 2003; Fehr and Gächter, 2000; Henrich et al., 2006; Janssen et al., 2010). However, we still cannot carry out experiments at the scale of real-world social groups that could involve hundreds to millions of individuals, or experiments to study effects over the course of decades to millennia. For such studies, computational modeling experiments, verified against the archaeological and historic records can play a vital role—especially since the coarse-grained nature of the archaeological record tends to emphasize processes occurring at larger spatial scales and long temporal scales, while processes happening in more familiar human time and space frames are often blurred, mixed, or impossible to distinguish (Shennan, 2002: 9–21; van der Leeuw, 2004).

Science is not simply the creation and use of technology, but some technologies can be transformational in science. The telescope transformed astronomy, the microscope transformed biology, and particle accelerators transformed physics. Social sciences have largely made little use of technologies—and in some cases have even appeared antitechnological. Modern archaeology is similarly a largely low-tech field. It is true that there are a few archaeologists who study bone isotopic chemistry, or use magnetometry or other geophysical methods in the field, but most data-collection and analysis methods have changed little in over a century. Surface data are collected in pedestrian walkovers; excavations are carried out by shovel and trowel or patishe and goofa; artifact classes are intuited and exemplars counted; analyses are statistically simple and largely can be done with a spreadsheet (or even by hand with a little more time). Computers are used most widely and intensively to write narratives. Traditional methods may suffice very well for some tasks, but there has been little impetus for technological development within archaeology. Rather, archaeologists are more often the (frequently reluctant) consumers of technologies developed by other fields. Astronomers and physicists, in contrast, well understand the importance of enabling technologies such as the Hubble space telescope and Large Hadron Collider and work to develop these technologies as they do other aspects of their science.

Computational model is a potentially transformative technology for the social sciences in general and archaeology in particular (Bankes, 2002; Cioffi-Revilla, 2009). It offers a new way to couple the rich and varied database of the archaeological record with field observations of societies today, microscale social experiments, and the complex dynamics of large-scale socioecological systems (Kohler and van der Leeuw, 2007; van der Leeuw, 2004). And although such modeling may give us a different window on ancient societies, its greater value lies in its potential to integrate the record of the past with a new science of social dynamics—a science we need to help us to face the challenges of the 21st century. To fully benefit from the potential of this technology, archaeology also will need to transform itself. Because the power of modeling lies in its potential to connect data, theory, and explanation in powerful ways and to carry out computational experiments in social process and dynamics, it must be used by archaeologists acting as social scientists; it cannot be relegated to a "black-box" type of specialist analysis such as radiocarbon dating. Currently, few social scientists, and even fewer archaeologists, use computational modeling or are more than vaguely aware of its existence. It will be necessary for archaeologists to intellectually retool—requiring investments of time and resources—and to establish new academic programs that teach computational or algorithmic thinking to the next generation of archaeologists, along with new methods in combination with the still important domain knowledge of the archaeological record. Likewise, archaeologists need to become actively involved with the development of technology such as computational modeling for social science. We cannot afford to leave this to others who are not trained in the social sciences. With a rich and unique database that is critical for building robust models of long-term social change, archaeologists can play a fundamental role in the development of an advanced and socially valued science of human society.

## Notes

1. Supported by National Science Foundation grants BNS-9115209, SBR-9904050, BCS-0075292, and BCS-0331583.
2. Supported by National Science Foundation grants BCS-410269, BCS-638879, and BCS-543848.

## References

Abu-Lugbod, L. 1991. Writing against culture, in Fox, R. G. (Ed.), *Recapturing Anthropology: Working in the Present*, pp. 137–62. Santa Fe, NM: SAR Press.

Bankes, S. C. 2002. Agent-based modeling: A revolution? *Proceedings of the National Academy of Sciences of the United States of America* 99(90003): 7199–200.

Bankes, S. C., Lempert, R., and Popper, S. 2002. Making computational social science effective: Epistemology, methodology, and technology, *Social Science Computer Review* 20(4): 377–88.

Barton, C. M. 1991. Retouched tools: Fact or fiction? Paradigms for interpreting chipped stone, in Clark, G. A. (Ed.), *Perspectives in Prehistory Paradigmatic Biases in Circum-Mediterranean Hunter-Gatherer Research*, pp. 143–63. Philadelphia: University of Pennsylvania Press.

———. 1997. Stone tools, style, and social identity: An evolutionary perspective on the archaeological record, in Barton, C. M., and Clark, G. A. (Eds.), *Rediscovering Darwin: Evolutionary Theory in Archaeological Explanation, Archaeological Papers of the American Anthropological Association*, Vol. 7, pp. 141–56. Washington, D.C.: American Anthropological Association.

Barton, C. M., Bernabeu Aubán, J., Aura Tortosa, J. E., and Garcia Puchol, O. 1999. Landscape dynamics and socioeconomic change: An example from the Polop Alto Valley, *American Antiquity* 64(4): 609–34.

Barton, C. M., Bernabeu Aubán, J., Aura Tortosa, J. E., Garcia, Puchol, O., and La Roca, N. 2002. Dynamic landscapes, artifact taphonomy, and landuse modeling in the western Mediterranean, *Geoarchaeology* 17(2): 155–90.

Barton, C. M., Bernabeu Aubán, J., Aura Tortosa, J. E., and Molina Balaguer, L. 2004a. Historical contingency, nonlinearity, and the neolithization of the western Mediterranean, in Wandsnider, L., and Athanassopoulos, E. (Eds.), *Current Issues in*

*Mediterranean Landscape Archaeology*, pp. 99–124. Philadelphia: University of Pennsylvania Press.

Barton, C. M., Bernabeu Aubán, J., Garcia Puchol, O., Schmich, S., and Molina Balaguer, L. 2004b. Long-term socioecology and contingent landscapes, *Journal of Archaeological Method and Theory* 11(3): 253–95.

Barton, C. M., and Clark, G. A. (Eds.). 1997. *Rediscovering Darwin: Evolutionary Theory in Archaeological Explanation*. Archaeological Papers of the American Anthropological Association (Vol. 7). Washington, D.C.: American Anthropological Association.

Barton, C. M., Ullah, I. I., and Bergin, S. 2010. Land use, water, and Mediterranean landscapes: Modelling long-term dynamics of complex socio-ecological systems, *Philosophical Transactions of the Royal Society A: Mathematical, Physical and Engineering Sciences*, 368(1931): 5275–97. doi:10.1098/rsta.2010.0193

Barton, C. M., Ullah, I., and Mitasova, H. 2010. Computational modeling and Neolithic socioecological dynamics: A case study from southwest Asia, *American Antiquity* 75(2): 364–86.

Bell, T., and Lock, G. 2000. Topographic and cultural influences on walking the Ridgeway in later prehistoric times, in Lock, G. R. (Ed.), *Beyond the Map: Archaeology and Spatial Technologies*, NATO Science Series, Series A, Life Sciences v. 321, pp. 85–100. Amsterdam: Washington, D.C., Tokyo: IOS Press.

Bernabeu Aubán, J., Barton, C. M., García, O., and La Roca, N. 1999. Prospecciones sistemáticas en el valle del Alcoi (Alicante): Primeros resultados, *Arqueología Espacial* 21: 29–64.

________. 2000. Systematic survey in Alicante, Spain: First results, *Türkyie Bilimer Akademisi Arkeoloji Dergisi* 3: 57–86.

Bernabeu Aubán, J., Guitart, I., and Pascual, L. 1989. Reflexiones en torno al patrón de asentamiento en el País Valenciano entre el Neolítico y la Edad del Bronce, *Saguntum* 22: 99–123.

Bernabeu Aubán, J., Molina Balaguer, L., Díez Castillo, A., and Orozco Köhler, T. 2006. Inequalities and power: Three millennia of prehistory in Mediterranean Spain (5600–2000 cal BC), in Díaz del Río, P., and García Sanjuan, L. L. (Eds.), *Social Inequality in Iberian Late Prehistory*, pp. 97–116. Oxford: Archaeopress, BAR International Series.

Bernabeu Aubán, J., and Orozco Köhler, T. 2005. Mas d'Is (Penàguila, Alicante): Un recinto monumental del VI milenio cal AC, in Arias, P., Ontañón, R., and García-Moncó, C. (Eds.), *III Congreso sobre el Neolítico en la Península Ibérica (Santander, 2003)*, pp. 485–95. Santander: Instituto Internacional de Investigaciones Prehistóricas de Cantabria.

Bernabeu Aubán, J., Orozco Köhler, T., Diez Castillo, A., and Gomez Puche, M. 2003. Mas d'Is (Penàguila, Alicante): Aldeas y recintos monumentales del Neolítico Antiguo en el Valle del Serpis, *Trabajos de Prehistoria* 60(2): 39–59.

Binford, L. R. 1962. Archaeology as anthropology, *American Antiquity* 28(2): 217–25.

Bleed, P. 2001. Trees or chains, links or branches: Conceptual alternative for consideration of stone tool production and other sequential activities, *Journal of Archaeological Method and Theory* 8(1): 101–27.

Bowles, S., and Gintis, H. 2003. The origins of human cooperation, in Hammerstein, P. (Ed.), *Genetic and Cultural Evolution of Cooperation*, pp. 429–43. Cambridge, MA: The MIT Press.

Buchanan, M. 2009. Economics: Meltdown modelling, *Nature* 460(7256): 680–82. doi:10.1038/460680a

Carey, C. J., Brown, T. G., Challis, K. C., Howard, A. J., and Cooper, L. 2006. Predictive modelling of multiperiod geoarchaeological resources at a river confluence: A case study from the Trent-Soar, UK, *Archaeological Prospection* 13(4): 241–50. doi:10.1002/arp.295

Cioffi-Revilla, C. 2009. Keynote address to the Computational Social Sciences Society, Arizona State University, Tempe.

Clifford, J. 1983. On ethnographic authority, *Representations* (2): 118–46.

Degani, A., Lewis, L. A., and Downing, B. B. 1979. Interactive computer simulation of the spatial process of soil erosion, *Professional Geographer* 31(2): 184–90.

Diamond, J. M. 1987. Soft sciences are often harder than hard sciences, *Discover* 8: 34–39.

Dunnell, R. C. 1982. Science, social science, and common sense: The agonizing dilemma of modern archaeology, *Journal Anthropology Research* 38: 1–25.

Ebert, J. I., Camilli, E. L., and Berman, M. J. 1996. GIS in the analysis of distributional archaeological data, in Maschner, H. D. (Ed.), *New Methods, Old Problems: Geographic Information Systems in Modern Archaeological Research*, Center for Archaeological Investigations Occasional Paper, Vol. 23, pp. 25–37. Carbondale: Southern Illinois University.

*The Economist* Staff. 2009. What went wrong with economics and how the discipline should change to avoid the mistakes of the past, *The Economist*, July 16th.

Farmer, J. D., and Foley, D. 2009. The economy needs agent-based modelling, *Nature* 460(7256): 685–86. doi:10.1038/460685a

Fehr, E., and Gächter, S. 2000. Cooperation and punishment in public goods experiments, *The American Economic Review* 90(4): 980–94.

Fisher, C. T., Hill, J. B., and Feinman, G. M. (Eds.). 2009. *The Archaeology of Environmental Change: Socionatural Legacies of Degradation and Resilience*. Tucson: University of Arizona Press.

Flannagan, D. C., Laflen, J. M., and Meyer, L. D. 2003. *Honoring the Universal Soil Loss Equation: Historic Landmark Dedication Pamphlet*. West Lafayette, IN: American Society of Agricultural Engineers and Purdue University, http://topsoil.nserl.purdue.edu/flanagan/usle-landmark/booklet.htm.

García Puchol, O., Barton, C. M., and Bernabeu Aubán, J. 2008. Aplicación de métodos de prospección geofísica en la detección de un gran foso del IV milenio cal ac en el yacimiento del Alt del Punxó (Muro de l´Alcoi, Alacant), *Trabajos de Prehistoria* 65(1): 143–54.

Geertz, C. 1973. *The Interpretation of Cultures: Selected Essays*. New York: Basic Books.

Gunderson, L. H., and Holling, C. S. (Eds.). 2001. *Panarchy: Understanding Transformations in Human and Natural Systems*. Washington, D.C.: Island Press.

Haan, C. T., Barfield, B. J., and Hayes, J. C. 1994. *Design Hydrology and Sedimentology for Small Catchments*. New York: Academic Press.

Hegmon, M. 1998. Technology, style, and social practices: Archaeological approaches, in Stark, M. T. (Ed.), *The Archaeology of Social Boundaries*, pp. 264–79. Washington, D.C.: Smithsonian Institution Press.

———. 2003. Setting theoretical egos aside: Issues and theory in North American archaeology, *American Antiquity* 68(2): 213–43.

Henrich, J., McElreath, R., Barr, A., Ensminger, J., Barrett, C., Bolyanatz, A., Cardenas, J. C., et al. 2006. Costly punishment across human societies, *Science* 312(5781): 1767–70. doi:10.1126/science.1127333

Hill, J. B. 1998. Ecological variability and agricultural specialization among the protohistoric Pueblos of central New Mexico, *Journal of Field Archaeology* 25: 275–94.

Hill, J. B., Clark, J. J., Doelle, W. H., and Lyons, P. D. 2004. Prehistoric demography in the Southwest: Migration, coalescence, and Hohokam population decline, *American Antiquity* 69(4): 689–716.

Indruszewski, G., and Barton, C. M. 2008. Cost surface DEM modelling of Viking Age seafaring in the Baltic Sea, in Frescher, B., and Dakouri-Hild, A. (Eds.), *Beyond Illustration: 2D and 3D Digital Technologies as Tools for Discovery in Archaeology, International Series*, pp. 56–64. Oxford: British Archaeological Reports.

Janssen, M. A., Holahan, R., Lee, A., and Ostrom, E. 2010. Lab experiments for the study of social-ecological systems, *Science* 328(5978): 613–17. doi:10.1126/science.1183532

Keesing, R. M. 1974. Theories of culture, *Annual Review of Anthropology* 3(1): 73–97. doi:10.1146/annurev.an.03.100174.000445

Killick, D. 2004. Social constructionist approaches to the study of technology, *World Archaeology* 36(4): 571–78. doi:10.1080/0043824042000303746

Kohler, T. A., and van der Leeuw, S. E. 2007. *The Model-Based Archaeology of Socionatural Systems*. Santa Fe, NM: School for Advanced Research Press.

Kuijt, I., and Goring-Morris, N. 2002. Foraging, farming, and social complexity in the pre-pottery Neolithic of the Southern Levant: A review and synthesis, *Journal of World Prehistory* 16(4): 361–440.

Kvamme, K. 2007. Integrating multiple geophysical datasets, in *Remote Sensing in Archaeology*, pp. 345–74, http://dx.doi.org.ezproxy1.lib.asu.edu/10.1007/0-387-44455-6_14.

Legge, T., and Harris, D. 1996. The beginnings of caprine domestication in Southwest Asia, in *The Origins and Spread of Agriculture and Pastoralism in Eurasia*, pp. 238–62. London: UCL Press.

Llobera, M. 2001. Building past landscape perception with GIS: Understanding topographic prominence, *Journal of Archaeological Science* 28: 1005–14.

Longacre, W. A. 1964. Archeology as anthropology: A case study, *Science* 144(3625): 1454–55.

Martin, L. 1999. Mammal remains from the eastern Jordanian Neolithic, and the nature of caprine herding in the steppe, *Paléorient* 25(2): 87–104.

Mayer, G. R., and Sarjoughian, H. S. 2007. Complexities of simulating a hybrid agent-landscape model using multi-formalism composability, in *Agent-Directed Simulation, Spring Simulation Multi-Conference, Norfolk, Virginia*, pp. 161–68.

________. 2008. A composable discrete-time cellular automaton formalism, in *First International Workshop on Social Computing, Behavioral Modeling, and Prediction*, pp. 187–96. Presented at the First International Workshop on Social Computing, Behavioral Modeling, and Prediction. Phoenix, AZ: Springer.

________. 2009. Composable cellular automata, *SIMULATION* 85(11–12): 735–49. doi:10.1177/0037549709106341

Mayer, G. R., Sarjoughian, H. S., Allen, E. K., Falconer, S. E., and Barton, C. M. 2006. Simulation modeling for human community and agricultural landuse, in

*Agent-Directed Simulation, Proceedings of the Agent-Directed Simulation Multi-Conference, Huntsville, Alabama*, pp. 65–72.

Mitasova, H., Mitas, L., and Brown, W. M. 2001. Multiscale simulation of land use impact on soil erosion and deposition patterns, in Stott, D. E., Mohtar, R. H., and Steinhardt, G. C. (Eds.), *Sustaining the Global Farm: Selected Papers from the 10th International Soil Conservation Organization Meeting, May 1999, Purdue University*, pp. 1163–69. Purdue, IN: USDA-ARS National Soil Erosion Research Laboratory.

Mitasova, H., Thaxton, C., Hofierka, J., McLaughlin, R., Moore, A., and Mitas, L. 2004. Path sampling method for modeling overland water flow, sediment transport, and short-term terrain evolution in Open Source GIS, in Miller, C. T., Farthing, M. W., Gray, V. G., and Pinder, G. F. (Eds.), *Proceedings of the XVth International Conference on Computational Methods in Water Resources (CMWR XV)*, pp. 1479–90. Chapel Hill, NC: Elsevier.

Moore, I. D., and Burch, G. J. 1986. Physical basis of the length-slope factor in the universal soil loss equation, *Soil Science Society of America Journal* 50(5): 1294–98.

Newton, D. 1981. The individual in ethnographic collections, *Annals of the New York Academy of Sciences* 376(1): 267–87. doi:10.1111/j.1749-6632.1981.tb28172.x

Pauketat, T. R., and Alt, S. M. 2005. Agency in a postmold? Physicality and the archaeology of culture-making, *Journal of Archaeological Method and Theory* 12(3): 213–37. doi:10.1007/s10816-005-6929-9

Peeples, M., Barton, C. M., and Schmich, S. 2006. Resilience lost: Intersecting landuse and landscape dynamics in the upland southwest, *Ecology and Society* 12(22).

Quintero, L. A., Wilke, P. J., and Rollefson, G. O. 2004. Highland towns and desert settlements: Origins of nomadic pastoralism in the Jordanian Neolithic, in *Central Settlements in Neolithic Jordan*, pp. 201–13. Berlin: ex oriente.

Redman, C. L. 2005. Resilience theory in archaeology, *American Anthropologist* 107(1): 70–77.

Redman, C. L., and Kinzig, A. P. 2003. Resilience of past landscapes: Resilience theory, society, and the Longue Durée, *Conservation Ecology* 7(1): 14.

Rollefson, G. O., and Kohler-Rollefson, I. 1992. Early Neolithic exploitation patterns in the Levant: Cultural impact on the environment, *Population and Environment* 13: 243–54.

Rosen, S. A. 2008. Desert pastoral nomadism in the Longue Durée: A case study from the Negev and the southern Levantine deserts, in *The Archaeology of Mobility: Old World and New World Nomadism*, pp. 115–40. Los Angeles: Cotsen Institute of Archaeology, University of California.

Schiffer, M. B. 1983. Towards the identification of formation processes, *American Antiquity* 48: 675–706.

Shennan, S. 2002. *Genes, Memes, and Human History: Darwinian Archaeology and Human Evolution*. London: Thames and Hudson.

Simmons, A. H. 2007. *The Neolithic Revolution in the Near East: Transforming the Human Landscape*. Tucson: University of Arizona Press.

Singh, R., and Phadke, V. S. 2006. Assessing soil loss by water erosion in Jamni River Basin, Bundelkhand region, India, adopting universal soil loss equation using GIS, *Current Science* 90(10): 1431–35.

Taylor, C. 1971. Interpretation and the sciences of man, *The Review of Metaphysics* 25(1): 3–51.

Tiffagom, M., Jordá Pardo, J. F., Barton, C. M., and Aura Tortosa, J. E. 2007. Le vent de l'Est: Essai de mise en contexte chronologique et paléogéographique de la pointe à cran solutréenne. Presented at the Le Solutréen: 40 ans après le Smith'66. Conference of the Union Internationale des Sciences Préhistoriques et Protohistoriques, Preuilly-sur-Claise, France.

Tucker, G. E., and Hancock, G. R. 2010. Modelling landscape evolution, *Earth Surface Processes and Landforms* 35(1): 28–50. doi:10.1002/esp.1952

Twiss, K. C. 2007. The Neolithic of the southern Levant, *Evolutionary Anthropology* 1: 24–35.

Ullah, I. I. 2011. A GIS method for assessing the zone of human-environmental impact around archaeological sites: A test case from the Late Neolithic of Wadi Ziqlâb, Jordan, *Journal of Archaeological Science* 38(3): 623–32. doi:10.1016/j.jas.2010.10.015

Ullah, I. I., and Bergin, S. 2012. Modeling the consequences of village site location: Least cost path modeling in a coupled GIS and agent-based model of village agropastoralism in eastern Spain, in White, D. A., and Surface-Evans, S. L. (Eds.), *Least Cost Analysis of Social Landscapes: Archaeological Case Studies for Beginners and Experts Alike*. Salt Lake City: University of Utah Press.

van der Leeuw, S. E. 2000. Land degradation as a socionatural process, in McIntosh, R. J., Tainter, J. A., and McIntosh, S. K. (Eds.), *The Way the Wind Blows: Climate, History, and Human Action*, pp. 357–84. New York: Columbia University Press.

________. 2004. Why model? *Cybernetics and Systems: An International Journal* 35(2): 117. doi:10.1080/01969720490426803

van der Leeuw, S. E., and Redman, C. L. 2002. Placing archaeology at the center of socio-natural studies, *American Antiquity* 67(4): 597–605.

Warren, S. D., Mitasova, H., Hohmann, M. G., Landsberger, S., Iskander, F. Y., Ruzycki, T. S., and Senseman, G. M. 2005. Validation of a 3-D enhancement of the Universal Soil Loss Equation for prediction of soil erosion and sediment deposition, *Catena* 64: 281–96.

Weatley, D. 1995. Cumulative viewshed analysis: A GIS-based method for investigating intervisibility and its archaeological application, in Lock, G., and Stančič, Z. (Eds.), *Archaeology and Geographical Information Systems: A European Perspective*, pp. 171–85. London: Taylor and Francis, Ltd.

Wilensky, U. 1999. *NetLogo*. Center for Connected Learning and Computer-Based Modeling. Evanston, IL: Northwestern University.

Williams, I., Limp, F., and Briuer, F. L. 1990. Using geographic information systems and exploratory data analysis for archaeological site classification and analysis, in Allen, K. M., Green, S. W., and Zubrow, E. B. (Eds.), *Interpreting Space GIS and Archaeology*, pp. 239–73. London: Taylor & Francis.

Wischmeier, W. H., Johnson, C. B., and Cross, B. V. 1971. A soil erodibility nomograph for farmland and construction sites, *Journal of Soil and Water Conservation* 26: 189–92.

Wischmeier, W. H., and Smith, D. D. 1978. Predicting rainfall-erosion losses: A guide to conservation planning, *USDA Agriculture Handbook* 282.

Wylie, A. 1992. The interplay of evidential constraints and political interests: Recent archaeological research on gender, *American Antiquity* 57(1): 15–35.

________. 2000. Questions of evidence, legitimacy, and the (dis)union of science, *American Antiquity* 65(2): 227–38.

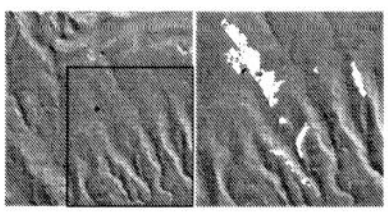

# 8 The Potential and Limits of Optimal Path Analysis

*Irmela Herzog*

## Aims of Least-Cost Path Calculations in Archaeology

In the last two decades an increasing number of archaeological least-cost path (LCP) studies have been published. In nearly all these publications, the aim has been to reconstruct ancient routes and route networks (for example, Fábrega Álvarez and Parcero Oubiña, 2007; Verhagen and Jeneson, 2012) or to identify the principal factors governing the construction of known roads or road segments (for instance, Bell and Lock, 2000; Kantner and Hobgood, 2003). Sometimes LCPs help to identify locations which hold a central position in a territory: in the example of Hare (2004), a site is at a crossroads of intra- and extra-valley LCPs and all north-south connections between major settlements pass through that site. LCP analysis recently has become part of the predictive modeling toolkit in American archaeology (Whitley and Burns, 2008). Such models are based on the assumptions that the initial travel arteries were created by migratory herbivore travel and that the early hunters followed herds of these animals; subsequently, subsistence and settlement patterns developed in connection with these first routes.

Bell and Lock (2000), as well as Batten (2007), proposed using LCP analysis for relative dating: in their view, the location of sites along routes between two other sites implies that the midpoint sites were founded later than the endpoints. But this does not apply to all roads: in medieval Britain, drove roads often avoided villages (Hindle, 2002: 16).

LCP analysis may also be applied to design modern routes that best uphold the visual impact of an important archaeological monument: Conolly and Lake (2006: 226) refer to a study by Batchelor published in 1999 trying to optimize the modern roads near the World Heritage site of Stonehenge so that the routes would have least visual impact. In fact, LCP analysis included in GIS software was not developed for the small community of archaeologists interested in this technology. Approaches to minimize travel time and fuel consumption while taking other factors such as the impact on the environment into account are important for modern highway planning, shipping companies, and the military. The LCP technology was developed mainly with focus on these purposes (Longley et al., 2005: 358).

Most archaeologists applying push-button LCP software are not aware of the limitations of these GIS functions. As Kantner notes: "GIS software has become reasonably accessible, with user-friendly interfaces that obscure underlying procedures, the details of which are hidden away in online manuals and white papers" (2012). The aim of this paper is to present a best-practice guide to LCP approaches and to outline possible improvements of the GIS software available for this purpose.

## The Model for LCP Analysis in Archaeology

LCP analysis is based on the assumption that people optimize the costs of routes that are taken frequently, which, over time, leads to the development of the real-world equivalent of an LCP. The route can be considered to have an optimal and a random component (Saerens et al., 2009). At first, the contribution of the random component is very high, but in the long run it is very small. This idea is the basis of genetic algorithms that have also been used for solving LCP problems in the presence of conflicting objectives (Zhang and Armstrong, 2005). Standard LCP implementations, however, calculate only the optimal route component.

Two small examples may help to explain, why LCPs are appropriate for prehistoric or medieval route reconstruction: a trader who is overtaken by a colleague on a faster route and delivers the goods later incurs more costs and will take the fast route next time to reduce his/her costs. When food is scarce, those people that minimize their energy expenditure while maximizing the energy gains resulting from their endeavors have a higher chance of survival. According to Surface-Evans and White, "LCA assumes that humans will tend to economize many aspects of their behavior" in general (2012: 2). At a small scale, people will optimize the route to the closest well or their preferred social venue, whereas at a larger scale, an LCP may connect two neighboring towns that exchange goods.

This model may not apply to once-in-a-lifetime journeys such as a mass exodus, a crusade, or some of the behavior of soldiers at war. Most probably, people on such journeys used existing roads, but, in spite of some guidance or advice by locals or experienced travelers in the area, the optimal route is rarely taken on these occasions. Similarly, tourists or newcomers to an island who are exploring an area may first climb a high peak to get an overview, although such a climb may require lots of energy. Alternatively, they may take long detours. Likewise, a shepherd walking with his or her flock will not follow the LCP connecting the start and the end points of the journey. For modeling movements of this kind, agent-based models are perhaps more appropriate (for example, Lake, 2000; Ratti, 2005); however, this chapter focuses on LCP calculations.

## LCP Calculation: An Overview

Several steps are necessary to calculate LCPs (Figure 8.1). First of all, the cost components should be identified that are to be included in the model. If topographic cost components are important, reconstruction of the ancient landscape may be necessary, especially if human impacts, such as bulk material extraction, or natural processes, such as erosion and accumulation, are present in the study area. Alternatively, the effects of landscape change may be analyzed in a later step by experiments with variations on the modern data. After choosing the appropriate cost function for each cost component, one must select a method to combine these functions, if several cost components are included in the model. Next, appropriate GIS software is required to calculate the LCP. In general, the software first creates an *accumulated cost surface (ACS)* that stores the accumulated costs for traveling outward from the origin and then traces the LCP from the target location back to the origin (for example, Conolly and Lake, 2006: 215–24, 252; de Smith, Goodchild, and Longley, 2007: 142–45; Wheatley and Gillings, 2002: 151–58).

Important aspects of ACS generation and of the backtracking step are implemented differently in the different LCP software currently available, and other relevant aspects (gray dot items in Figure 8.1) are at present not supported in any of the popular GIS packages. Owing to these different implementations in GIS software, the outcome of LCP analysis varies significantly, and it is often not possible to re-create the results of one software package with another (Gietl, Doneus, and Fera, 2008). For this reason, the software user should be aware of the advantages and disadvantages of different LCP implementations. This guideline applies to calculation of a single LCP and even more so to networks, because different network models result in widely varying least-cost route networks.

The LCP analysis should not stop once the first optimal routes were generated. Instead, the reliability of the results should be investigated by comparing them to archaeological evidence. If ancient route sections are known, the LCPs should be compared to these known route locations. In the absence of

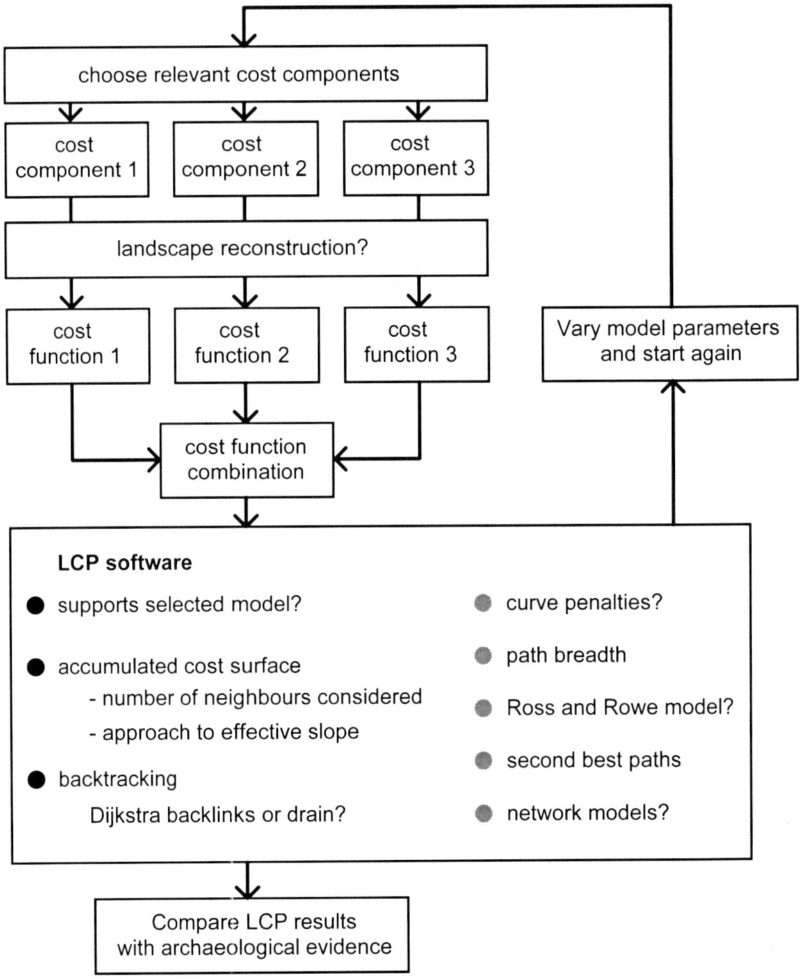


**Figure 8.1** The steps of LCP generation: First, the relevant cost components are identified (three cost components in this example); palaeoenvironmental studies might be required to reconstruct the topography of the past period studied. Second, for each cost component, an appropriate cost function is selected. Third, LCP software is chosen on the basis that it should support the desired combination of cost functions; moreover, the user should be aware of the drawbacks of this software implementation and of aspects not yet supported in current GIS packages (gray dot items on right). Several models should be tested and compared with the archaeological evidence.

such information, LCPs can be analyzed with respect to the distribution of known sites, features, or finds of the period considered. Several studies indicate (for instance, Andresen, Iversen, and Jensen, 2008; Kvamme, 1992) that finds of certain periods are associated with ancient roads.

Several computations with different cost components and/or different weights for the cost components are necessary to investigate how adequately each model fits archaeological reality and to assess the impact of different cost factors on route construction. Moreover, the stability of the result should be investigated, for example by experimenting with slightly varying departure and destination points for the paths (Helbing, Keltsch, and Molnár, 1997). In the simulation study by Helbing and his coauthors, trail bundles appeared. Such trail bundles have also been observed in the archaeological record and in modern times.

If we introduce small deviations of the model parameters, we can reconstruct alternative routes, which often form bundles, but sometimes small variations will result in widely differing LCPs (for an example, see Herzog and Posluschny, 2011). In fact, some historic accounts record two alternative routes between A and B that were taken depending on weather conditions or other factors such as inundation. An example is given by Hindle (2002: 12), who describes a situation in the medieval Forest of Dartmoor, England, where a detour had to be taken during bad weather conditions that was twice as long as the regular route.

## The Development of LCP Analysis in Archaeology

Already in 1977, Bintliff (112–13) underlined the importance of walking time rather than map distances and described catchment areas delimited by a given walking time threshold. A first step toward computer-assisted LCP calculations in archaeology was the publication by Ericson and Goldstein in 1980. They suggested analyzing the relative efficiencies in energy expenditure among identical or substitute resources and evaluating the uniqueness of a path between two loci by comparing the ancient route to the path of least effort. The study by Gorenflo and Gale (1990) is another precursor for modern archaeological LCP calculations. They present a popular slope-dependent cost curve, the Tobler hiking function, for the first time in an archaeological context; however, their approach to model a cost surface is not as exact as that created by modern GIS software.

The developments of LCP analysis and cost-surface analysis in archaeology are closely connected, because, as mentioned, most LCP methods calculate an ACS as a first step. With the advent of efficient raster GIS procedures, some archaeologists felt that it became feasible to calculate fairly realistic cost surfaces (for instance, Gaffney and Stančič, 1992). The first cost-surface analysis with GIS presented at a Computer Application in Archaeology conference (CAA) was by Gaffney and Stančič in 1991 (published in 1992) in their study of the island of Hvar. A more detailed version of this study (Gaffney and Stančič, 1998) includes an LCP between two locations on the island. Shortly thereafter and in the same conference series, van Leusen (1993) suggested a method for determining territorial boundaries using cost-surface techniques. The method is outlined in general, no cost function is given, and the section on cost surfaces ends with the statement that the task of finding the optimal path between two sites "is yet to be explored."

In 2000 the book *Beyond the Map* (Lock, 2000) appeared containing two LCP studies that are quoted widely: The cost function in Llobera's contribution has received much attention; another even more important point is that he combined costs of the landscape and attraction to cultural features in one model. The chapter by Bell and Lock in the same book introduces the significant concept of effective slope.

A new development is the use of LCP analysis for understanding human movement in built-up areas (for example, Branting, 2004; Haciguzeller, 2008; Paliou, 2008; Stöger, 2008). The LCP for an individual traveler on a natural landscape differs significantly from the traffic flows within an urban landscape, any other built-up area, or within a larger building: pedestrian movement is restricted to streets and squares, and congestion may play a role (for instance, Batty, 2003; Longley et al., 2005: 373–74).

LCP analysis in rural areas is often considered a well-established task in modern computer applications in archaeology (for example, Rahn, 2005, 2007), so that some authors even skip naming the method or cost function used (for instance, Moreno Martín, 2008; cf. Kantner, 2012). The case studies in the book by White and Surface-Evans (2012) mainly rely on the software available in current GIS software. Eight out of eleven case studies apply the ArcGIS path distance function, GRASS r.walk, and two different custom applications are used in the other studies. Although various cost functions and cost factors are discussed in some of the contributions, most case studies rely on the LCP results of the GIS software, and some derive new methods on this basis. However, calculating LCPs in rural areas is by no means trivial; the relevant issues are discussed in the next sections.

## Cost Components for LCPs: An Overview

Two types of costs are defined in publications on cost paths: isotropic and anisotropic (for example, Conolly and Lake, 2006: 215; Husdal, 2000: 13–14; Surface-Evans and White, 2012; Wheatley and Gillings, 2002: 151). *Isotropic costs* are independent of the direction of movement; typical isotropic costs are those based on vegetation, soil properties, lakes, barriers such as large rivers without fords, taboos, and social attraction. In general, a friction raster map is used to store the isotropic costs for traversing each cell. If only isotropic costs are considered, such friction maps form the basis of the ACS calculation. *Anisotropic costs* are dependent on the direction of movement. Slope is anisotropic, and slope is so important that sometimes the term *anisotropic costs* actually means costs based on slope (Conolly and Lake, 2006: 253). Another type of anisotropic cost is traveling by boat on a river—upstream is slower than downstream. Taking the prevailing wind direction and the wind velocity into account when sailing on the sea or a lake is somewhat anisotropic; that is, in this case anisotropy has one favored direction. Implementation of isotropic cost calculations in a raster grid is fairly straightforward, whereas anisotropic models require more sophisticated approaches.

An anisotropic cost function can be converted easily to an isotropic function by assuming that the same route was chosen for returning back to the source location. However, no appropriate isotropic cost grid can be derived from this isotropic function, because the cost of traversing the cell still depends on the direction of movement. This aspect is discussed in more detail in the next section.

## Slope-Dependent Cost Functions

Nearly all archaeological LCP studies are based on slope, often combined with some other component (for a lengthier treatment, see Herzog, 2010). The following cost function for modeling pedestrian movement can be recommended, drawing on physiological data

$$\text{cost}(s) = 1337.8\, s^6 + 278.19\, s^5 - 517.39\, s^4 - 78.199\, s^3 + 93.419\, s^2 + 19.825\, s + 1.64$$

where $s$ is mathematical slope (rise over run). The function is based on data presented by Minetti and associates (2002). The authors of that study proposed a fifth-degree polynomial curve fitting that data, but this function is unsatisfactory, because it produces unrealistic negative values for steep downhill slopes. Another alternative is the fourth-degree polynomial function discussed by Llobera and Sluckin (2007); yet the estimates given by this function deviate significantly from the Minetti data for steep downhill slopes and flat terrain. The estimates provided by the sixth-degree polynomial shown here are close to the Minetti data, and the function does not produce unintuitive negative values.

Gorenflo and Gale (1990) introduced the Tobler hiking function, used in many archaeological LCP studies (for example, Batten, 2007; Fábrega Álvarez and Parcero Oubiña, 2007; Parslow, 2009; Verhagen and Jeneson, 2012; four case studies in White and Surface-Evans, 2012). This function measures time rather than energy expenditure and is a rule of thumb, not based on scientific experiments.

A wide variety of pack animals have been found useful for transport, and it is possible to suggest differing cost functions depending on the animal species involved (although a detailed review of the studies available on the energy expenditure of selected species is beyond the scope of this chapter). However, in general, one might argue that pack animals are led by walkers who choose the direction of movement, with walkers adjusting their paths to the needs of pack animals only if the animals are less resourceful than humans. Studies carried out with burros (Yousef, Dill, and Freeland, 1972) suggest that these animals experience less energy expenditure in walking—that is, they can follow a human walker easily. Most probably the cost function for human pedestrians is therefore appropriate even for walkers followed by pack animals, if the pack animals considered can cope with more difficult terrain and are faster than walkers.

However, another cost function is appropriate for wheeled vehicles, since these cannot climb steep slopes as easily as walkers can. For wheeled transport, a cost function depending on a *critical slope* parameter (that is, the transition where switchbacks become more effective than direct uphill or downhill paths) has been defined based on ideas presented by Llobera and Sluckin (2007):

$$\mathrm{cost}(s) = 1 + (s \,/\, \check{s})^2$$

where both $s$ and $\check{s}$ are percent slope values. Critical slope is given by $\check{s}$ and typically is in the range of 8 to 16. This cost curve is symmetric and can be used if a route was taken in both directions.

## Other Topographic Cost Components

Pedestrian and wheeled vehicle speeds and energy requirements also depend on vegetation cover, wetlands, and other soil properties. However, land cover may have changed considerably with time; some examples are given by Bell, Wilson, and Wickham (2002): swamps can be drained, forests can be felled, and scrubland cleared. So it might be inappropriate to include values for different vegetation cover directly in LCP calculations, because finally the path was there and progress on the path was faster than in the surrounding area. Still, for travelers to create a path by repeatedly and regularly walking on a given route is very difficult given some vegetation types and fairly easy with others. For this reason, vegetation cover should be considered primarily in the context of path formation and not for travelers using an existing route. Even if a path exists, wetness and snow or ice matter, except if extensive construction work has been carried out. Different path properties were preferred in different periods: for example, according to Hindle (2002: 34–35), medieval travelers in England and Wales shunned the hard surface of surviving Roman roads for the softer ground alongside.

Archaeological LCP studies that deal with traveling on water are based on cost estimates rather than physiological experiments or data calibrated on the scene (for example, Bevan, Forthcoming; Indruszewski and Barton, 2007; Livingood, 2012). A model accounting for varying winds during the year in the Mediterranean Sea was published by Arcenas (n.d.). In general, the traveling speed and costs depend on the boat, the mode of propulsion, the type of water (creek, river, lake, sea), currents, winds, weather, and many other factors. Owing to this multitude of factors, modeling the costs of movement on water is quite difficult. With mixed means of transport, the model must ensure that appropriate costs are paid when changing the mode of transportation.

## Social and Cultural Cost Components

Archaeological LCP analysis has been criticized for being environmentally deterministic, although cultural or social variables can be included in the LCP calculation. But it is difficult to model social and cultural cost contributions, because they often cannot be derived in a straightforward way from geographic parameters, and different meanings and values may be attached to one location by different people (cf. Lock, 2003: 176–77). Cost functions to model attraction or repulsion to given localities in the landscape have been suggested (for instance, Llobera, 2000).

Van Leusen (2002: 6.8) notes that "it is unlikely that 'social' costs can be established with any degree of objectivity," and this assertion is also supported by Bell, Wilson, and Wickham (2002), who point out that political boundaries, social divisions, cultural taboos, or attractions leave hardly any mark in the archaeological record and therefore can be difficult or impossible to reconstruct.

The route layout depends also on the purpose of the path. For example, civil and military roads have some aspects in common; that is, quick progress is the main aim, but on a military road high observability is often avoided, whereas it increases safety on a civil road (Lee and Stucky, 1998). In fact, the only social or cultural cost factors that have so far been regularly considered in LCP calculations are connected with visibility issues (for example, Bell and Lock, 2000; Lock and Pouncett, 2010; Madry and Rakos, 1996; Rahn, 2005, 2007; Verhagen and Jeneson, 2012).

Different cost models are appropriate for the reconstruction of formal and informal routes as described by Thrombold (1991) and Earle (1991): formal routes show evidence of planning and purposeful construction such as roadbeds, pavement, or bridges, whereas informal routes involve minimal or no labor directed to their creation or maintenance. Formal routes require some engineering skill and organization (for instance, the construction and maintenance of the Roman road network). Route layouts designed mainly for social prestige or religious purposes may be detected only by comparing LCPs with remains of ancient routes. To my knowledge, specific slope-dependent cost functions for formal or informal route reconstruction do not exist. Instead, in a formal context, a complex model is required that allows contour line roads and is able to reconstruct tunnels and bridges (for example, Yu, Lee, and Munro-Stasiuk, 2003), whereas the model for informal route reconstruction should avoid contour line roads.

### Dealing with Several Cost Components

Batten (2007) advocates starting with the simplest model, which, in rugged terrain, is based mainly on slope, and to refine this model if necessary after comparing the paths based on the simple model with the known or expected paths. Similarly, Bevan (Forthcoming) argues that hyper-real models (that is, those with a greater number of inputs) "can become so bewilderingly complex that they lose all explanatory strength," and he suggests comparing the results computed on the basis of a simple model to archaeological reality. Bell, Wilson, and Wickham (2002) argue along the same lines: in their view, topographic features are the most static and reliable elements of the landscape, and they propose using a topographic model as a foundation on which additional models can be constructed. Different methods have been used by archaeologists to combine several cost components. Herzog (2010) discusses these methods and recommends multiplying isotropic and anisotropic costs, first because multipliers are published in studies on energy expenditure depending on different terrain, and second because this definitely avoids zero or negative costs.

## Least-Cost Path Algorithms

### Properties of Algorithms Currently Implemented in GIS Software

The implementation of LCP calculations in most GIS software is based on an algorithm published by Dijkstra in 1959 (Cormen et al., 2001: 595–99). This algorithm identifies the optimal LCP for weighted graphs with positive weights. A *weighted graph* is a network of nodes (locations) and weighted links (connections between the nodes, where the weights are the costs for traveling along the link). If the centers of the grid cells are considered as graph nodes and the moves to the neighboring cells are the links, with the weights proportional to the costs required for the moves, then the vector algorithm can be applied to raster data, and the optimal cost path is found (Collischonn and Pilar, 2000; Husdal, 2000: 13; Llobera, 2000; Yu, Lee, and Munro-Stasiuk, 2003). Positive weights are required for Dijkstra's algorithm, a cost function that sets the cost of flat terrain to 0 or includes negative values (for example, Bell and Lock, 2000) does not fulfill this requirement.

Conolly and Lake (2006: 254) point out that "another objection to least-cost paths is that they often do not appear to follow the globally optimal route." However, they present Tomlin's algorithm (221–23) for creating the ACS, which is not equivalent to Dijkstra's approach. Even if Dijkstra's algorithm is implemented to generate the ACS grid, the second step of tracing the LCP from the target back to the origin on the basis of a backlink raster is sometimes replaced by a drainage algorithm (Yu, Lee, and Munro-Stasiuk, 2003). The drainage algorithm identifies the route of steepest cost reduction, which does not necessarily coincide with the optimal path (de Smith, Goodchild, and Longley, 2007: 146). Maybe the additional memory requirements for the backlink raster storing for each cell the backlink to the predecessor cell prevented the full implementation of the algorithm in some GIS packages. The GRASS commands r.walk and r.cost introduced the backlink option in version 6.5 (which is in the "testing" status in November 2012). A correct implementation of Dijkstra's algorithm identifies the globally optimal path, but unfortunately some archaeologists are not aware of this fact (Kantner, 2012; Lock and Pouncett, 2010; van Leusen, 2002: 6.9).

Rowe and Ross (1990) note that one of the disadvantages of raster-based LCP calculations is the fact that much computation time is wasted because the search for the goal is mostly blind. The class of informed search algorithms, notably the A* search algorithm, address this issue. The A* algorithm is a modification of Dijkstra's approach, which reduces the computational load in most situations considerably by taking at each step the distance to the goal into account (for instance, Husdal, 2000: 12; Surface-Evans and White, 2012).

## Some Issues in Raster-to-Graph Conversion

The accuracy of the LCP calculation depends also on how the raster data is converted to the graph—that is, the constraints imposed by links restricted to the immediate neighbors. Figure 8.2 shows a 13 × 13 environment of a specific grid cell, marked by O. All cells that can be reached by successive Queen's moves in one direction are marked by a Q. In chess games a Queen is able to move any number of squares vertically, horizontally, or diagonally. A Knight can move either two squares horizontally and one square vertically, or two squares vertically and one square horizontally. If a direct path in one direction using Knight's moves exists to a cell, this cell is marked by a K. Finally, A- and B-moves are defined, which are comparable to Knight's moves, but whereas Knight's moves consist of 2-1 steps, A-moves are 3-1 steps, and B-moves are defined by 3-2 steps. Cells marked with A or B, respectively, can be reached by successive A- or B-moves in one direction; 64 out of 169 cells remain unmarked, and, owing to the raster-to-graph conversion, paths from the origin cell O to unmarked cells are longer than the corresponding straight line. At present, most LCP implementations in GIS software support only 8 neighbors—that is, in this case the unmarked cells as well as the cells marked by K, A, or B cannot be reached directly.

In a terrain with uniform isotropic costs, the optimal path between any two points is a straight line. However, if the raster was converted to a graph as just described, the LCP calculated by Dijkstra's algorithm is not the shortest path but deviates from the straight line (Figure 8.3). Figure 8.3a depicts a situation where only 8 neighbors are considered. In this case the lengths of two LCPs

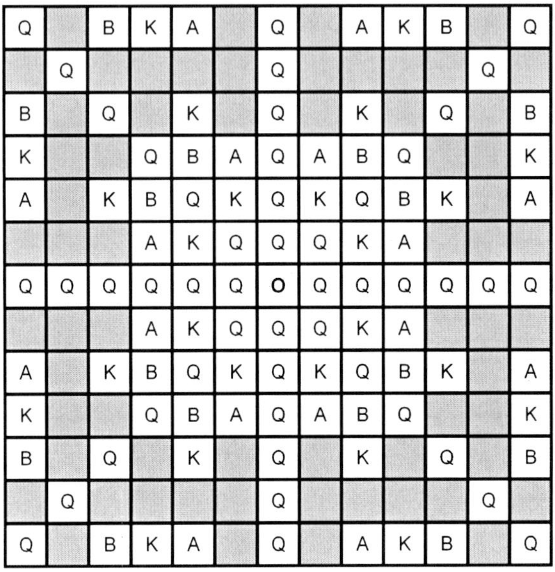


**Figure 8.2** Raster cells that can be reached directly from origin O by moves in 8 directions are marked by Q; moves in 24 directions can directly reach all cells marked by Q and K. If 48 directions are supported, all cells with white background can be reached by straight-line moves.

with minimal costs (dashed black and dotted gray) exceed the length of the optimal path by about 8%. Because of this distortion, the ACS for constant costs is an octagon (Adriaensen et al., 2003; Wheatley and Gillings, 2002: 158, Figure 7.6). Or, to put the same fact differently, if we change the orientation of the grid, the length of the LCP may increase by 8%. In Figure 8.3b the dashed and the alternative dotted path are based on a neighborhood consisting of 24 cells, which reduces the error in length to about 3%. For this reason the corresponding ACS for constant costs is a regular polygon with 24 sides, fairly closely resembling a circle. By increasing the neighborhood size to 48, we reduce the maximum elongation error to about 1% (Figure 8.3c; see also Harris, 2000, citing Huber and Church, 1985), and the resulting ACS is very close to a circle.

Huber and Church (1985) recommend using a neighborhood of 24 cells, which "appears to provide the most satisfactory trade-off between accuracy and computational burden." The elongation errors introduced via inaccuracies in the input data (for example, the elevation model), the modeling of intangible variables, and the choice of cost function probably exceed those introduced by the grid to vector conversion with 24 neighbors. However, for most archaeological path reconstructions, the maximum distance between calculated path and optimal path is more important than is the error in path length. Indeed, even if a 48-cell neighborhood is considered, in the worst-case scenario in uniform landscape the maximum distance between the calculated cost path and the optimal path is 8% of the optimal path length. To illustrate the effect of a small number of neighbors, examples of paths in a raster grid with constant costs are shown in standard GIS books for archaeologists (Conolly and Lake, 2006: 252–53, Figure 11.16; Wheatley and Gillings, 2002: 157, Figure 7.5). In these images the moves are depicted close to the optimal path, as the gray dotted lines in Figure 8.3a and b. But the black dashed line shows an alternative LCP incurring the same costs. Dijkstra's algorithm selects one optimal path in the presence of several alternatives. A standard implementation of the algorithm will test the neighbors in a fixed sequence so that the worst-case LCP will result.

The examples in Figure 8.3 show the following maximum distances between the calculated path and the optimal path in a uniform environment: 20% of the total optimal path length for 8 neighbors, reduced to 11% for 15 neighbors, 8% for 48 neighbors, still as large as 6% with 80 neighbors, and 4.6%

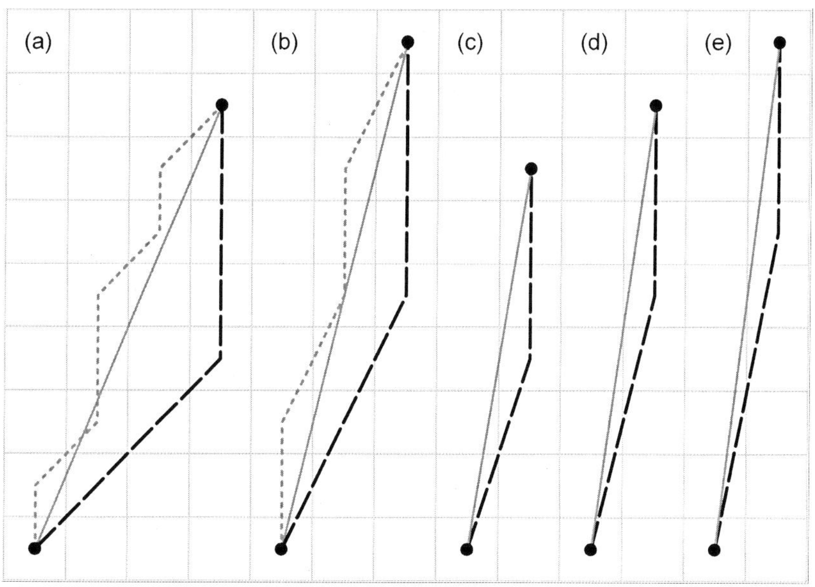


**Figure 8.3** Worst-case scenarios for different neighborhood sizes, when a straight line is optimal but LCPs deviate from the straight line owing to the raster-to-graph conversion. Neighborhood sizes: (a) 3 × 3, (b) 5 × 5, (c) 7 × 7, (d) 9 × 9, (e) 11 × 11. The maximum distance of the worst-case LCP (dashed black) to the true optimal path (gray straight line) decreases with increasing neighborhood sizes. The dotted gray polylines are LCPs incurring the same cost as the worst-case LCPs.

with 120 neighbors. These examples suggest that for site catchment analysis 48 neighbors are fine, but even with 120 neighbors, cost paths will deviate from the true optimal path by up to 4.6%. The disadvantages of using only 8 neighbors are apparent in the LCPs calculated by Nolan and Cook (2012): the LCPs start at a given central location and represent the most effective paths in one of 8 cardinal directions. Nearly all LCPs start with a straight-line section in a vertical or horizontal direction and end with a diagonal section (or vice versa), and this situation does not appear very realistic.

Van Leusen (2002: 16.14–16.16, Figure 13) points out that high cost barriers in the cost surface are skipped by diagonal moves if the barrier consists of corner-connected cells. He uses a neighborhood of 8 cells, and, in this case, increasing the breadth of the barrier to two cells ensured that no move could cross the barrier without paying the appropriate costs. This problem and the workaround by "magnification" of the barrier were also presented and discussed by Adriaensen and associates (2003: 233–47). Moreover, Adriaensen and coauthors note that this problem does not occur in the case of facilitating linear elements, since the grid-based algorithm will create LCPs following 1-cell wide corridors even if the cells merely touch by corners. Similarly, the problem does not arise in the presence of a line-shaped attractor.

Wheatley and Gillings (2002: 156–57) present a model of how to include a feature such as a navigable river as both a barrier and a transportation opportunity. They suggest representing the river as a central corridor with fairly low costs but surrounding it by a thin buffer of high cost. This method ensures that crossing the river is expensive and changing the means of transportation is costly as well, but, after the change, moving on the river is inexpensive. If such a construction is used and A- and B-moves are allowed, the cost values of 12 successive cells must be changed to ensure that none of the 3 bands are crossed without paying the due costs. However, by extending the breadth of the 3 bands, valuable information concerning the neighboring cost landscape is lost. In addition, Knight's, A- or B-moves might skip over small local barriers unnoticed. In fact, grid resolution is increased by allowing simple long moves (Rowe and Ross, 1990).

To avoid this problem, the author's LCP software subdivides the Knight's move so that it consists of two submoves of equal length (Figure 8.4). For the first submove the cost of the end point is the average of the costs of the two neighboring cells of this end point. Similarly, the A- and B-moves are subdivided into three smaller submoves, and the costs of reaching the intermediate points are calculated by taking the distances to each of the 2 neighboring cells into account.

Huber and Church (1985) suggest another way of calculating the costs of these long moves, which works fine in an isotropic cost grid but is not appropriate in an anisotropic situation: for each cell passed they calculate the length (or area for roads with a specific width) of the path segment in this cell and

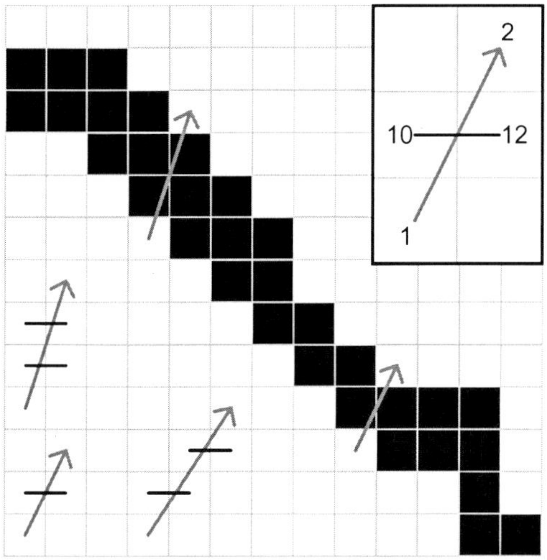


**Figure 8.4** When Knight's move or A- and B-moves cross barriers (black cells), care must be taken that the costs of the barrier are taken into account. This is done by subdividing each long move into submoves; short black horizontal lines indicate the cuts. The costs at the cut points are calculated from the costs of the neighboring cells. The top right box shows an example of subdividing a Knight's move (in an isotropic grid): the move from the cell with cost 1 to the cell with cost 2 consists of two submoves, and the cost of the cut point is 11 = (10 + 12)/2. So the cost total of the move is (*L*/2)*((1 + 11)/2 + (11 + 2)/2), where *L* is the length of the Knight's move.

multiply the cell costs by the segment length; finally the costs accumulated in each cell passed are added.

## Some Slope and Curve Issues

The arguments in the previous section focused on isotropic cost grids. This topic is now extended to cost calculations involving slope: the slope map, which can be calculated easily by most GIS programs (but depends on the accuracy and resolution of the elevation data—see Herzog and Posluschny, 2011; and on the method used for slope calculation, see Lock and Pouncett, 2010; Warren et al., 2004)—normally records only the slope value of the direction of steepest descent (Conolly and Lake, 2006: 217–18). The slope of a path crossing the surface is dependent on the direction of the path relative to the underlying surface. So the slope of the path, called the *effective slope*, is always less than or equal to the slope of the surface (de Smith, Goodchild, and Longley, 2007: 259). However, some of the archaeological LCP studies based on slope cost do not take the direction of movement into account (for example, Batten, 2007; Kantner and Hobgood, 2003; van Leusen, 2002: 16.12–16.16).

## Approaches to Dealing with Effective Slope

Several unsuitable approaches have been proposed to solve this problem of direction-dependence in calculations of effective slope; applying a cost function that averages the costs of upslope and downslope movement to an isotropic slope raster map is one possibility, but it does not take movements perpendicular to the direction of maximum slope into account. The approach of Hare (2004) is an extension of this idea; he creates an isotropic slope grid by averaging the costs of traversal to the four nearest neighbor cells for each grid cell.

Stephen Stead (pers. comm.) suggested at the CAA conference in 1995 creating eight slope grids corresponding to eight directions and to select the appropriate slope value for each move depending on the direction. Apart from additional storage requirements, the main drawback is that again only eight neighbor cells are considered for each move.

The developers of IDRISI introduced a method to deal with anisotropic costs that has received repeated mention (Conolly and Lake, 2006: 221; de Smith, Goodchild, and Longley, 2007: 146; Wheatley and Gillings, 2002: 152–53) and was implemented in other GIS software as well (Olaya, 2004: 161–62).

According to Eastman (2006: 264), IDRISI's standard anisotropic cost functions are constructed in such a way that for all parameter choices, downhill directions accumulate less expenditure than level ground does. This is not appropriate for steep downhill movement, because walking down an extremely steep slope is more costly than walking on level ground. To address this issue, Bell and Lock (2000) as well as Bell, Wilson, and Wickham (2002) reclassified the slope data so that extremely steep slopes were given proportionally high cost, but downhill movement on steep slopes is still considered less costly than walking perpendicular to the slope. Moreover, substituting the standard anisotropic cost functions of the IDRISI-approach by a user-defined function (Eastman, 2006: 264) still does not ensure that small downhill slopes are considered as forces, whereas steep slopes incur higher costs than does movement on level ground.

Another complex approach to effective slope has been implemented in ESRI's path distance procedure, with eight parameters controlling the result, including the isotropic friction grid, the DEM, and an optional table providing values of the slope-dependent cost function. Additional friction grids and cost parameters can be used to model anisotropic costs such as wind directions. This approach is extremely versatile but also hard to understand, and the ESRI implementation is restricted to movements in eight directions only. It seems that this GIS function has gained popularity in recent years (Parslow 2009: 61; Verhagen and Jeneson, 2012; eight case studies in White and Surface-Evans, 2012), probably because Tripcevich (2009) published a tutorial including an example on how this procedure can be used for effective slope calculations with the popular Tobler cost function.

Equations involving some trigonometry for calculating effective slope based on the slope and the reversed aspect raster map are presented by Conolly and Lake (2006: 217–18). No reference to an implementation is given.

A more straightforward method has been implemented in the author's software. This approach was also outlined by Llobera (2000) and described by Yu, Lee, and Munro-Stasiuk (2003) in more detail. The raster data is converted to a bidirectional graph, and the weights of the links are calculated from the DEM directly and combined with the isotropic cost. Slope values are computed from the DEM by constructing a right triangle, with the baseline connecting the *from* cell to the *to* cell, and the height equals the elevation differences of the two cells. The surface distance is calculated using the Pythagorean theorem, and percent slope is the height of the triangle divided by the baseline length, multiplied by 100. This slope estimation for LCP sections is equivalent to that resulting from bilinear interpolation of the DEM grid. The disadvantage of bilinear interpolation is that sharp edges may occur; other interpolation methods create smooth surfaces. The advantages of bilinear interpolation are that undershoots or overshoots are avoided and that it is extremely fast.

However, even a narrow footpath needs a minimum breadth, and this is why informal paths tend to avoid routes along a contour line of a steep slope. Some construction work is necessary to build level paths along steep slopes, and this involves removing material from the uphill inclination or constructing some support toward the valley; that is, such paths are formal (Figure 8.5). Bell and colleagues (Bell and Lock, 2000; Bell, Wilson, and Wickham, 2002) discuss LCP implementations where the direct traverse of the hillside, that is, moving perpendicular to the force, simply incurs the cost for movement on level ground, and they call this an *apparent discrepancy*. But once a comfortable formal road is constructed along the steep slope, the energy expended on this road and on flat terrain is about the same. For informal paths, the method proposed by Gonçalves (2010) for modeling wide paths adequately addresses the problem, if the resolution of the DEM is higher than the width of the desired path. Even with a DEM of less detail, this approach effectively avoids contour line routes, but the computational load of this method remains quite heavy. An alternative approach is to multiply the cost of the intended move by the sum of the costs of the two moves perpendicular to the intended move, multiplied by some constant. The choice of the constant controls whether a formal (low value) or an informal (high value) route reconstruction is modeled.

Discussing past routes in the Bergische Land, Nicke (2001: 13) points out that the Romans were able to build contour line roads requiring paving or a tar cover, whereas no such roads were constructed in medieval or early modern times. Olaya (2004: 162) recommends discarding cells with slopes exceeding a certain threshold because it is impossible to climb steep slopes and descent is dangerous. This recommendation is appropriate for informal routes, whereas formal routes may be found along the contour lines of steep slopes, although

**Figure 8.5** Modern path along a contour line of a steep slope in southern Spain. Some construction effort was necessary to create the track (photograph by the author).

they involve high construction costs. It is plausible that the extra costs for building a formal route lateral to a slope is proportional to the amount of material removed from the uphill inclination and the material deposited as a support toward the valley. Estimating the weight of the extra cost is a difficult task, because the weight depends on the breadth of the route and many other factors, such as the number of travelers per day.

To check if contemporary LCP software deals with effective slope at all, an artificial "roof"-shaped DEM was created with small level bands at both ends. For a cost function with a given critical slope, LCPs with switchbacks are expected, if the roof's slope exceeds the critical slope. Neither ESRI's cost distance nor IDRISI's anisotropic procedures created hairpin bends, whereas the author's implementation resulted in small-scale zigzags (Figure 8.6, next page). The r.walk function of GRASS GIS was not submitted to the test because the method does not support the quadratic cost function used.

## Case Study

The LCPs in this case study were created with the author's software, which allows neighborhood sizes up to 48 and dividing moves into submoves to prevent loss of resolution. The program, written in Borland Delphi 5, generates LCPs in the MapInfo Interchange format. In this case study LCPs connect the settlements Wipperfürth and Gimborn covering a straight-line distance of

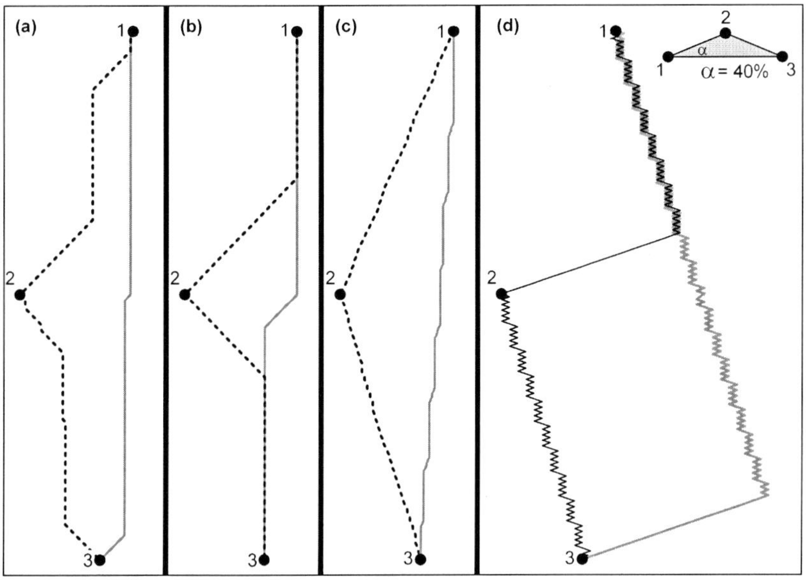


**Figure 8.6** Test of different LCP software with an artificial roof-shaped landscape; points 1 and 3 are on the gutters, point 2 is located on top of the roof. A quadratic cost function was used with a critical slope of 10%. LCPs from 1 to 2 and 2 to 3 are marked by black dotted lines in diagrams (a) to (c), and thin black lines in diagram (d); the LCP connecting points 1 and 3 are depicted in gray. (a) Result of ESRI's cost distance, which is independent of the slope of the roof. (b) LCPs created by IDRISI with default anisotropic parameters; the LCPs follow the same course on roofs with different slopes. (c) Result of the author's software on a roof with a slope of 10% and (d) on a roof with a slope of 40%.

about 13 km (Figure 8.7). Wipperfürth is located about 40 km northeast of Cologne, Germany, at a crossroads of several ancient trade routes and is the first settlement in the region that received town privileges (in 1217). Gimborn was first mentioned in historical sources dating back about 1180 C.E. An early map finished in 1719 C.E. is available for this region, which is less exact for the Gimborn area because of political reasons of that time. The main roads depicted on this old map were transferred to a GIS.

In previous case studies focusing on a larger area, it was found that slope and avoiding wet areas are important components for reconstructing known ancient trade routes. Several LCPs were created; one of these does not avoid wet soils but runs through a wet valley southeast of Wipperfürth (Figure 8.7). For all other LCPs slope-dependent costs were multiplied by 5 in areas of wet soils. In this case study the Tobler LCPs and the LCPs created with a critical slope of 12% for wheeled vehicles do not agree at all. The LCPs resulting from the 6th degree polynomial previously mentioned are fairly close to the

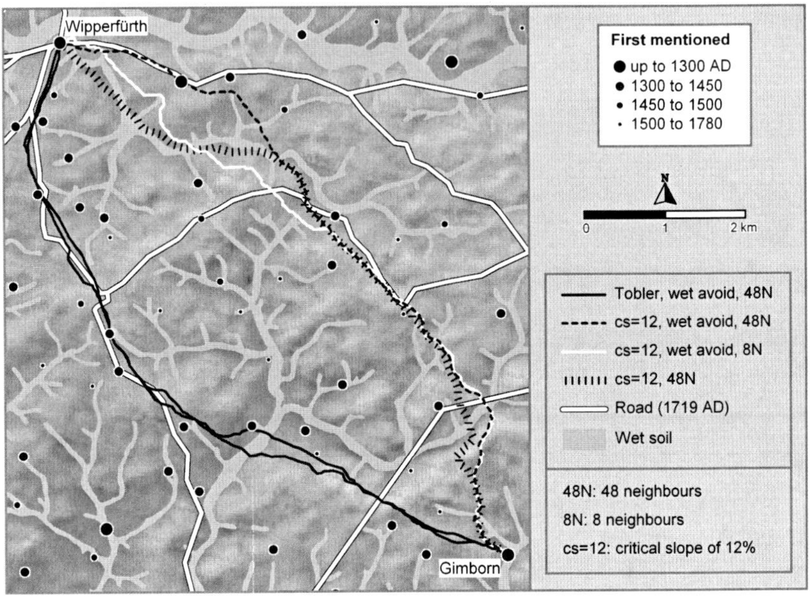


**Figure 8.7** LCPs connecting Gimborn and Wipperfürth. The ASTER DEM created by NASA and METI was used for slope calculations. Wet soil areas were extracted from a digital soil map provided by the Geologischer Dienst, Krefeld, Germany.

LCPs for wheeled vehicles (not shown). The Tobler function is anisotropic; therefore, the return LCP differs from the initial LCP. The LCP for wheeled vehicles taking 48 neighbors into account is closer to the road on the historic map than that based on 8 neighbors only.

Landscape reconstruction attempts are beyond the scope of this small case study, and the ASTER DEM with a grid resolution of about 30 m does show some errors. In spite of these drawbacks the LCPs agree quite well with some sections of the old main roads.

A case study focusing on a small study area is not able to exhibit all issues concerning LCP calculations. The LCP software was applied to route reconstruction in different countries (Germany, Ecuador, Italy). According to the mostly unpublished results, in cases where only one obvious route choice is available, most LCPs coincide irrespective of the parameter choice. Sometimes very small deviations in the model parameters or varying the start locations result in widely differing LCPs.

## Modeling the Cost of Curves

Some authors (Conolly and Lake, 2006: 253; Gietl, Doneus, and Fera, 2008; Husdal, 2000: 19–20; Llobera and Sluckin, 2007) express regret that LCP algorithms do not create realistic mountain roads for ascending a steep slope but at best small-scale zigzags instead of larger-scale switchbacks (Figure 8.6d). Without doubt, the small-scale zigzags are due to the fact that the costs of sharp turns are not included in the cost calculations. The first carts and wagons could be turned only with difficulty; therefore "turning penalties" should be higher for these vehicles than for later, more sophisticated models including an axle suitable for steering. But even the curves of Roman roads in northern Greece have a radius of 100 m or more; only in mountainous areas was the curve radius sometimes reduced in the range of 10 to 20 m (Xeidakis and Varagouli, 1997). For modern forest roads, a minimum radius of 7 to 9 m is required (Cretu and Rusnac, 1998). Even pedestrians do not like to change their direction, because doing so involves some energy costs and most likely some decrease in comfort (Llobera and Sluckin, 2007), so the LCP should follow a *turn-hostile route*, a term introduced by Rowe (1997). However, Rowe goes way beyond the objective by concluding that "turn-hostility says that optimal paths cannot zigzag." At least on fairly flat terrain, roads and paths often disregard slope in favor of straightness (Pingel, 2009).

Collischonn and Pilar (2000) proposed incorporating a "turning cost" or "turning penalty" into the cost function. This proposal is quite difficult, because the calculations depend not only on the vehicle used (if any) but also on the resolution of the geographic raster data. For example, if the minimum turning radius is 20 m, and grid data with 5 m cells are used, a sharp turn of 45° should be modeled by successive moves to neighboring cells, whereas with a resolution of 50 m or more, a single move to a neighboring cell is appropriate. Another improvement is to replace the polyline segments by arcs for a realistic curve representation (Husdal, 2000: 15; Longley et al., 2005: 369), although this replacement requires additional computation time. But to the author's knowledge, incorporating curve costs has not been implemented in standard GIS packages, and no archaeological application of such an approach has been published.

## The Model by Rowe and Ross

An even more complex model was introduced by Rowe and Ross (1990), refined by Rowe (1997). This model takes the energy needed for braking into consideration, as well as the dangers of loss of traction and of catastrophic overturn. A vehicle on a steep slope has to stay away from three impermissible zones: (1) sideslopes on both sides (to avoid the risk of overturn), (2) steep upslope, if the forces of the vehicle do not allow to manage such a steep upslope movement, (3) the braking zone is in the opposite direction (Figure 8.8).

This model was designed for robots but applies to vehicles as well. For walkers, the model might be valid, too, since walkers also tend to avoid these

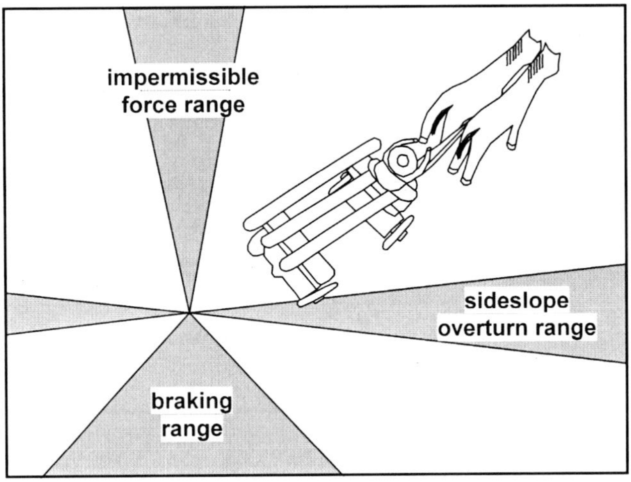


**Figure 8.8** The model of Rowe and Ross (1990) for traversing a slope in informal conditions. The range of "impermissible force" includes all uphill directions that are too steep for vehicle progress. For downhill movements the braking range is generally avoided, because the braking power of the vehicle is insufficient to maintain a safe speed in this zone. The term *sideslope overturn range* refers to those cross-slope directions that cannot be negotiated owing to risk of overturning.

impermissible zones. If the zones can be identified by a set of rules, implementing this model seems fairly easy. On converting the raster grid to the graph, omit all moves into the problematic zones. When one is considering a neighborhood of only 8 cells, the number of choices may reduce to none, whereas a larger neighborhood may still uphold some choices—which is yet another reason for including more than 8 neighboring cells into the LCP calculations. But this implementation allows borderline impermissible routes for long sections of the path (Rowe, 1997). Although this model was published more than 20 years ago, it seems that this approach has not had any effect on GIS development or archaeological LCP calculations.

## Networks

The simplest reconstruction of a least-cost network is by repeated application of a method for calculating a single LCP (Figure 8.9a). This is appropriate only if the aim is to minimize the travel costs between any two locations in the network. Waugh (2000: 615) introduces the term *least-cost network to the user* in this context. According to Earle (1991), such paths develop in an informal context, and the resulting network is highly redundant and without centrality. With this model, LCPs between different sites coincide only when avoiding

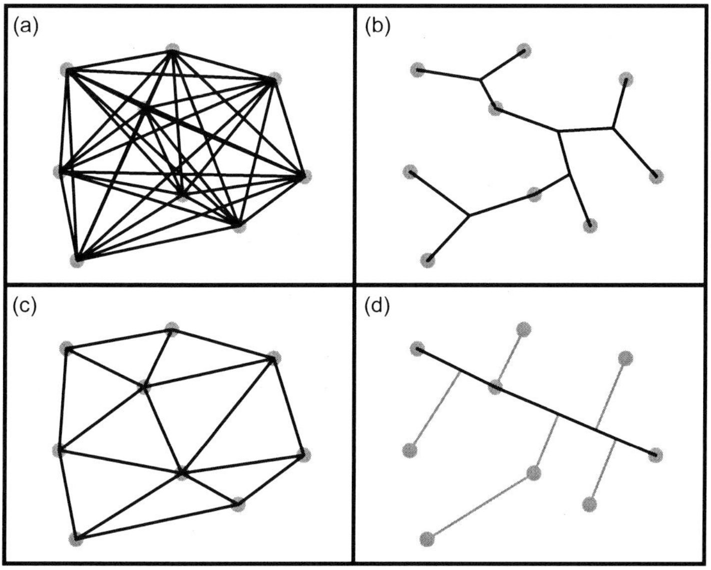


**Figure 8.9** Different network models in a terrain with constant friction: (a) least-cost network to the user; (b) least-cost network to the builder; (c) Delauney triangulation; (d) main route and later subsidiary pathways.

high-friction areas—that is, if topography forces LCPs along certain routes, or if a new site was constructed close to an existing route. The complementary objective is to optimize the road network so that it covers only a minimum amount of space of the valuable land used for agrarian purposes (for example, Husdal, 2000: 17–19). The term *least-cost network to the builder* is used by Waugh (2000: 615) for route systems with the shortest overall length (Figure 8.9b). With uniform friction, Steiner trees connecting all relevant locations form the least-cost network to the builder (cf. paper presented by Jens Andresen at CAA, 2008). In a simulation study, Helbing, Keltsch, and Molnár (1997) create networks that look, to the user and to the builder, like least-cost networks as well as intermediate solutions because of their varying the parameter controlling the attractiveness of existing tracks.

Another compromise between a builder- and a user-optimized network might focus on the immediate neighbors, whereas locations at a distance can be reached only via intermediate stops at neighboring settlements. In general, exchange with the immediate neighbors is more important than with more distant locations. Moreover, intermediate stops at other settlements on a long journey help to keep the traveler on track and to ensure supply with food and water. Least-cost extensions of traditional Delauney triangulation approaches (for example, Wheatley and Gillings, 2002: 190–92) are probably appropriate for such models (Figure 8.9c).

However, the three models mentioned here give equal weight to each location to be connected and are based on the assumption that all locations were set up at the same time. A network study with a known central origin of exploration for the study area considered is presented by Fábrega Álvarez and Parcero Oubiña (2007). The model for a main travel route with later subsidiary paths (Figure 8.9d) is described in the study of Whitley and Burns (2008). These approaches are discussed in more detail in Herzog (2012) and applied to a study region east of Cologne, Germany, where several ancient trade routes are known.

Many archaeological LCP studies try to reconstruct a network of roads (for instance, Bell, Wilson, and Wickham, 2002; Kantner and Hobgood, 2003). Generally, push-button LCP software does not support different network models, and hardly any information can be found in the documentation of the software concerning the methods used for calculating optimal networks. Consequently, most archaeologists applying this software are not aware that different models exist and that the software might not be appropriate for the model they have in mind.

## Beyond Least-Cost Paths

The ACS raster can serve as starting point for several other least-cost methods and concepts (Figure 8.10). For these concepts symmetric cost functions are required; that is, the cost of one step is the average of the costs of the forward and the backward steps. Wheatley and Gillings (2002: 162) use the term *return to base* in this context.

On the basis of a cost surface, isolines may be generated so that for all points on the isoline the energy expenditure or the time required to reach this point from the origin is constant. If the cost function estimates time requirements, the isolines are also called isochrones (for example, Lock, 2003: 172, reprinting the results of a study by Stančič and Gaffney first published in 1991). The site catchment area is the territory included by a given cost isoline and can also be generated by reclassifying the ACS (Conolly and Lake, 2006: 224; Wheatley and Gillings, 2002: 162). Already in 1980 Ericson and Goldstein proposed creating site catchments on the basis of equal work or energy expenditure, and they defined the term *isoergoic contours*. However, site catchments do not take into account that a field within a 10-minute walking distance requires less total effort in terms of time and energy consumption than a field on similar soil at a larger distance. For modeling the decreasing effects of the site on the landscape, site catchments with increasing cost limits could be added, so that the area close to the site center is assigned a higher weight or importance than areas at a distance. To get a smooth result without arbitrary steps, least-cost kernel density estimation (KDE) can be applied (Herzog and Yépez, 2010). The term *kernel* is used for the function that models the effect

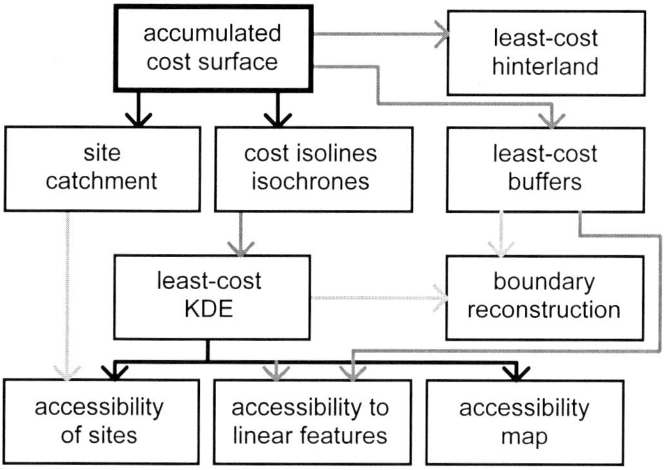


**Figure 8.10** The concept of an ACS is related to several other least-cost concepts. Black arrows indicate strong relationships, and light gray arrows connect weakly related concepts. If concept A forms the basis of concept B, the arrow direction is from A to B. For example, site catchments can be derived directly from corresponding ACSs; therefore the black arrow connects the ACS box with the site catchment box.

of the site on its environment. If the catchment area is subdivided into adjacent bands delimited by equidistant isolines, a kernel can be defined as a weighted sum of these bands, with weights decreasing with increasing cost distance from the central location.

Several approaches for least-cost accessibility concepts can be derived from least-cost KDE [Herzog, n.d.]. The basic dataset for calculating accessibility consists of opportunities with given size and location. A typical opportunity in prehistoric times is a source of water, of flint stone, or other raw material. Accessibility to point-shaped sources can be modeled by choosing the size of the kernel proportional to the importance of the source and by adding the kernels of all sources. For line-shaped sources like streams this concept can be readily extended (based on least-cost buffers, which are explained shortly) so that the impact of the source decreases with increasing cost distance. A special variant of accessibility is sometimes calculated in connection with predictive modeling when one is trying to identify the routes or the corridor on which man first explored a region (Whitley and Burns, 2008). In this case all points in a raster grid of the study area may be considered as opportunities of equal weight. For this variant the term *general accessibility* is used. A general accessibility map can be created by calculating the least-cost KDE result for cell centers of a high-resolution raster covering the study area (Herzog and Yépez, 2010; Herzog, n.d.).

Creating buffers or single-distance bands is a standard procedure in GIS analysis (Conolly and Lake, 2006: 209–11; Longley et al., 2005: 329–30;

Wheatley and Gillings, 2002: 148–49)—for example, to establish the relationship of settlements of a certain period to fresh water supply. The outer bound of a cost-distance band for a single point can be calculated from a cost surface; it is an isoline, as explained. For other geometric objects covering a set of (connected) cells, a cost-buffer can be generated by starting the spreading process from all seed cells; spreading is stopped when the user-supplied cost threshold is exceeded.

Another least-cost method requiring ACSs allocates supply areas to a set of demand centers. This *hinterland* procedure (Eastman, 2006: 86) is based on a spreading process that moves outward based on some distance measure, and cost distance seems to be appropriate in this case so that the spreading process will stop at barriers. This approach may serve as an alternative to site catchment analysis. The user may stop the spreading process when the supply area allows sustainable living of the people in the settlement considered. Afterward the sizes of the hinterlands created in this way could be compared. However, to my knowledge this concept has not yet been implemented.

Boundary reconstruction is often based on Theissen polygons. Least-cost Theissen polygons (Hare, 2004) can be generated easily by starting a least-cost buffer process at the source locations and stopping the spreading process when two (or more) buffers meet (Herzog, 2012). Another approach is the *Xtent* algorithm, which is derived from interaction models. Hare (2004) presents results of the least-cost Xtent method, where cost distance is used in the formula for the political influence of a central position on a given location, although he does not give any details on how this method was implemented in software. Least-cost KDE is another alternative for modeling the impact of given sites on the surrounding landscape, allowing to take the importance of the sites into account by varying the height of the kernel.

## Conclusions, Discussion, and Future Work

In 1998 Gaffney and Stančič pointed out that success of a particular method can be claimed only when the majority of field and academic archaeologists have access to such techniques on a day-to-day basis. While this point has become true for GIS in general, the present chapter shows that standard GIS software lacks some features that are important in archaeological optimal path calculations. The author's programs address some of these issues and are also able to create site catchments, least-cost KDE rasters, and some of the network models referred to. Raster grids such as site catchments and accessibility maps are output in the ESRI ASCII grid format. (The software will be available on the author's website after the user interface has been improved.)

Issues requiring additional research include methods for generating costs both on the basis of effective slope and the maximum slope of the cell, for avoiding the risks of overturn and for particularly steep uphill or downhill movements, and for modeling the cost of bends. Moreover, reliable data for estimating the costs of movement for pack animals as well as prehistoric and historic vehicles on the road and on water are not readily available. Calibrating the models and methods on the basis of historic maps and remains of ancient routes ensures realistic results.

With the LCP software currently available, LCP calculations can be applied tentatively to identify the factors governing the route layout connecting two points in ancient times.

It is recommended to start with a simple model first and to refine this model as required. Good practice includes testing the stability of the results by slightly varying the model parameters. A systematic approach for introducing such variations was proposed by Saerens and associates (2009), who present a theoretical framework for randomized optimal paths given an expected level of entropy. However, the algorithms proposed are computationally demanding, do not apply to networks, and are at present not implemented in GIS software.

When employing LCP software for archaeological route reconstruction, keep in mind the limitations of software, the deficiencies in both models, and underlying geographic data. With the tools currently available, using GIS-based software to reconstruct prehistoric routes without validating the results on the basis of archaeological evidence or historic sources is mere guesswork.

After establishing sound methods for calculating LCPs that connect two locations, constructing route networks based on different models is possible. Reliable least-cost calculations are also required for the kinds of boundary reconstruction methods discussed by Hare (2004) as well as for attempts to analyze spatial developments in a region—for example, by creating accessibility maps (Herzog, n.d.) or by reconstructing spreading processes.

## Acknowledgments

Many people facilitated my research on LCPs. First of all, Axel Posluschny, who initiated my interest in this topic, invited me to Frankfurt to test LCPs with his commercial GIS software, contributed many valuable comments, and provided information on the ArcGIS 9.1 documentation and some other references. I thank Andrew Bevan for inviting me to London, for fruitful mail discussions, and for providing me with references. Other colleagues who discussed these issues with me or sent references relevant for this publication include Mark Lake, Benjamin Ducke, Karsten Lambers, Martin Sauerbier, Ingo Eichfeld, and Petra Dittmar.

### References

Adriaensen, F., Chardon, J. P., De Blust, G., Swinnen, E., Villalba, S., Gulinck, H., and Matthysen, E. 2003. The application of least-cost modelling as a functional landscape model, *Landscape and Urban Planning* 64: 233–47.

Andresen, J., Iversen, R. B., and Jensen, P. 2008. On the war-path: Terrestrial military organisation in prehistoric Denmark, in Posluschny, A., Lambers, K., and Herzog, I. (Eds.), *Layers of Perception, Proceedings of the 35th International Conference on Computer Applications and Quantitative Methods in Archaeology (CAA), Berlin, Germany, April 2–6, 2007*, p. 283. Bonn: Dr. Rudolf Habelt GmbH, Kolloquien zur Vor- und Frühgeschichte 10.

Arcenas, S. n.d. *ORBIS and the Sea: A Model for Maritime Transportation under the Roman Empire*, http://orbis.stanford.edu (select: Applying ORBIS); accessed November 24, 2012.

Batten, D. C. 2007. Least-cost pathways, exchange routes, and settlement patterns in late prehistoric East-Central New Mexico, in Clark, J. T., and Hagemeister, E. M. (Eds.), *Digital Discovery, Exploring New Frontiers in Human Heritage, CAA 2006, Computer Applications and Quantitative Methods in Archaeology, Proceedings of the 34th Conference, Fargo, North Dakota, April 2006*, pp.167–74.

Batty, M. 2003. Agent-based pedestrian modelling, in Longley, P. A., and Batty, M. (Eds.), *Advanced Spatial Analysis*, pp. 81–106. Redlands, CA: ESRI Press.

Bell, T., and Lock, G. 2000. Topographic and cultural influences on walking the Ridgeway in later prehistoric times, in Lock, G. (Ed.), *Beyond the Map: Archaeology and Spatial Technologies*, pp. 85–100. Amsterdam: IOS Press.

Bell, T., Wilson, A., and Wickham, A. 2002. Tracking the Samnites: Landscape and communications routes in the Sangro Valley, Italy, *American Journal of Archaeology* 106: 169–86.

Bevan, A. Forthcoming. Computational models for understanding movement and territory, in Mayoral, V., and Celestino, S. (Eds.), *Sistemas de Información Geográfica y Análisis Arquelógico del Territorio. V Simposio Internacional de Arqueología de Mérida, Anejos de Archivo Español de Arqueología*, www.homepages.ucl.ac.uk/%7Etcrnahb/downloads/Bevan08_manuscript.pdf; accessed September 13, 2009.

Bintliff, J. L. 1977. *Natural Environment and Human Settlement in Prehistoric Greece*, Oxford: Archeopress, British Archaeological Reports International Series 28.

Branting, S. 2004. *Iron Age Pedestrians at Kerkenes Dag: An Archaeological GIS-T Approach to Movement and Transportation*. Unpublished Ph.D. dissertation. Buffalo: The State University of New York.

Collischonn, W., and Pilar, J. V. 2000. A direction-dependent least-cost path algorithm for roads and canals, *International Journal of Geographical Information Science* 14(4): 397–406.

Conolly, J., and Lake, M. 2006. *Geographical Information Systems in Archaeology*, Cambridge Manuals in Archaeology. Cambridge: Cambridge University Press.

Cormen, T. H., Leiserson, C. E., Rivest, R. L., and Stein, C. 2001. *Introduction to Algorithms* (2nd ed.). Cambridge, MA: MIT Press.

Cretu, O., and Rusnac, C. 1998. Forest roads in Romania: Planning and design, in Food and Agriculture Organization of the United Nations (Ed.), *Proceedings of the Seminar on Environmentally Sound Forest Roads and Wood Transport*. Rome, www.fao.org/docrep/X0622E/x0622e02.htm; accessed November 13, 2010.

de Smith, M. J., Goodchild, M. F., and Longley, P. A. 2007. *Geospatial Analysis: A Comprehensive Guide to Principles, Techniques, and Software Tools*. Leicester: Matador.

Earle, T. 1991. Paths and roads in evolutionary perspective, in Thrombold, C. D. (Ed.), *Ancient Road Networks and Settlement Hierarchies in the New World*, pp. 10–16. Cambridge: Cambridge University Press.

Eastman, J. R. 2006. *IDRISI Andes. Guide to GIS and Image Processing*. Manual Version 15.00. Worcester, MA: Clark University.

Ericson, J. E., and Goldstein, R. 1980. Work space: A new approach to the analysis of energy expenditure within site catchments, *Anthropology UCLA* 10(1&2): 21–30.

Fábrega Álvarez, P., and Parcero Oubiña, C. 2007. Proposals for an archaeological analysis of pathways and movement, *Archeologia e Calcolatori* 18: 121–40.

Gaffney, V., and Stančič, Z. 1992. Diodorus Siculus and the island of Hvar, Dalmatia: Testing the text with GIS, in Lock, G., and Moffet, J. (Eds.), *Computer Applications and Quantitative Methods in Archaeology*, pp. 113–25. Oxford: Archeopress, British Archaeological Reports International Series 577.

———. 1998. *GIS Approaches to Regional Analysis: A Case Study of the Island of Hvar*, http://acl.arts.usyd.edu.au/VISTA/gaffney_stancic/index.html; accessed November 6, 2010.

Gietl, R., Doneus, M., and Fera, M. 2008. Cost-distance analysis in an Alpine environment: Comparison of different cost-surface modules, in Posluschny, A., Lambers, K., and Herzog, I. (Eds.), *Layers of Perception, Proceedings of the 35th International Conference on Computer Applications and Quantitative Methods in Archaeology (CAA), Berlin, Germany, April 2–6, 2007*, p. 342. Bonn: Dr. Rudolf Habelt GmbH, *Kolloquien zur Vor- und Frühgeschichte 10*.

Gonçalves, A. B. 2010. An extension of GIS-based least-cost path modelling to the location of wide paths, *International Journal of Geographical Information Science* 24(7): 983–96.

Gorenflo, L. J., and Gale, N. 1990. Mapping regional settlement in information space, *Journal of Anthropological Archaeology* 9: 240–74.

Haciguzeller, P. 2008. Modeling human circulation in the Minoan palace at Malia, in Posluschny, A., Lambers, K., and Herzog, I. (Eds.), *Layers of Perception, Proceedings of the 35th International Conference on Computer Applications and Quantitative Methods in Archaeology (CAA), Berlin, Germany, April 2–6, 2007*, pp. 336–41. Bonn: Dr. Rudolf Habelt GmbH, *Kolloquien zur Vor- und Frühgeschichte 10*.

Hare, T. S. 2004. Using measures of cost distance in the estimation of polity boundaries in the Postclassic Yautepec valley, Mexico, *Journal of Archaeological Science* 31: 799–814.

Harris, T. 2000. Moving GIS: Exploring movements within prehistoric cultural landscapes using GIS, in Lock, G. (Ed.), *Beyond the Map: Archaeology and Spatial Technologies*, pp. 116–23. Amsterdam: IOS Press.

Helbing, D., Keltsch, J., and Molnár, P. 1997. Modelling the evolution of human trail systems, *Nature* 388: 47–50.

Herzog, I. 2010. Theory and practice of cost functions, in Melero, F. J., Cano, P., and Revelles, J. (Eds.), *Fusion of Cultures*. Abstracts of the XXXVIII Conference on Computer Applications and Quantitative Methods in Archaeology, CAA 2010, Granada, pp. 431–34.

Herzog, I. 2012. Least-cost networks, in Earl, G., Sly, T., Chrysanthi, A., Murrieta-Flores, P., Papadopoulos, C., Romanowska, I., and Wheatley, D. (Eds.), *Proceedings of the 40th Conference on Computer Applications and Quantitative Methods in Archaeology, CAA 2012, Southampton*, pp. 240–51 (paper in press).

Herzog, I. n.d. Calculating accessibility, *Proceedings of the 40th Conference on Computer Applications and Quantitative Methods in Archaeology, CAA 2012, Southampton* (paper in review).

Herzog, I., and Posluschny, A. 2011. Tilt–slope-dependent least-cost path calculations revisited, in Jerem, E., Redo, F., and Szeverényi, V. (Eds.), *On the Road to Reconstructing the Past, Proceedings of the 36th CAA Conference, Budapest, April 2–6 2008*, pp. 212–18.

Herzog, I., and Yépez, A. 2010. Least-cost kernel density estimation and interpolation-based density analysis applied to survey data, in Melero, F. J., Cano, P., and Revelles, J. (Eds.), *Fusion of Cultures, Abstracts of the XXXVIII Conference on Computer Applications and Quantitative Methods in Archaeology, CAA 2010*, Granada, pp. 447–50.

Hindle, P. 2002. *Medieval Roads and Tracks* (3rd ed.), Shire Archaeology 26. Princes Risborough: Shire Publications.

Huber, D. L., and Church, R. L. 1985. Transmission corridor location modeling, *Journal of Transportation Engineering* 111(2): 114–30.

Husdal, J. 2000. *How to Make a Straight Line Square: Network Analysis in Raster GIS with Time-Dependent Cost Variables*. Unpublished master's thesis. Leicester, www.husdal.com/2000/09/01/how-to-make-a-straight-line-square; accessed November 6, 2010.

Indruszewski, G., and Barton, C. M. 2007. Simulating sea surfaces for modeling Viking age seafaring in the Baltic Sea, in Clark, D. and Hagemeister, E. M. (Eds.), *Digital Discovery, Exploring New Frontiers in Human Heritage, CAA 2006, Computer Applications and Quantitative Methods in Archaeology, Proceedings of the 34th Conference, Fargo, North Dakota, April 2006*, pp. 616–29.

Kantner, J. 2012. Realism, reality, and routes: Evaluating cost-surface and cost-path algorithms, in White, D., and Surface-Evans, S. (Eds.), *Least-Cost Analysis of Social Landscapes: Archaeological Case Studies*, pp. 225–38. Salt Lake City: The University of Utah Press.

Kantner, J., and Hobgood, R. 2003. Digital technologies and prehistoric landscapes in the American Southwest, in Forte, M., and Williams, P. R. (Eds.), *The Reconstruction of Archaeological Landscapes through Digital Technologies, Proceedings of the First Italy-United States Workshop, Boston, November 1–3*, pp. 117–24. Oxford: Archeopress, British Archaeological Reports International Series 1151.

Kvamme, K. L. 1992. Geographic information systems and archaeology, in Lock, G., and Moffet, J. (Eds.), *CAA 91: Computer Applications and Quantitative Methods in Archaeology 1991*, pp. 77–84. Oxford: Archeopress, British Archaeological Reports International Series S577.

Lake, M. W. 2000. MAGICAL computer simulation of mesolithic foraging, in Kohler, T. A., and Gumerman, G. J. (Eds.), *Dynamics in Human and Primate Societies: Agent-Based Modeling of Social and Spatial Processes*, pp. 107–43. New York: Oxford University Press.

Lee, J., and Stucky, D. 1998. On applying viewshed analysis for determining least-cost paths on digital elevation models, *International Journal of Geographical Information Science* 12(8): 891–905.

Livingood, P. 2012. No crows made mounds, in White, D., and Surface-Evans, S. (Eds.), *Least-Cost Analysis of Social Landscapes: Archaeological Case Studies*, pp. 174–87. Salt Lake City: The University of Utah Press.

Llobera, M. 2000. Understanding movement: A pilot model towards the sociology of movement, in Lock, G. (Ed.), *Beyond the Map: Archaeology and Spatial Technologies*, pp. 65–84. Amsterdam: IOS Press.

Llobera, M., and Sluckin, T. J. 2007. Zigzagging: Theoretical insights on climbing strategies, *Journal of Theoretical Biology* 249: 206–17.

Lock, G. (Ed.). 2000. *Beyond the Map: Archaeology and Spatial Technologies*. Amsterdam: IOS Press.

________. 2003. *Using Computers in Archaeology: Towards Virtual Pasts*. London: Routledge, Taylor & Francis Group.

Lock, G., and Pouncett, J. 2010. Walking the Ridgeway revisited: The methodological and theoretical implications of scale dependency for the derivation of slope and the calculation of least-cost pathways, in Frischer, B., Webb Crawford, J., and Koller, D. (Eds.), *Making History Interactive, Computer Applications and Quantitative Methods in Archaeology (CAA), Proceedings of the 37th International Conference, Williamsburg, Virginia, March 22–26, 2009*, pp. 192–203. Oxford: Archeopress, British Archaeological Reports International Series S2079.

Longley, P. A., Goodchild, M. F., Maguire, D. J., and Rhind, D.W. 2005. *Geographic Information Systems and Science* (2nd ed.). Chichester: John Wiley & Sons.

Madry, S., and Rakos, L. 1996. Line-of-sight and cost-surface techniques for regional archaeological research in the Arroux River Valley, in Maschner, H. D. G. (Ed.), *New Methods, Old Problems: Geographic Information Systems in Modern Archaeological Research*, pp. 104–26. Carbondale: Southern Illinois University Center for Archaeological Investigations.

Melero, F. J., Cano, P., and Revelles, J. (Eds.). *Fusion of Cultures*, Abstracts of the XXXVIII Conference on Computer Applications and Quantitative Methods in Archaeology, CAA 2010, Granada.

Minetti, A. E., Moia, C., Roi, G. S., Susta, D., and Ferretti, G. 2002. Energy cost of walking and running at extreme uphill and downhill slopes, *Journal of Applied Physiology* 93: 1039–46.

Moreno Martín, A. 2008. Using GIS to deconstruct Iberian Iron Age landscapes: The territory of Kelin between 6th and 5th centuries BC (La Plana d'Utiel, València), in Posluschny, A., Lambers, K., and Herzog, I. (Eds.), Layers of perception, *Proceedings of the 35th International Conference on Computer Applications and Quantitative Methods in Archaeology (CAA), Berlin, Germany, April 2–6, 2007*, p. 279. Bonn: Dr. Rudolf Habelt GmbH, Kolloquien zur Vor- und Frühgeschichte 10.

Nicke, H. 2001. *Vergessene Wege: Das historische Fernwegenetz zwischen Rhein, Weser, Hellweg und Westerwald, seine Schutzanlagen und Knotenpunkte*. Nümbrecht: Martina Galunder-Verlag.

Nolan, K., and Cook, R. 2012. A method for multiple cost-surface evaluation of a model of Fort Ancient interaction, in White, D., and Surface-Evans, S. (Eds.), *Least-Cost Analysis of Social Landscapes: Archaeological Case Studies,*, pp. 67–93. Salt Lake City: The University of Utah Press.

Olaya, V. 2004 (rev.). *A Gentle Introduction to SAGA GIS*, Edition 1.1, http://switch.dl.sourceforge.net/project/saga-gis/SAGA%20-%20Documentation/SAGA%20 Documents/SagaManual.pdf; accessed November 6, 2010.

Paliou, E. 2008. An autonomous agent approach to the investigation of intra-site movement and visibility: The visual consumption of Theran murals from the public spaces of LBA Akrotiri (Thera, Greece), in Posluschny, A., Lambers, K., and Herzog, I. (Eds.), *Layers of Perception, Proceedings of the 35th International Conference on Computer Applications and Quantitative Methods in Archaeology (CAA), Berlin, April 2–6, 2007*, pp. 328–35. Bonn: Dr. Rudolf Habelt GmbH, Kolloquien zur Vor- und Frühgeschichte 10.

Parslow, C. A. 2009. *Social Interaction in the Prehistoric Natufian: Generating an Interactive Agency Model Using GIS*. Oxford: Archeopress, British Archaeological Reports International Series 1916.

Pingel, T. 2009. *Modeling Slope as a Contributor to Route Selection in Mountainous Areas*, www.geog.ucsb.edu/~pingel/documents/Pingel-Modeling.Slope.as.a.Contributor.To.Route.Selection.in.Mountainous.Areas.2009.pdf; accessed April 4, 2010.

Posluschny, A., Lambers, K., and Herzog, I. (Eds.). 2008. Layers of perception, *Proceedings of the 35th International Conference on Computer Applications and Quantitative Methods in Archaeology (CAA), Berlin, Germany, April 2–6, 2007*. Bonn: Dr. Rudolf Habelt GmbH, Kolloquien zur Vor- und Frühgeschichte 10.

Rahn, R. B. 2005. Cost-surface-derived least-cost paths: A case study from Iron Age Orkney, *Internet Archaeology*, http://intarch.ac.uk/journal/issue19/rahn_index.html; accessed November 6, 2010.

________. 2007. Praise the sea, on land remain? GIS analysis of travel routes in an Iron Age island environment, in Figueiredo, A., and Velho, G. (Eds.), *The World Is in Your Eyes*, Proceedings of CAA 2005, Tomar, Portugal, pp. 391–96.

Ratti, C. 2005. The lineage of the line: Space syntax parameters from the analysis of urban DEMs, *Environment and Planning B: Planning and Design* 32: 547–66.

Rowe, N. C. 1997. Obtaining optimal mobile-robot paths with nonsmooth anisotropic cost functions using qualitative-state reasoning, *The International Journal of Robotics Research* 16: 375–99.

Rowe, N. C., and Ross, R. S. 1990. Optimal grid-free path planning across arbitrarily contoured terrain with anisotropic friction and gravity effects, *IEEE Transactions on Robotics and Automation* 6(5): 540–53.

Saerens, M., Achbany, Y., Fouss, F., and Yen, L. 2009. Randomized shortest-path problems: Two related models, *Neural Computation* 21: 2363–404.

Stöger, H. 2008. Roman Ostia: Space syntax and the domestication of space, in Posluschny, A., Lambers, K., and Herzog, I. (Eds.), *Layers of Perception, Proceedings of the 35th International Conference on Computer Applications and Quantitative Methods in Archaeology (CAA), Berlin, Germany, April 2–6, 2007*, pp. 322–27. Bonn: Dr. Rudolf Habelt GmbH, Kolloquien zur Vor- und Frühgeschichte 10.

Surface-Evans, S., and White, D. 2012. An introduction to the least-cost analysis of social landscapes, in White, D., and Surface-Evans, S. (Eds.), *Least-Cost Analysis of Social Landscapes: Archaeological Case Studies*, pp. 1–7. Salt Lake City: The University of Utah Press.

Thrombold, C. D. 1991. An introduction to the study of ancient New World road networks, in Thrombold, C. D. (Ed.), *Ancient Road Networks and Settlement Hierarchies in the New World*, pp. 1–9. Cambridge: Cambridge University Press.

Tripcevich, N. 2009. *Workshop No. 1: Viewshed and Cost Distance*, http://mapaspects.org/book/export/html/3743; accessed November 24, 2012.

van Leusen, M. 1993. Cartographic modelling in a cell-based GIS, in Andresen, J., Madsen, T., and Scollar, I. (Eds.), *CAA 1992: Computer Applications and Quantitative Methods in Archaeology*, pp.105–24. Aarhus: Aarhus University Press.

———. 2002. *Pattern to Process: Methodological Investigations into the Formation and Interpretation of Spatial Patterns in Archaeological Landscapes*. Ph.D. dissertation. Groningen: Rijksuniversiteit Groningen, http://irs.ub.rug.nl/ppn/239009177; accessed November 6, 2010.

Verhagen, P., and Jeneson, K. 2012. A Roman puzzle: Trying to find the Via Belgica with GIS, in Chrysanthi, A., Flores, P. M., and Papadopoulos, C. (Eds.), *Thinking beyond the Tool: Archaeological Computing and the Interpretive Process*, pp. 123–30. Oxford: Archeopress, British Archaeological Reports International Series 2344.

Warren, S. D., Hohmann, M. G., Auerswald, K., and Mitasova, H. 2004. An evaluation of methods to determine slope using digital elevation data, *Catena* 58: 215–33.

Waugh, D. 2000. *Geography. An Integrated Approach*. Cheltenham: Nelson Thornes.

Wheatley, D., and Gillings, M. 2002. *Spatial Technology and Archaeology: The Archaeological Application of GIS*. London: Taylor & Francis.

White, D., and Surface-Evans, S. (Eds.). 2012. *Least-Cost Analysis of Social Landscapes, Archaeological Case Studies*, pp. 1–7. Salt Lake City: The University of Utah Press.

Whitley, T. G., and Burns, G. 2008. Conditional GIS surfaces and their potential for archaeological predictive modelling, in Posluschny, A., Lambers, K., and Herzog, I. (Eds.), *Layers of Perception, Proceedings of the 35th International Conference on Computer Applications and Quantitative Methods in Archaeology (CAA), Berlin, Germany, April 2–6, 2007*, pp. 292–98. Bonn: Dr. Rudolf Habelt GmbH, Kolloquien zur Vor- und Frühgeschichte 10.

Xeidakis, G. S., and Varagouli, E. G. 1997. Design and construction of Roman roads: The case of Via Egnatia in the Aegean Thrace, northern Greece, *Environmental and Engineering Geoscience* 3(1): 123–32.

Yousef, M. K., Dill, D. B., and Freeland, D. V. 1972. Energetic cost of grade walking in man and burro, Equus asinus: Desert and mountain, *Journal of Applied Physiology* 33(3): 337–40.

Yu, C., Lee, J., and Munro-Stasiuk, M. J. 2003. Extensions to least-cost path algorithms for roadway planning, *International Journal of Geographical Information Science* 17(4): 361–76.

Zhang, X., and Armstrong, M. P. 2005. Using a genetic algorithm to generate alternatives for multiobjective corridor location problems, in Yichun Xie and Brown, D. G. (Eds.), GeoComputation CD-ROM, *Proceedings of the 8th International Conference on GeoComputation, University of Michigan, July 31–August 3*, www.geocomputation.org/2005/ZhangX.pdf; accessed March 7, 2011.

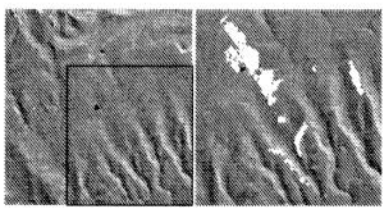

# 9 Compute-Intensive GIS Visibility Analysis of the Settings of Prehistoric Stone Circles

*Mark Lake and Damon Ortega*

## The Past and Future of GIS-Based Visibility Analysis

For 20 years visibility analysis has been one of the most popular archaeological applications of geographical information systems (GIS) for interpretive purposes. In 2003 Lake and Woodman provided a detailed account of the various forms that GIS-based visibility analysis had taken up to that time. They argued that such analyses could be divided into those that were predominantly informal, statistical, or humanistic and, furthermore, that this tripartite division recapitulated—albeit over a compressed timescale—theoretically driven developments in non-GIS visibility studies. Nearly 10 years on it is probably safe to say that all three forms of GIS-based "viewshed analysis" have lost their novelty value. Thus informal viewshed analyses, those that lack statistical or theoretical sophistication and adopt a largely common-sense approach to inference (Lake and Woodman, 2003), are no longer found in methodological literature but are scattered through the relevant subject literature. More interesting is the lack of evidence that more sophisticated statistical or humanistic analyses routinely contribute to archaeological explanation/interpretation (but see Gillings, 2009, for a recent exception). We suspect that the increasing use of multicore processors—and the power of modern desktop computers more generally—will lead to a resurgence of interest in GIS-based visibility analysis. To see why such resurgence might occur, we revisit the distinction between statistical and humanistic GIS-based visibility analyses

*Computational Approaches to Archaeological Spaces* by Andrew Bevan and Mark Lake, Eds., 213–242 

and consider how the failure of either approach to become a routine part of landscape archaeology was in no small part due to computational limitations.

Statistical studies of visibility, those characterized by an explicit concern with quantification and inferential rigor, predate the archaeological application of GIS and may in a general sense be considered examples of processual archaeology (Lake and Woodman, 2003: 690). In an European context, the influence of the New Geography, as expounded in Haggett's (1965) *Locational Analysis in Human Geography*, is clearly seen in Fraser's later (1988) work on study of visibility from Orcadian Neolithic chambered cairns and in a number of other studies (for example, Barnatt and Pierpont, 1983; Bradley, Harding, and Mathews, 1993; Bradley et al.,1993; Ruggles, Martlew, and Hinge, 1991), which, although not always explicitly subscribing to, nevertheless supported the prospects for a so-called cognitive-processual archaeology (Renfrew, 1982, 1994). What all these studies share with subsequent GIS-based statistical visibility analysis (for instance, Fisher et al., 1997; Lake and Woodman, 2000; Ruggles and Medyckyj-Scott, 1996; Wheatley, 1995, 1996; Woodman, 2000) is the use of a control sample to ascertain whether the viewshed properties of the archaeological sites in question could have occurred by chance alone or were more likely to reflect the intentions of past people (Lake and Woodman, 2003: 693). The most sophisticated of these studies (for example, Fisher et al., 1997; Woodman, 2000) used Monte Carlo simulation and stratified random sampling in an attempt to distinguish association from causation; in other words, they sought to ensure that apparently significant viewshed properties of archaeological sites were more than simply a byproduct of other locational choices such as elevation.

The humanistic turn in archaeology (see, for example, Hodder, 1986, and Shanks and Tilley, 1987) produced a number of visibility studies that frequently share the cognitive-processual interest in ideology and cognition but that often place greater emphasis on the nondiscursive knowledge of past people. These studies are typically characterized by a concern with situated visibility—the changing field of view as one moves through a landscape or an archaeological site—and, perhaps inevitably, they usually focus on the particular rather than the general. The phenomenological approach of Tilley and his coworkers (Bender, Hamilton, and Tilley, 1997, 2007; Tilley, 1994, 2004) has been particularly influential, but there are others, such as Barrett (1994: 15–17) and Thomas (1993: 42), who have shown greater interest in visibility as a resource for the maintenance of power relations and social reproduction. Much archaeological writing in a humanistic vein has explicitly questioned conventional scientific reasoning (see Edmonds et al., 1990; also Thomas, 2004), but recently there has been renewed interest in the development of methodology to permit a degree of repeatability (Hamilton et al., 2006). The mid-1990s witnessed a bout of soul searching in which leading proponents (for example, Wheatley, 1993: 133) of archaeological GIS expressed their fear that the uptake of GIS either wittingly or unwittingly encouraged a functionalist approach to archaeological explanation that they felt had otherwise been largely rejected as part of the humanistic critique of processual archaeology. In many respects the ensuing critique of archaeological GIS reflected the wider post-positivist critique of GIS in geography, incorporating both ontological (see, for instance, Thomas, 1993; c.f. Sui, 1994) and epistemological (such as Gaffney, Stančič, and Watson, 1996: 132; c.f. Wright, Goodchild, and Proctor, 1997) concerns. Lake and Woodman (2003: 694–95) provide a detailed account of the charges laid against GIS-based visibility studies. For present purposes, note that the response was ultimately a focus on "perception," whether by attempting to compute more nuanced forms of viewshed, such as so-called Higuchi viewsheds (Wheatley and Gillings, 2000), or to compute visual landscape "affordances" *sensu* Gibson's (1986) ecological theory of perception (primarily the work of Llobera [2001, 2003, 2007]), or the call for greater awareness of the interconnectedness of sensory experience (Freeman and Gillings, 2007) and possibly the use of virtual reality (Gillings and Goodrick, 1996).

By the turn of the millennium the state of the art in GIS-based visibility analyses was such that any substantive contribution to a real archaeological problem (as opposed to a case study intended only as an illustration of method) would require substantial computing resources. This fact was equally true of both the statistical approach, with its use of control samples and in particular Monte Carlo simulation, and the humanistic approach, whose emphasis on visual affordances would require the computation of viewsheds for entire landscapes (so-called total viewsheds; see Llobera, 2003) or the construction of virtual reality models. We suspect that the increased computational requirement goes some way to explaining the relative paucity of sophisticated and inferentially successful GIS-based visibility analysis over the past 10 years. The point is not so much whether the necessary computational resources could be found at all—Exon and associates' (2000) intensive study of the intervisibility of barrows in the vicinity of Stonehenge and Llobera and colleagues (Llobera et al., 2010) use of the Condor high throughput computing framework demonstrate the possibility of such studies—but whether they were or were perceived to be available in the context of primarily subject-focused archaeological projects. It is our contention that the power of desktop computers has now reached the point where some of the more sophisticated forms of analysis proposed as many as 10 years ago can now be more routinely used "in anger"—that is, to draw inferences from real datasets rather than simply to illustrate methods on "toy" datasets. Coupled with the development of a more eclectic and less sharply polarized theoretical climate (Pearce, 2011) this will lead to a renewal of interest in, on the one hand, large-scale generalizing of comparative visibility analyses where the lack of subtlety of traditional GIS-based visibility is less problematic and, on the other hand, the merging of GIS-based visibility analysis with the use of augmented reality (Eve, 2012)—

in this case precisely to overcome the lack of subtlety required for particularizing studies. In the remainder of this chapter we provide an example of the former approach, made possible by the use of adaptive parameterization of the viewshed calculation.

## The Visual Setting of Stone Circles

The particular problem we address in this chapter concerns the visual setting of stone circles. The prehistoric stone circles found across many parts of Great Britain and northern France were constructed between the Middle Neolithic and the Late Bronze Age (c. 3500 B.C.E.–1000 B.C.E.), although the form reemerges in Scandinavia and north east Europe from the Late Iron Age (500+ B.C.E.). Here we are concerned with British stone circles, which, despite disagreement about their exact definition (see Barnatt, 1989: 505 contra Burl, 2000: 317 and Bradley, 1993: 55, in the case of Avebury) basically comprise a number of large stones or boulders (orthostats) arranged as an ellipse, although their size and form vary considerably with time and from region to region; the diameters of the British stone circles studied in this chapter range from 1.5 m to 350.6 m and the estimated original numbers of stones from 4 to 99 (data from Barnatt, 1989).

Stone circles are among the most enigmatic of prehistoric monuments, and their purpose is not certain. First described in detail by the early Antiquarians Stukeley (1687–1765) and Tolland (1670–1722), who thought them to be Celtic Druidic temples (Burl, 2000: 16), they have more recently been explained as places of assembly (Harding and Lee, 1987) and/or instruments for calculating and predicting the seasons or the movement of the sun or the moon (Hawkins, 1966). Their form has likewise been attributed to various forces, ranging from the punishment of wrongdoers resulting in circles of petrified beings (folk tales reported in Burl, 2000: 69), to their careful layout employing a standard unit of measurement known as the megalithic yard (Thom, 1967). Explanations for the location of stone circles are equally varied and include coincidence with the path of ley-lines forming a telepathic network across the landscape (Williams, 1968); the actions of prehistoric water diviners (Underwood, 1969); centrality within hierarchically organized territories (Renfrew, 1973); orientation with respect to the movement of celestial bodies (Ruggles, 1999; Thom, 1971); and most recently the symbolism of the setting (Bradley, 1998; Richards, 1996). It is attempts to explain location in terms of visibility that interest us here.

The idea that stone circles functioned as high-precision astronomical instruments has been largely discredited by a number of studies (Barnatt, 1989; Barnatt and Pierpont, 1983; Ruggles, 1984; Ruggles and Burl, 1985; Ruggles, Martlew, and Hinge, 1991), which have, nevertheless, confirmed that many stone circles are indeed oriented with reference to astronomical events. The finding of these broadly cognitive-processual studies that astronomical alignments—although real—were of rather lower precision than once supposed has informed a humanistic strand of interpretation, one that takes a more experiential approach to stone circles (Watson, 2001: 307). At their most ambitious such interpretations combine the solar and lunar orientation of stone circles with insights from ethnographic analogy to posit that stone circles such as Stonehenge and Avebury were part of complexes of monuments that ritualized an understanding of the human lifecycle as one involving life, death, and rebirth (Bradley, Arrowsmith, and Society of Antiquaries of Scotland, 2005; Pearson and Ramilisonina, 1998).

In a less ambitious approach that focuses more on terrestrial visibility, it has been noted that stone circles are not necessarily prominent in the landscape and may have been difficult to locate from both nearby and far away (Bradley, 2002: 75), especially when compared with burial monuments, which were often built of materials providing high visual contrast with the immediate environment (Burl, 1988: 47–50). Consequently, some have argued for the importance of concealment as a means of effecting social differentiation in the experience of rituals taking place within stone circles (Barrett, 1994: 15–18; Bradley, 1993: 53; Thomas, 1993: 42). This concern with visibility looking in toward stone circles is mirrored by an interest in the terrestrial view out, notably in the work of Bradley (1998) and Richards (1996), both of whom have examined prehistoric monuments in relation to their landscape setting. Bradley characterizes the locations of certain stone circles as embodying a "circular perception of space" (1998: 122), citing as examples Castlerigg stone circle, which is situated with "a facade of standing stones confronting a chain of mountains"; Long Meg and Her Daughters, which commands a "virtually continuous horizon of hills and mountains"; and Avebury, which he describes as being "ringed by a horizon of hills" (1998: 122–23). Richards notes that, although being almost surrounded by water owing to its location on an isthmus, the Ring of Brodgar is nevertheless "enclosed by the encircling hills" (1996: 203). Thus both Bradley and Richards suggest that the forms of certain stone circles "echo the characteristic features" of their topographic setting and so provide a "metaphor" for the wider landscape (Bradley, 1998: 122–23). Indeed, Richards goes so far as to suggest that by creating a "a microcosm of landscape" stone circles provided the "physical and cultural centers of the world" for the people who used them (1996: 203).

The specific cases cited by Bradley and Richards may well support their argument that stone circles in some sense recapitulate their landscape setting, but it is not clear how far one can generalize this insight, even within one region. As already noted, there is substantial variability in the form and size of stone circles, and it may or may not be coincidental that three of the four stone circles cited by Bradley and Richards are among the very largest in Britain (Avebury, Long Meg and Her Daughters, and the Ring of Brodgar). In addition, there is the question of intentionality: one can envisage situations where the nature of the terrain is such that it would have been difficult to locate a stone circle so that it did not appear to be enclosed by encircling hills. Of course, discovery that the locational properties of stone circles are common in the landscape at large would not prove that the builders of stone circles were not attentive to the visual properties of the circle's setting, but identifying cases where stone circles were built in locations with settings that were not characteristic of the local norm would certainly strengthen the argument.

In the remainder of this chapter we demonstrate how it is possible to harness the power of contemporary desktop computers to investigate the generality of claims about the visual settings of stone circles and the extent to which they may have been deliberately selected.

## A GIS-Based Methodology

Our method for investigating patterning in the visual settings of stone circles has three components: (1) a quantitative measure of the visual properties of landscape setting; (2) a statistically robust inferential framework; and (3) a means of making the first two components computationally tractable. We outline each in turn.

### Measures of Setting

The particular quality of a visual setting that interests us here is one obtained by an observer placed at a specified viewpoint (in this case one located within a stone circle) and that Bradley termed a "circular perception of space" (Bradley, 1998: 122)—in other words the sense of being located within a topographic basin (Richards, 1996). We measure several factors that contribute to this experience.

*Viewshed Size* The size of the area visible from a specified viewpoint (the viewshed) does not directly measure the circular perception of space, since viewsheds of equal size can be very different in shape. Nevertheless, viewshed size may provide a rough first proxy for the basin-like feel of a setting to the extent that the notion of a basin presupposes a bounded area and thus a less than maximal viewshed size.

*Fragmentation* More subtly, we understand the notion of a basin to imply not just a rim—in this case a clearly defined far horizon—but also reasonably uninterrupted visibility within it; in other words, a continuous rather than fragmented viewshed. Even if we cannot exactly calibrate the amount of fragmentation required to disrupt the basin-like feel of a setting, with a suitable measure we can at least compare the setting of stone circles in this respect. We have devised a measure of viewshed fragmentation based on the ratio of the visible area of a given stone circle's viewshed to that of a hypothetical viewshed with the same perimeter but that filled a perfect circle. The measure of fragmentation $F$ for a given stone circle $c$, is calculated as

$$F = V_c / V_h$$

where $V_c$ is the area of the stone circle's actual viewshed and $V_h$ is the area of the hypothetical circular viewshed of equivalent perimeter. $V_h$ is calculated thus:

$$V_h = \pi(P_c / 2\pi)^2$$

where $P_c$ is the perimeter of the stone circle's actual viewshed.

Given that for same perimeter any shape will have a smaller area than that of a circle, the measure of fragmentation, $F$, will range from $F = 1$ for a perfectly circular unfragmented viewshed to $F \rightarrow 0$ for an irregular and fragmented viewshed.

*Properties of the Horizon* We have already seen how Bradley and Richards both place great emphasis on the horizon—the rim of the basin—in their discussion of the circular perception of space. Lake and Woodman (2003) explored how the properties of the horizon contribute to the circular perception of space from a viewpoint. They noted that while the far horizon may be perceived as circular, no aspect of the topography need actually demonstrate patterning on a fixed radius (see Figure 9.1 for the very irregular and fragmented viewshed of Long Meg and Her Daughters). In particular, the inclination (vertical angle of view) at which the horizon is seen is a function of (1) the horizontal distance to the horizon and (2) the difference in elevation between the land on the horizon and the viewpoint. Consequently, a viewshed of reasonably constant radius need not offer a constant line of horizon (inclination) if the land on the horizon is very variable in elevation. Conversely, a viewshed of variable radius could in fact offer a constant line of horizon if elevation of land on the horizon happened to vary in the right way. In their study of Scottish recumbent stone circles, Lake and Woodman (2003) demonstrated that this effect is not simply hypothetical, so we follow them in rejecting measures dependent on specification of a fixed radius (such as concavity; see Yokoyama, Shirasawa, and Pike, 2002) and instead adopt their technique of computing the distance to, elevation of, and inclination of land on the far horizon at azimuths from 1 degree through 360 degrees. For the purposes of statistical testing, we summarize the properties of the horizon of each stone circle in terms of the minimum, maximum, mean, and standard deviation of each of the three attributes just mentioned.

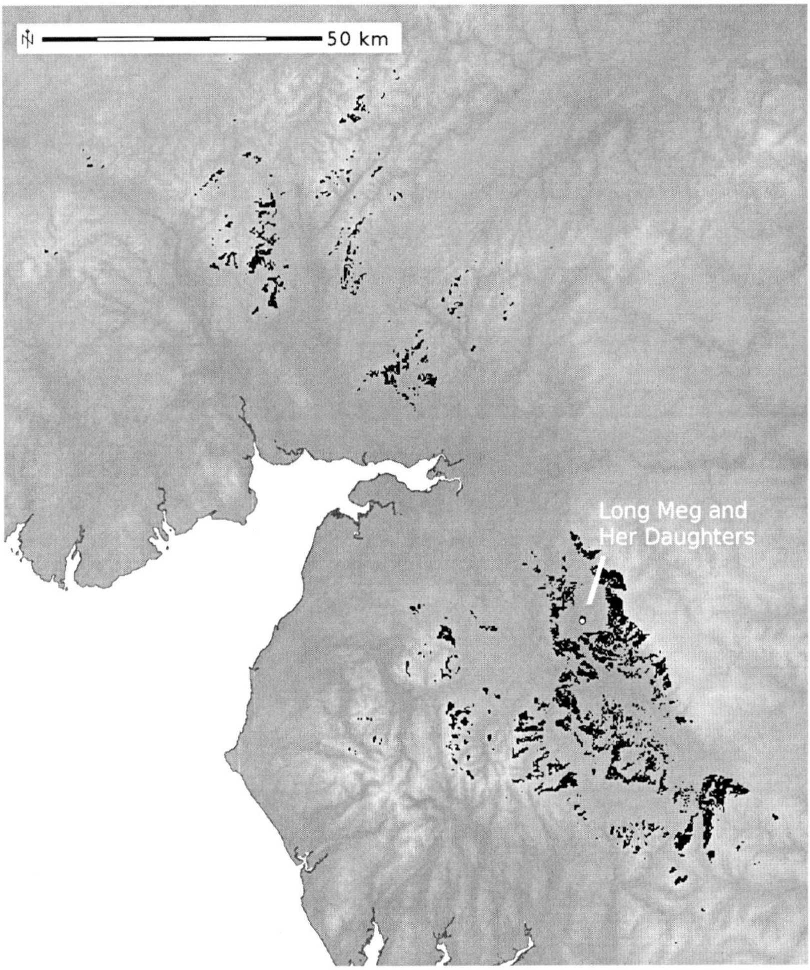


**Figure 9.1** The fragmented viewshed of Long Meg and Her Daughters (derived from elevation data © Crown Copyright/database right 2012, an Ordnance Survey/EDINA supplied service)

## Inferential Framework

Having outlined how we attempt to measure Bradley's "circular perception of space" in a quantitative framework, we now turn to the issue of how we hope to learn from those measurements. We emphasize that we do not suppose that our quantitative measures can replicate the experience of actually standing in a stone circle, so to that extent we acknowledge that they do not overcome the well-rehearsed criticism that GIS-based analysis fails to capture the subtlety of real-world visibility (see Wheatley and Gillings, 2000, for an overview)—although the measurement of fragmentation and the inclination of the horizon go somewhat farther in this direction than most other published studies. Rather, given that it has been demonstrated that such measures do broadly correlate with variability experienced in the field (Lake and Woodman, 2003: 701–03), we focus on one of the strengths of a GIS-based approach, which is to provide a large-scale comparative framework within which regional or other patterning might be observed. As already noted, our aim here is to test the generality of claims about the visual settings of stone circles and the extent to which it is possible to support the argument that the settings were intentionally selected for their topographic properties.

*Generality* We address the question of generality by computing the viewshed size, fragmentation, and horizon properties of 529 stone circles distributed across Great Britain. We took the spatial coordinates from a keyhole markup language (KML) file available for download from the website of *The Modern Antiquarian* (www.themodernantiquarian.com), having first rejected sites whose status as a stone circle is uncertain or whose accuracy of coordinates was doubted in the light of comparison with other sources (primarily PastScape and CANMORE; the national monuments records of England and Wales, and Scotland; and inspection in the Google Earth viewer). Barnatt (1989) assembled a gazetteer of British stone circles that records a number of their attributes, including his own taxonomic grouping, the maximum diameter of the circle, and the likely original number of stones. We were able to cross-reference 306 of the stone circles listed in Barnatt's gazetteer with those that we deemed to have credible coordinates in *The Modern Antiquarian* dataset (Barnatt provided only low-precision coordinates for a significant number of entries in his gazetteer). In the long run it would clearly be desirable to build a definitive spatial database of British stone circles, but for present purposes even the smaller subset of 306 stone circles for which we have both high-precision spatial coordinates and attribute data does at least provide numerous cases in all areas of Great Britain where stone circles are present (predominantly the west and the north).

*Intentionality* We adopt the standard approach from classical inferential statistics, which is to establish the probability that the relevant properties of the topographic setting of each stone circle could be obtained by a random draw from the background population of possible settings—that is to say, if the stone circles had been located without reference to those properties. In archaeological GIS-based analysis this approach was pioneered by Kvamme (1985, 1988, 1990), and the inferential logic has been thoroughly discussed by Fisher and associates (1997), Lake and Woodman (2000), and Woodman (2000), with further examples of its application in a number of studies, including Wheatley's well-known investigation of intervisibility between southern British Neolithic long barrows (1995, 1996). There are three points worth elaborating here:

1. This technique cannot ultimately disprove that the builders of a stone circle intentionally chose its location for the nature of its viewshed, but what it can do is to increase or decrease our confidence that this was the case according to whether or not we can reliably reject the null hypothesis of a random draw.
2. In line with much of the work just cited, we use Monte Carlo simulation to compute the significance level at which we can reject the null hypothesis. Specifically, we compute the relevant measures (viewshed size, fragmentation, and the summary horizon properties listed earlier) for 55 control sample locations around each stone circle and then compute the (one-tailed) significance level at which we can reject the null hypothesis, *p*, as $p = R_c/N$, where $R_c$ is the rank of the measure obtained for the stone circle in question among the *N* control sample locations (see Fisher et al., 1997, and Robert and Casella, 2004, for more detail). Figure 9.2 provides some examples of the results of this process.

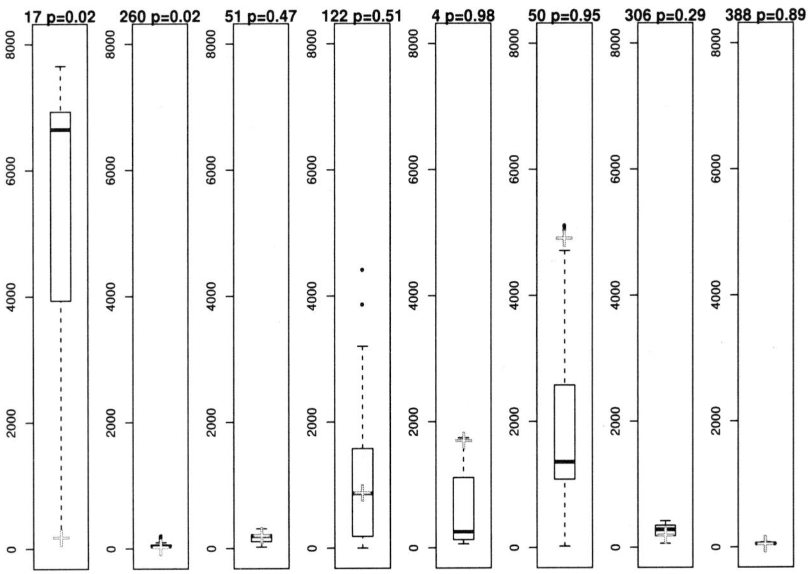


**Figure 9.2** Eight examples from the Monte Carlo simulation of viewshed area. Stone circles 17 and 260 have viewsheds (open cross) that are smaller than expected of viewpoints in the locality; 51 and 122 have viewsheds that are typical; 4 and 50 have viewsheds that are larger than expected. Stone circles 306 and 388 are Long Meg and Her Daughters and the Ring of Brodgar, respectively.

3. Note the importance of distinguishing mere patterning from causation. Both Fisher and associates (1997) and Woodman (2000) provide useful discussion of this problem, which, in this case, is how to be reasonably confident that a viewshed property for which one can reject the null hypothesis of random location was in fact the subject of conscious choice rather than the inadvertent consequence of the choice of some other locational attribute with which it happens to correlate. We attempt to limit the likelihood of such confounding variables by restricting the control sample for each stone circle to points within 500 m of that site. In this way we accept that there may have been myriad other reasons for the rough location of stone circles (such as placement within a territory, proximity to or distance from settlement, and so on) and concentrate on testing whether there is statistical support for the hypothesis that, within these constraints, people sought out locations which afforded particular visual properties. This method also has the additional benefit over global control sample of being less likely to produce spurious significance owing to the fact that, for example, no stone circles (setting aside the timber Holme I and Holme II) are found in the flatlands of northern East Anglia.

## Computational Tractability

The GIS-based methodology just outlined required the computation of the viewsheds of each stone circle, plus the points in the control samples, followed by postprocessing of the computed viewsheds to extract the horizons and subsequently compute the fragmentation and horizon measures. The viewsheds were computed using the British Ordnance Survey's Land-Form Panorama digital terrain model, which describes the terrestrial elevation of Great Britain on a 50 m grid.

All computation was carried out using GRASS GIS software (GRASS Development Team, 2012) running under Linux. We used the r.horizon GRASS module written by Lake (see Lake and Woodman, 2003: 697 for details) to extract the viewshed horizons from previously computed viewshed maps, and then Unix bash shell and R statistical programming language (R Development Core Team, 2012) scripts written by Ortega to compute the fragmentation and other summary statistics from the horizon maps. Since postprocessing the viewshed contributed only a small fraction of the total processing time we focus here on describing the method that allowed us compute the viewsheds of the stone circles and their associated control points.

The challenge we faced was to compute some $529 \times 56 = 29{,}624$ viewsheds, each potentially extending in excess of 100 km from the viewpoint and covering some 12.5 million raster map cells. Our solution to this challenge was to embed an efficient viewshed module in a shell script, which effected adaptive parameterization. The shell script ultimately invokes the r.viewshed

GRASS module to compute the viewshed of each stone circle/control point; this module utilizes a fast algorithm written by Laura Toma and colleagues (Fishman, Haverkort, and Toma, 2009). Before doing so, however, the script limits analysis to the smallest possible geographical region consistent with the need to compute a far horizon truncated only by the curvature of the earth or intervening land, but not by an arbitrary maximum viewing distance. Computation of the necessary region is achieved in three steps:

1. The maximum possible distance to the horizon $D$ for a viewpoint in Great Britain was calculated using the formula $D = \sqrt{(2Rh)}$ (Lodge and Muirhead, 1924) where $R$ is the average radius of the earth (6,371,009 m) and $h$ is the maximum possible height differential between a viewpoint and another location that might be visible from it.
2. For each stone circle/control point $D$ is used to calculate the boundary of a geographical region marking the maximum extent of visibility around that stone circle as if it happened to have a viewshed with the maximum distance to the horizon that is possible anywhere in Great Britain. Since that is unlikely, this region is searched for the maximum height differential with the stone circle in question, and that value is then used to calculate a further reduced geographical region as per step 1.
3. For each stone circle/control point the second smaller geographical region is used to constrain the area that must be examined when one is computing the viewshed. This approach has two advantages: first it ensures that r.viewshed need sweep only the minimum collection of raster map cells necessary to locate the true maximum extent of the viewshed; second it ensures that the minimum necessary amount of data is loaded into computer memory, thereby reducing the likelihood of needing to use swap-space on disk, which is of course orders of magnitude slower than accessing physical memory (for example, we found that r.viewshed was nearly four times slower when arbitrarily limited to 512 MB of physical memory than when it was configured so as to load the entire geographical region into physical memory).

This technique allowed us to compute the necessary 29,624 viewsheds in approximately 425 hours (18 days) using an Intel Core i5-2500–based desktop computer offering 4 cores and 4 logical processors, each running at 3.3 GHz. (Note that GRASS GIS software is not currently thread-safe, so we launched parallel processes to benefit from the availability of multiple processors.)

## Statistical Analysis

Armed with 29,624 viewsheds we have undertaken a variety of statistical analyses of their size and fragmentation, as well as the distance to, elevation of, and inclination of the locations falling on their horizons. In keeping with the inferential framework described in the previous section, we use spatial plots, scatterplots, boxplots, and cluster analyses to explore patterning in viewshed properties and Monte Carlo simulation to address the issue of intentionality. There is insufficient space here to reproduce every graphical output from our analyses, but we do attempt to illustrate the principal findings, measure by measure.

## Viewshed Size (Area)

The viewsheds of our larger sample of 529 stone circles range in area from 21,011.25 km$^2$ to 0.98 km$^2$. The numerical distribution is strongly skewed toward smaller sizes, such that the median (137.06 km$^2$) is substantially smaller than the mean (1,599.92 km$^2$) and only 23% of stone circles have viewsheds that exceed the mean. Viewshed area does not exhibit a convincing linear or low-order polynomial correlation with the elevation of the corresponding stone circle, although it is the case that the largest viewsheds occur at lower elevations (only 7 of the 55 stone circles with viewsheds in excess of 5,000 km$^2$ are located at elevations above 300 m). This result is interesting because it is commonly supposed that higher viewpoints have large viewsheds. There is similarly no easily modeled relationship between viewshed area and either slope or aspect, although the stone circles with the very largest viewsheds are most often located on slight slopes and land facing south through west to north. Among the 306 stone circles for which we have attribute data there is no simply modeled relationship between viewshed area and the maximum diameter of the circle, although the largest circles (with diameters in excess of 60 m) all have small viewsheds (all below the mean viewshed area and all except 2 below the median).

There is very pronounced geographical patterning in the spatial distribution of viewshed area (Figure 9.3). In particular, the central Scottish stone circles have small viewsheds, the west Cumbrian and most Pennine stone circles have moderate sized viewsheds, and the southwestern stone circles (Devon and Cornwall) have large viewsheds. Among the Scottish recumbent stone circles (Grampian) there is a clearly defined gradation from small viewsheds in the west to large viewsheds in the east. In general (the central Pennines excepted), stone circles with larger viewsheds are located in closer proximity to the sea.

Monte Carlo simulation allows us to address the question of whether the observed numerical and/or spatial distributions of viewshed area imply intentional choices by those who built stone circles. We calculate that 25.8% of the smaller sample of 306 stone circles have viewsheds that are smaller or larger than would be expected by chance alone ($p = 0.05$), and this increases to 39.2% at a more relaxed significance level of $p = 0.1$. This result does not outwardly appear very promising, but there is an interesting relationship between whether viewshed sizes are significantly different from those expected by chance alone and their absolute size. Specifically, Figure 9.4 demonstrates

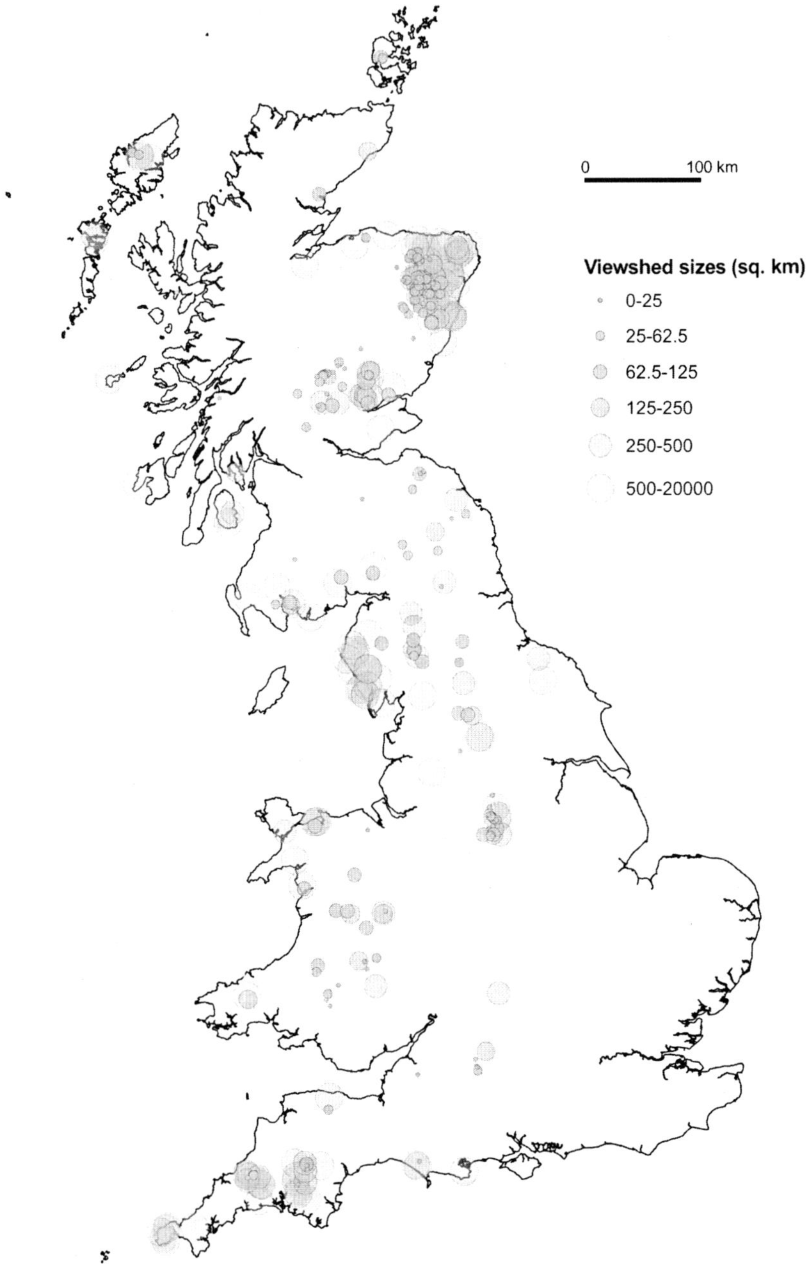


**Figure 9.3** The geographical distribution of the stone circles studied, showing their viewshed area

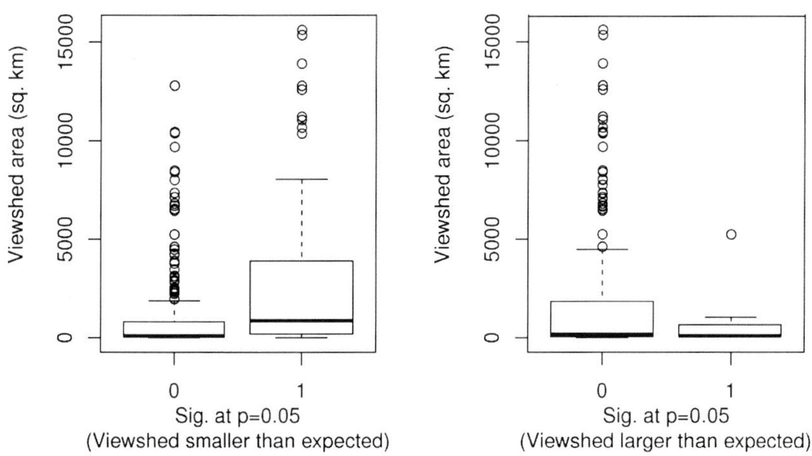


**Figure 9.4** The relationship between the statistical significance of viewshed areas and their absolute values

that most of the largest viewsheds are smaller than expected and that none of the very large viewsheds are larger than expected. Furthermore, the median of viewshed sizes that are smaller than expected is greater than the median of those that are not statistically distinguishable from the background population. We return to what this might mean in our concluding comments.

Turning to the spatial distribution of viewshed sizes, we found no evidence of robust spatial patterning whether viewshed sizes are smaller or larger than expected. Thus, although it may not be true of any particular stone circle, the general tendency for larger viewsheds to occur in closer proximity to the sea is one that would be expected without the exercise of intentional choice.

## Viewshed Fragmentation

The viewshed fragmentation of our larger sample of 529 stone circles ranges from 0.502 to 0.0000766. The numerical distribution is massively skewed toward smaller sizes, such that the median (0.00476) is substantially smaller than the mean (0.0146), and only 22% of stone circles have viewsheds that exceed the mean. Given that the fragmentation statistic, $F$, ranges from $F \rightarrow 0$ for very high fragmentation to $F = 1$ for a perfectly circular and continuous viewshed, it is clear that the vast majority of stone circles have very fragmentary viewsheds, something that is borne out by examination of viewshed maps. Viewshed fragmentation does not exhibit a convincing linear or low-order polynomial correlation with the elevation, slope, or aspect of the corresponding stone circle; nevertheless, 12 of the 15 stone circles with the least fragmented viewsheds (that is, $F > 0.1$) are located below 200 m above sea level, and 13 of them on slopes of less than 5 degrees. Among the 306 stone circles for which we have attribute data there is no simply modeled relationship between fragmentation and the maximum diameter of the circle. The largest circles (with diameters in excess of 60 m) all have very fragmented viewsheds ($F < 0.0180$), although the distribution of fragmentation values is so massively skewed that this is also true of 82% of all stone circles for which we have attribute data. However, considered another way, 90% of stone circles whose viewsheds are less fragmented than the mean are 30 m or less in diameter.

There is some geographical patterning in the spatial distribution of viewshed fragmentation (Figure 9.5). Most notably, stone circles nearer the sea are more likely to have less fragmented viewsheds, as might be expected if those viewsheds include a significant area of sea. Inland circles have more highly fragmented viewsheds. This geographical cline from low to high fragmentation is especially clear as one moves from east to west through the Scottish recumbent stone circles.

The results of the Monte Carlo simulations indicate that 22.6% of the smaller sample of 306 stone circles have viewsheds that are more or less fragmented than would be expected by chance alone ($p = 0.05$), and this fraction increases to 36.6% at a more relaxed significance level of $p = 0.1$. Unlike with viewshed size, there is no discernible pattern in either the numerical or geographical distribution of statistically significant cases. It is worth noting, however, that viewsheds are three times more likely to be significantly ($p = 0.05$) less fragmented than expected by chance than they are to be more fragmented than expected.

## Horizon Properties

As noted, we follow Lake and Woodman (2003) in computing for each stone circle the distance to, elevation of, and inclination of land on the far horizon at aziumuths from 1 degree through to 360 degrees. However, whereas Lake and Woodman studied 19 stone circles, we are dealing with a maximum of 529 sites plus an additional 29,095 control points and for this reason do not analyze the horizon measures by azimuth but instead summarize them in terms of the minimum, maximum, mean, and standard deviation per stone circle/control point. We discuss distance, elevation, and inclination in turn but before doing so must note that these measures show very little correlation with one another per stone circle/control point, which is unsurprising given that multiple combinations of distance and elevation can produce the same inclination. With just one exception, we also find no convincing correlation between any of the three horizon measures and the elevation, slope, or aspect of the stone circle itself. The exception is a strong positive linear relationship between the minimum elevation of the horizon and the elevation of the stone circle; again, this result is as expected.

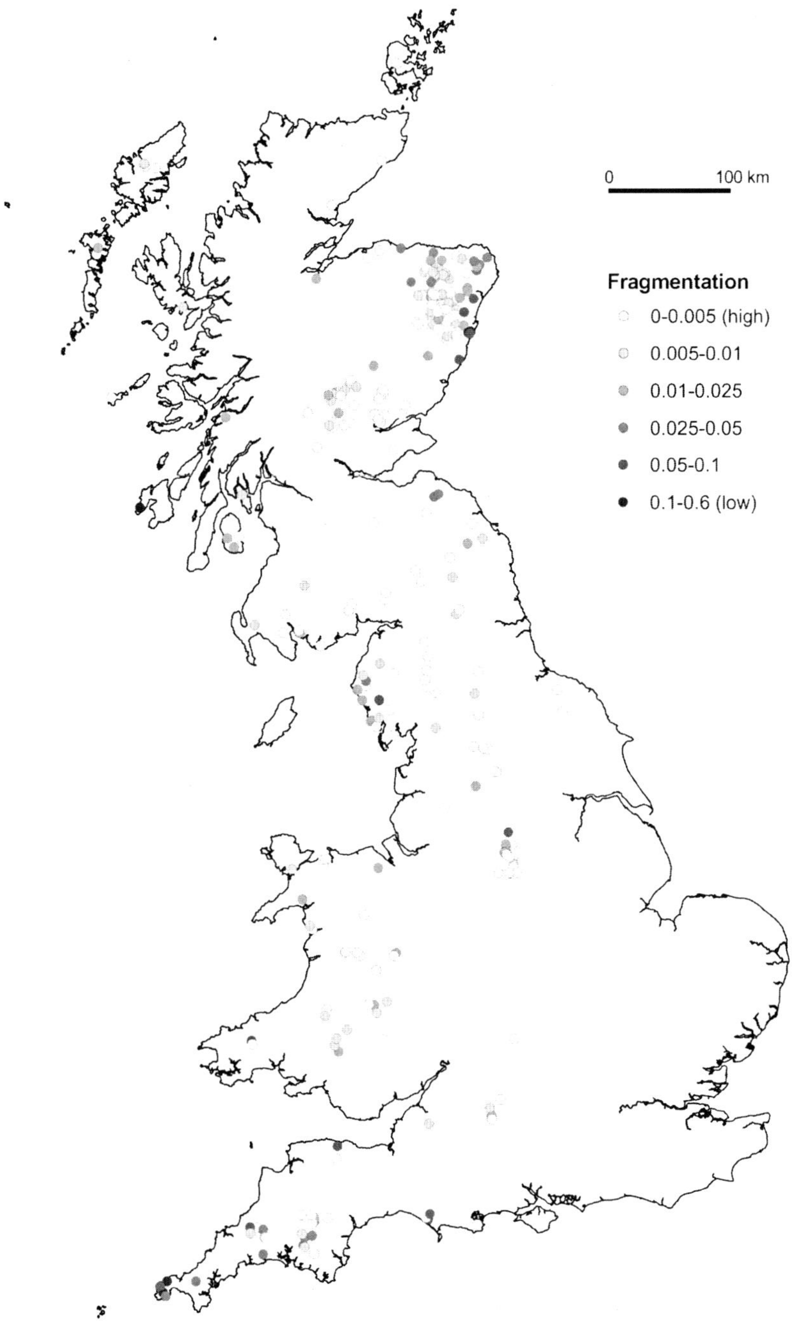


**Figure 9.5** The geographical distribution of the stone circles studied, showing the fragmentation of their viewsheds

**Table 9.1** Elevation of the Horizon (meters)

| | *Min. Elevation* | *Mean Elevation* | *Max. Elevation* | *Std. Dev.* |
|---|---|---|---|---|
| Min. | 0.0 | 1 | 136 | 11.4 |
| Max. | 555 | 822 | 1,309 | 406.6 |
| Mean | 110.4 | 350.0 | 766.6 | 158.3 |
| Std. Dev. | 118.7 | 175.7 | 282.4 | 82.3 |

*Elevation of the Horizon* The elevation of the horizon is variable both within and between stone circles, as can be seen from the summary statistics presented in Table 9.1. In this table the columns refer to per-circle summaries, whereas the actual numerical values presented in rows measure the variability in per-circle summaries across all 529 stone circles.

Although the minimum horizon elevation at any given stone circle varies from 0 to 555 m, Figure 9.6 shows that the minimum at any stone circle with a large viewshed (in terms of area) is always 0, confirming that large viewsheds include the sea. The maximum horizon elevation does not correlate with viewshed size. Figure 9.6 also shows that high mean horizon elevation is, as one might expect, associated with small viewsheds. There is no correlation between horizon elevation and the maximum diameter of stone circles.

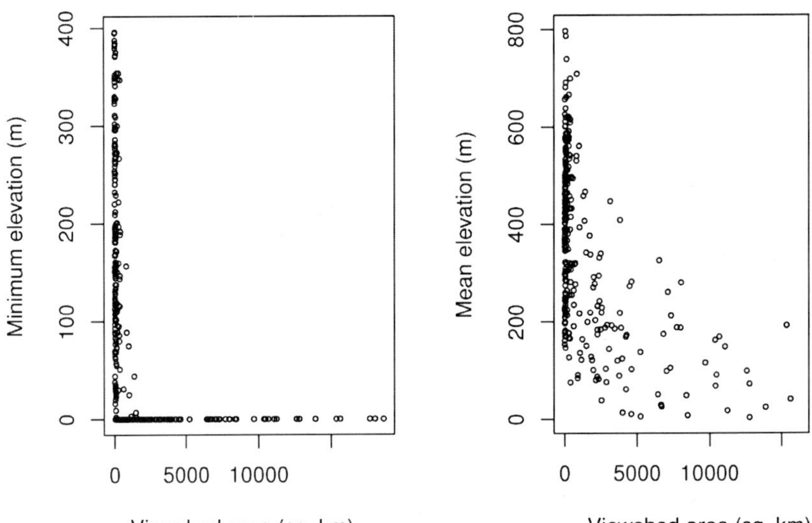


**Figure 9.6** The relationship between the elevation of the horizon and viewshed area (Avebury omitted)

The Monte Carlo simulations suggest that between 19% and 25% of the 306 stone circles for which we have attribute data have horizon elevations characterized by summary statistics that are different from those expected by chance alone (at $p = 0.05$). In these cases the minimum horizon elevation is equally likely to be lower or higher than expected by chance, but the maximum, mean, and standard deviation of horizon elevation are variously 2–3 times more likely to be lower than would be expected by chance. It is interesting to note that when the maximum and mean horizon elevations are greater than expected they are also low in absolute terms, and the converse is also true (Figure 9.7). We return to the possible significance of this finding in our concluding comments.

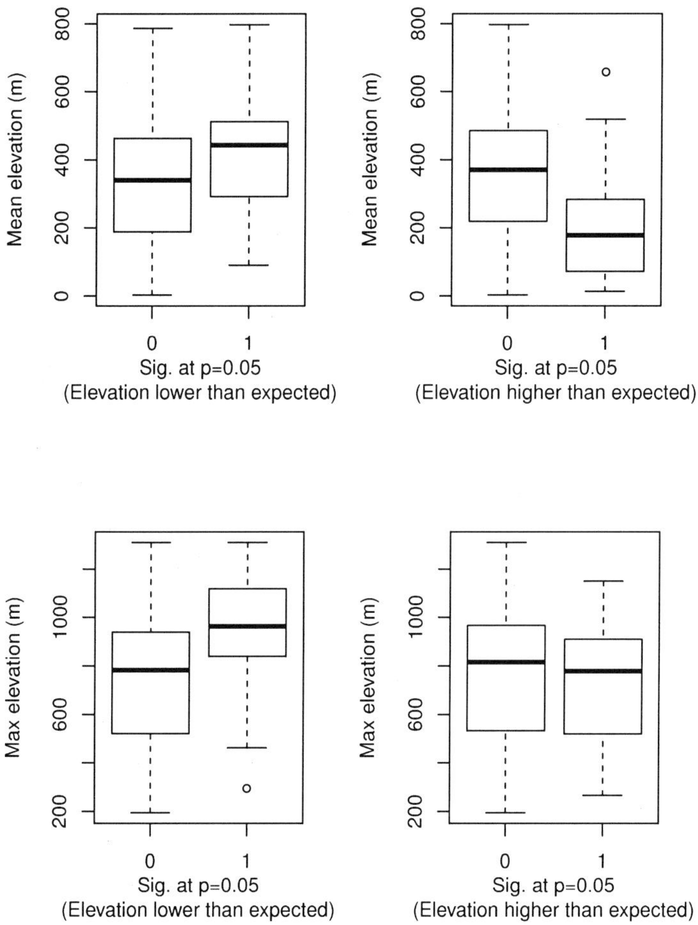


**Figure 9.7** Relationship between the statistical significance of horizon elevation and its absolute value

**Table 9.2** Distance to the Horizon (kilometers)

| | *Min. Distance* | *Mean Distance* | *Max. Distance* | *Std. Dev. Distance* |
|---|---|---|---|---|
| Min. | 0.05 | 1.34 | 2.45 | 0.54 |
| Max. | 6.76 | 124.75 | 129.88 | 21.58 |
| Mean | 0.32 | 41.36 | 72.95 | 61.53 |
| Std. Dev. | 0.50 | 30.73 | 38.95 | 14.10 |

*Distance to the Horizon* The variability in distance to the horizon is documented in Table 9.2, which is laid out in the same way as Table 9.1. The minimum distance to the horizon at any given stone circle does not correlate with viewshed area, but there is a clear relationship between both mean and maximum distance to the horizon and viewshed size. Although viewsheds with large mean and maximum distance to the horizon can be small in terms of their area (because they are highly fragmented), increasingly large viewsheds are, as one might expect, associated with larger mean and maximum distances to the horizon (Figure 9.8). There is no correlation between distance to the horizon and the maximum diameter of stone circles.

The Monte Carlo simulations suggest that the per-circle maximum, mean, and standard deviation of distance to the horizon are smaller or larger than would be expected by chance ($p = 0.05$) at 44%, 38%, and 39% of stone circles (the sample of 306), respectively. It is interesting that in these cases the measures are 5.8–6.5 times more likely to be smaller than expected by chance than

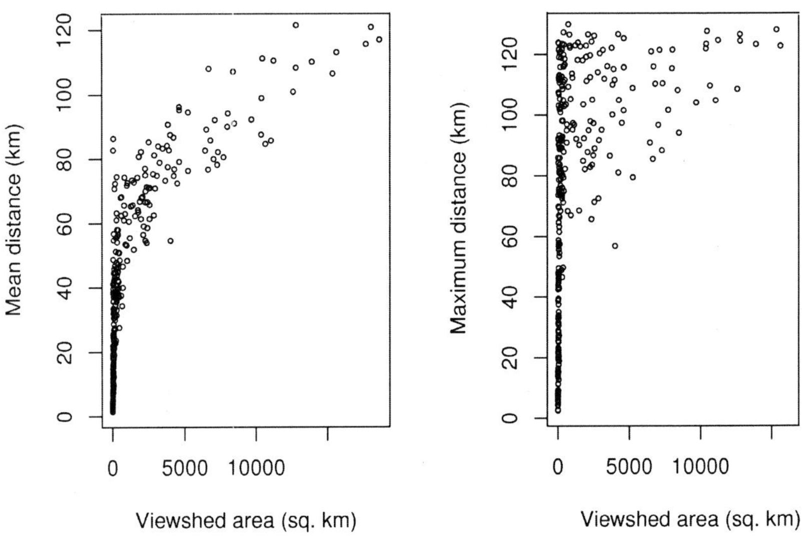


**Figure 9.8** Relationship between the distance to the horizon and viewshed area (Avebury omitted)

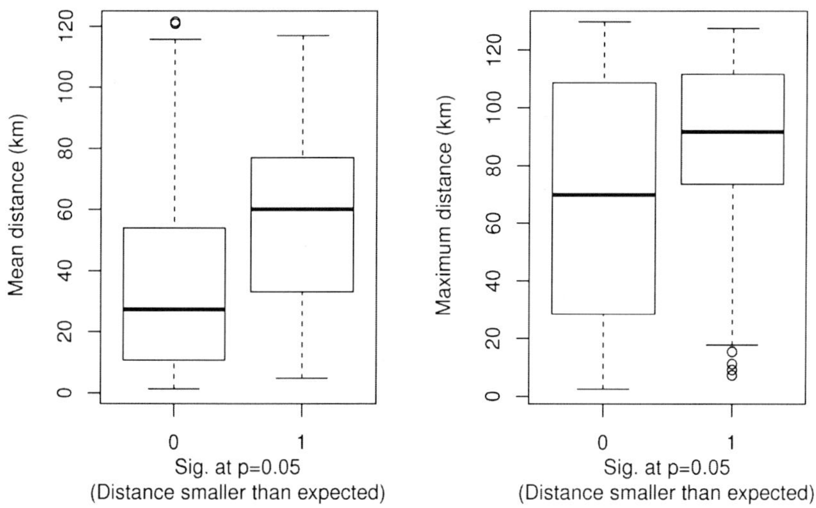


**Figure 9.9** Relationship between the statistical significance of the distance to the horizon and its absolute value (Avebury omitted)

they are to be larger than expected. Furthermore, Figure 9.9 shows that the values of smaller than expected measures of distance to the horizon are typically high in absolute terms.

The latter finding also shows spatial patterning. A majority of the eastern Scottish recumbent stone circles have viewsheds with a smaller than expected maximum distance to the horizon, whereas this is not so at the more westerly recumbent stone circles. Approximately 40%–50% of southwestern and Welsh stone circles also have viewsheds with a smaller than expected maximum distance to the horizon. In our concluding comments we discuss how this finding relates to the tendency for these same viewsheds to be large.

*Inclination of the Horizon* Table 9.3 documents the variability in the inclination of the horizon and is laid out in the same way as Table 9.1. The

**Table 9.3** Inclination of the Horizon (decimal degrees)

| | *Min. Inclination* | *Mean Inclination* | *Max. Inclination* | *Std. Dev. Inclination* |
|---|---|---|---|---|
| Min. | 84.55 | 89.76 | 90.70 | 0.11 |
| Max. | 94.78 | 99.49 | 118.00 | 5.61 |
| Median | 89.94 | 90.70 | 95.00 | 0.72 |
| Mean | 89.97 | 91.07 | 96.26 | 0.99 |
| Std. Dev. | 0.65 | 1.30 | 4.21 | 0.81 |

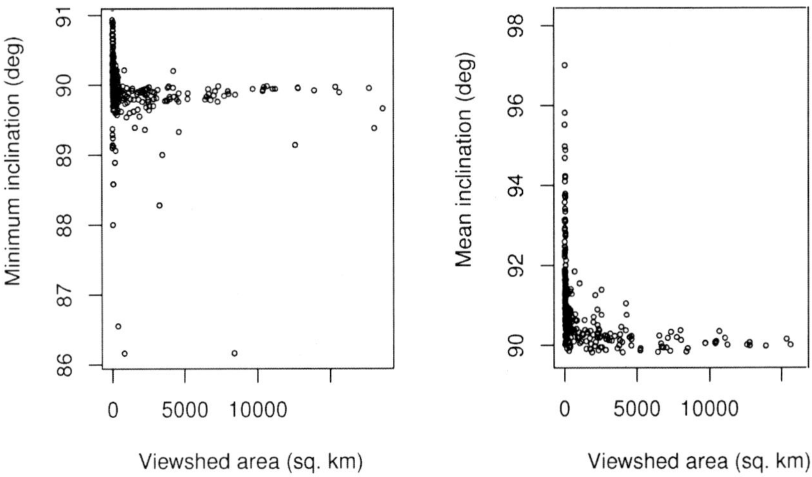


**Figure 9.10** Relationship between the inclination of the horizon and viewshed area (Avebury omitted)

distribution of the minimum horizon inclination across the sample of 529 stone circles is approximately normal, centered on a line of sight to the horizon that is close to horizontal. In general, stone circles at which the minimum inclination falls more than 0.5 degree either side of horizontal have small viewsheds (Figure 9.10). The 84% of stone circles at which the minimum inclination falls with this band have viewsheds ranging from very small to very large. The association of large viewsheds with near horizontal minimum inclination is an expected outcome of them also having distant horizons, as is their association with low standard deviation of inclination. There is a similar relationship between mean horizon inclination and viewshed size, except that in this case 92% of stone circles have a mean line of site to the horizon that is horizontal or above. The relationship between large viewsheds and mean minimum inclination, low mean inclination and low standard deviation of inclination, is largely repeated for the maximum diameter of stone circles (Figure 9.11), at least in the 306 cases for which we have that information; nevertheless, the 5 very largest stone circles in our sample (Avebury, Long Meg and Her Daughters, Stanton Drew Central, the Ring of Brodgar, The Twelve Apostles) actually have viewshed horizons with mean inclinations in the range 90.70–91.34 degrees, making their horizons appear on average higher than at 53% of stone circles in the sample, although in none of these 5 cases is the horizon inclination higher than expected (at $p = 0.05$ or $p = 0.1$). We return to this point in our concluding comments.

The Monte Carlo simulations suggest that only 9% and 6% of stone circles have horizons characterized by a minimum or maximum inclination,

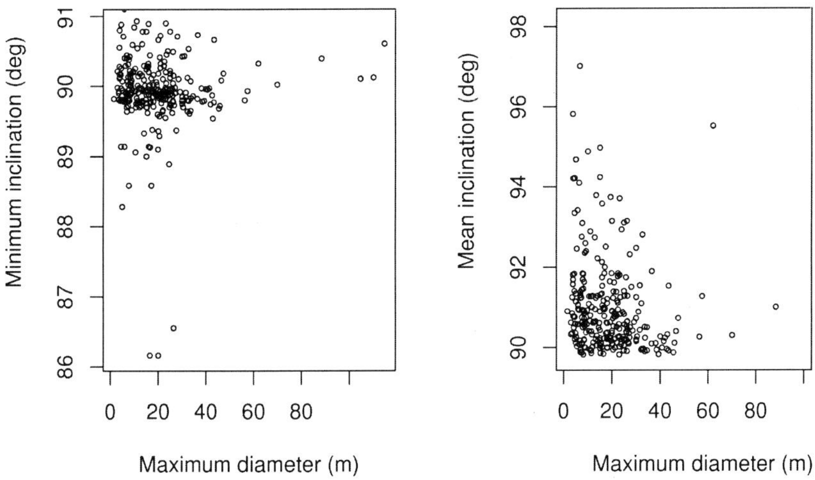


**Figure 9.11** Relationship between the inclination of the horizon and the maximum diameter of the stone circle (Avebury omitted)

respectively—that is, either smaller or larger than would be expected by chance alone (at $p = 0.05$). It is notable, however, that where significantly different from the control sample, the maximum horizon inclination is over five times more likely to be higher rather than lower than expected. This fact might be taken as tentative evidence for the intentional choice of locations with horizons providing a pronounced "rim," especially since only 7% of these stone circles have a smaller than expected mean horizon inclination. This inference does not, however, fit so well with the fact that slightly more (by a factor of 1.8) stone circles have horizons characterized by a mean inclination that is lower rather than higher than expected; but in this case it may be that their builders were first and foremost seeking a reasonably constant line of horizon, since only 7% of these stone circles have greater than expected variability in horizon inclination (standard deviation), whereas 44% have lower than expected variability.

## Conclusion

We computed the viewsheds of some 529 stone circles (approximately 60% of known circles) to explore the generality of Bradley and Richards' suggestions that these prehistoric monuments were built in locations that offered a "circular perception of space" (Bradley, 1998: 122–23). We also used Monte Carlo simulation in an attempt to find statistical evidence for intentional placement of stone circles in settings with particular visual properties. Our principal findings are outlined in the following sections.

## Geographical Distribution of Viewshed Properties

The stone circles with the largest viewsheds are found in southwest England, western Cumbria, and among the more easterly of the Scottish recumbent stone circles. The stone circles with the smallest viewsheds are generally found in central southern England, the southern half of Wales, the eastern Scottish border counties, central Scotland, and among the more westerly Scottish recumbent stone circles. The Monte Carlo simulations provide no evidence that this global variability in viewshed size reflects anything other than broad differences in local terrain form and, in particular, the likelihood of at least part of the viewshed, including the sea. The impact of the latter is also seen in the more coastal distribution of the least fragmented viewsheds.

Although there is no evidence that regional variability in the size and fragmentation of viewsheds is the result of different choices by the builders of stone circles, that lack does not alter the fact that those building and using them in different parts of Great Britain may have experienced them differently in terms of their relationship with their topographic setting. However, if the Scottish recumbent stone circles were built and used by related groups with similar cultural values, then perhaps the west-east geographical cline in the size and fragmentation of the viewsheds of those circles is evidence that viewshed size and fragmentation were largely irrelevant to these people, at least relative to, say, lunar orientation Bradley, Arrowsmith, and Society of Antiquaries of Scotland, 2005.

## Intentional Selection of Basin-Like Settings

One might intuitively suppose that a perfectly basin-like setting would have a relatively unfragmented small or moderately sized viewshed. The Monte Carlo simulation analyses of viewshed area and fragmentation provide evidence that the builders of at least some (22% at $p = 0.05$) stone circles may have sought topographic settings that offered smaller viewsheds than those typical of the locality. The importance of restricting the viewshed in some way is supported by the finding that most of the very largest viewsheds are actually smaller than those typical of other viewpoints in the locality, and, crucially, there is no evidence for the choice of larger than typical viewsheds. A similar result was obtained for viewshed fragmentation: stone circle viewsheds with atypical fragmentation are three times more likely to be less rather than more fragmented than is typical of other viewpoints in the locality.

Interesting though these results are, viewshed area and fragmentation may not be reliable indicators of Bradley's "perception of circularity." One of his examples of this phenomenon, Long Meg and Her Daughters, has a viewshed that is small only because it is also highly fragmented. The viewshed from Long Meg and Her Daughters does, however, have relatively high mean horizon inclination and low standard deviation of inclination, supporting Lake and Woodman's (2003) claim that the properties of the horizon (especially the inclination) are likely to provide a more direct measure of the impression of circularity. The Monte Carlo simulation analysis of horizon elevation was inconclusive with respect to whether the builders of stone circles sought locations with particularly high or low horizon elevations; nevertheless, for the 24% of stone circles where there is evidence for locally atypical horizon elevation, one might tentatively offer evidence for the operation of some kind of norm in the fact that their builders seem to have chosen locations with atypically high horizon elevations when these were low in absolute terms and, conversely, chosen locations with atypically low horizon elevations when these were high in absolute terms. Some 38% of stone circles have viewsheds with a smaller or larger mean distance to the horizon than is typical of other viewpoints in the locality, and these are around six times more likely to be smaller than larger, again pointing to a desire to "contain" the viewshed. Finally, as discussed in the results given in this chapter, the Monte Carlo simulation analysis of horizon inclination provides tentative support that the builders of at least some stone circles sought to minimize the variability in the line of the horizon (its inclination). Overall then, our analysis provides some evidence that the desire for a contained and possibly basin-like viewshed may have been a factor in the siting of up 25% of the stone circles in our sample.

## Large Stone Circles

As discussed in the results, the largest stone circles in our sample (that is, those in excess of 60 m maximum diameter) have relatively small viewsheds (in terms of area) that are highly fragmented; nevertheless, those viewsheds are also characterized by moderately high mean horizon inclination and low variability in the horizon inclination. This fact suggests that the viewsheds of these very large stone circles combine two properties: on the one hand, they may offer a "circular perception of space" (Long Meg and Her Daughters and the Ring of Brodgar are both cited by Bradley and/or Richards in this regard), but, on the other hand, their highly fragmented viewsheds provide ample scope for the control of the visual encounter of these sites, along the lines suggested by Barrett (1994) and Thomas (1993) for Avebury. Despite this tantalizing prospect, the Monte Carlo simulation analyses provide no evidence that any of these very large stone circles were intentionally built in locations offering these particular properties, although it may be that in these cases the control points should have been drawn from a wider area, on the grounds that larger monuments were built and/or used by larger groups of people spread over a larger area.

In light of these results—tentative though they are—we hope this chapter demonstrates that carefully designed experimental methods in combination with contemporary desktop computing power now make it possible to deploy GIS-based viewshed analysis in a manner that appeals to its strengths rather than weaknesses—that is, the comparative analysis of large numbers of sites at a regional or larger scale.

## References


Barnatt, J. 1989. *Stone Circles of Britain: Taxonomic and Distributional Analyses and a Catalogue of Sites in England, Scotland and Wales*. Oxford: Archeopress, British Archaeological Reports 215.

Barnatt, J., and Pierpont, S. 1983. Stone circles: Observatories or ceremonial centres? *Scottish Archaeological Review* 2: 101–15.

Barrett, J. C. 1994. *Fragments from Antiquity: An Archaeology of Social Life in Britain*, 2900–1200 B.C. Oxford: Blackwell.

Bender, B., Hamilton, S., and Tilley, C. 1997. Leskernick: Stone worlds; alternative narratives; nested landscapes, *Proceedings of the Prehistoric Society* 63: 147–78.

________. 2007. *Stone Worlds: Narrative and Reflexivity in Landscape Archaeology*. Walnut Creek, CA: Left Coast Press.

Bradley, R. 1993. *Altering the Earth*, Vol. 8, Society of Antiquaries of Scotland Monograph Series. Edinburgh: Society of Antiquaries of Scotland.

________. 1998. *The Significance of Monuments: On the Shaping of Human Experience in Neolithic and Bronze Age Europe*. London: Routledge.

________. 2002. *The Past in Prehistoric Societies*. Oxford: Taylor & Francis.

Bradley, R., Arrowsmith, S., and Society of Antiquaries of Scotland. 2005. *The Moon and the Bonfire: An Investigation of Three Stone Circles in North-East Scotland*. Edinburgh: Society of Antiquaries of Scotland.

Bradley, R., Harding, J., and Mathews, M. 1993. The siting of prehistoric rock art in Galloway, southwest Scotland, *Proceedings of the Prehistoric Society* 59: 269–83.

Bradley, R., Harding, J., Rippon, S., and Mathews, M. 1993. A field method for investigating the distribution of rock art, *Oxford Journal of Archaeology* 12: 129–45.

Burl, A. 1988. *Four-Posters: Bronze Age Stone Circles of Western Europe*. Oxford: Archeopress, British Archaeological Reports 195.

________. 2000. *The Stone Circles of Britain, Ireland, and Brittany*. New Haven, CT: Yale University Press.

Edmonds, M., Thomas, J., et al. 1990. Papers from the Theoretical Archaeology Group (TAG) Conference, Newcastle, 1989, *Scottish Archaeological Review* 7: 1–36.

Eve, S. 2012. Augmenting phenomenology: Using augmented reality to aid archaeological phenomenology in the landscape, *Journal of Archaeological Method and Theory* 19: 582–600.

Exon, S., Gaffney, V., Woodward, A., and Yorston, R. 2000. *Stonehenge Landscapes: Journeys through Real-and-Imagined Worlds*. Oxford: Oxbow.

Fisher, P. F., Farrelly, C., Maddocks, A., and Ruggles, C. 1997. Spatial analysis of visible areas from the Bronze Age cairns of Mull, *Journal of Archaeological Science* 24: 581–92.

Fishman, J., Haverkort, H., and Toma, L. 2009. Improved visibility computation on massive grid terrains, *Proceedings of the 17th ACM SIGSPATIAL International Conference on Advances in Geographic Information Systems*, pp. 121–30. New York: ACM.

Fraser, D. 1988. The orientation of visibility from the chambered cairns of Eday, Orkney, in Ruggles, C. L. N. (Ed.), *Records in Stone*, pp. 325–37. Cambridge: Cambridge University Press.

Freeman, C., and Gillings, M. 2007. Seeing is perceiving? *World Archaeology* 39: 4–16.

Gaffney, V., Stančič, Z., and Watson, H. 1996. Moving from catchments to cognition: Tentative steps toward a larger archaeological context for GIS, in Aldenderfer, M., and Maschner, H. D. G. (Eds.), *Anthropology, Space, and Geographic Information Systems*, pp. 132–54. Oxford: Oxford University Press.

Gibson, J. J. 1986. *The Ecological Approach to Perception*. Hillsdale, NJ: Lawrence Erlbaum Associates.

Gillings, M. 2009. Visual affordance, landscape, and the megaliths of Alderney, *Oxford Journal of Archaeology*, 28: 335–56.

Gillings, M., and Goodrick, G. T. 1996. Sensuous and reflexive GIS: Exploring visualisation and VRML, *Internet Archaeology*, 1, http://intarch.ac.uk/journal/issue1/gillings_index.html; accessed August 30, 2004.

GRASS Development Team. 2012. *Geographic Resources Analysis Support System (GRASS GIS) Software*. Beaverton, OR: Open Source Geospatial (OSGeo) Foundation.

Haggett, P. 1965. *Locational Analysis in Human Geography* (1st ed.). New York: John Wiley.

Hamilton, S., Whitehouse, R., Brown, K., Combes, P., Herring, E., and Seager-Thomas, M. 2006. Phenomenology in practice: Towards a methodology for a "subjective" approach, *European Journal of Archaeology* 9: 31–71.

Harding, A. F., and Lee, G. E. 1987. *Henge monuments and related sites of Great Britain: Air Photographic Evidence and Catalogue*. Oxford: Archeopress, British Archaeological Reports 175.

Hawkins, G. S. 1966. *Stonehenge Decoded*. London: Souvenir Press.

Hodder, I. 1986. *Reading the Past*. Cambridge: Cambridge University Press.

Kvamme, K. L. 1985. Determining relationships between the natural environment and prehistoric site locations: A hunter-gatherer example, in Carr, C. (Ed.), *For Concordance in Archaeological Analysis: Bridging Data Structure, Quantitative Technique and Theory*, pp. 208–38. Kansas City: Westport Publishers.

________. 1988. Development and testing of quantitative models, in Judge, W. J., and Sebastian, L. (Eds.), *Quantifying the Present and Predicting the Past: Theory, Method, and Application of Archaeological Predictive Modeling*, pp. 325–427. Washington, D.C.: U.S. Bureau of Land Management, Department of Interior, U.S. Government Printing Office.

________. 1990. One-sample tests in regional archaeological analysis: New possibilities through computer technology, *American Antiquity* 55: 367–81.

Lake, M. W., and Woodman, P. E. 2000. Viewshed analysis of site location on Islay, in Mithen, S. J. (Ed.), *Hunter-Gatherer Landscape Archaeology: The Southern Hebrides Mesolithic Project, 1988–98. Volume 2: Archaeological Fieldwork on Colonsay, Computer Modelling, Experimental Archaeology, and Final Interpretations*, pp. 497–503. Cambridge: The McDonald Institute for Archaeological Research.

Lake, M. W., and Woodman, P. E. 2003. Visibility studies in archaeology: A review and case study, *Environment and Planning B: Planning and Design* 30: 689–707.

Llobera, M. 2001. Building past landscape perception with GIS: Understanding topographic prominence, *Journal of Archaeological Science* 28: 1005–14.

________. 2003. Extending GIS-based visual analysis: The concept of "visualscapes," *International Journal of Geographical Information Science* 17: 25–48.

________. 2007. Reconstructing visual landscapes, *World Archaeology* 39: 51–69.

Llobera, M., Wheatley, D., Steele, J., Cox, S., and Parchment, O. 2010. Calculating the inherent visual structure of a landscape (total viewshed) using high-throughput computing, in *Beyond the Artefact: Digital Interpretation of the Past, CAA 2004*, pp. 146–51. Budapest: Archaeolingua.

Lodge, A., and Muirhead, R. F. 1924. The Stroud system, *The Mathematical Gazette* 12(172): 207–08.

Pearce, M. 2011. Have rumours of the "death of theory" been exaggerated? in Bintliff, J., and Pearce, M. (Eds.), *The Death of Archaeological Theory?* pp. 80–89. Oxford: Oxbow Books.

Pearson, M. P., and Ramilisonina. 1998. Stonehenge for the ancestors: The stones pass on the message, *Antiquity* 72: 308–26.

R Development Core Team. 2012. *R: A Language and Environment for Statistical Computing*. Vienna: R Foundation for Statistical Computing.

Renfrew, A. C. 1982. *Towards an Archaeology of Mind* (Inaugural Lecture to the University of Cambridge). Cambridge: Cambridge University Press.

Renfrew, C. 1973. Monuments, mobilisation and social organisation in Neolithic Wessex, in Renfrew, C. (Ed.), *The Explanation of Culture Change*, pp. 539–58. London: Duckworth.

________. 1994. Towards a cognitive archaeology, in Renfrew, C., and Zubrow, E. B. W. (Eds.), *The Ancient Mind: Elements of a Cognitive Archaeology*, pp. 3–12. Cambridge: Cambridge University Press.

Richards, C. 1996. Monuments as landscape: Creating the centre of the world in late Neolithic Orkney, *World Archaeology* 28: 190–208.

Robert, C. P., and Casella, G. 2004. *Monte Carlo Statistical Methods* (2nd ed.). New York: Springer-Verlag.

Ruggles, C. 1999. *Astronomy in Prehistoric Britain and Ireland*. New Haven, CT: Yale University Press.

Ruggles, C. L. N. 1984. A new study of the Aberdeenshire recumbent stone circles, 1: Site data, *Archaeoastronomy* 6: 55–79.

Ruggles, C. L. N., and Burl, H. A. W. 1985. A new study of the Aberdeenshire recumbent stone circles, 2: Interpretation, *Archaeoastronomy* 8: 25–60.

Ruggles, C. L. N., Martlew, R., and Hinge, P. 1991. The North Mull Project (2): The wider astronomical significance of the sites, *Archaeoastronomy* 16: 51–75.

Ruggles, C. L. N., and Medyckyj-Scott, D. J. 1996. Site location, landscape visibility, and symbolic astronomy: A Scottish case study, in Maschner, H. D. G. (Ed.), *New Methods, Old Problems: Geographic Information Systems in Modern Archaeological Research*, number 23 in Occasional Paper, pp. 127–46. Carbondale: Southern Illinois University Center for Archaeological Investigations.

Shanks, M., and Tilley, C. 1987. *Re-Constructing Archaeology*. Cambridge: Cambridge University Press.

Sui, D. Z. 1994. GIS and urban studies: Positivism, post-positivism, and beyond, *Urban Geography* 15: 258–78.

Thom, A. 1967. *Megalithic Sites in Britain*. Oxford: Clarendon Press.

________. 1971. *Megalithic Lunar Observatories*. Oxford: Oxford University Press.

Thomas, J. 1993. The politics of vision and the archaeologies of landscape, in Bender, B. (Ed.), *Landscape: Politics and Perspectives*, pp. 19–48. Oxford: Berg.

________. 2004. *Archaeology and Modernity*. London: Routledge.

Tilley, C. 1994. *A Phenomenology of Landscape: Paths, Places, and Monuments*. Oxford: Berg.

________. 2004. Round barrows and dykes as landscape metaphors, *Cambridge Archaeological Journal* 14: 185–203.

Underwood, G. 1969. *The Pattern of the Past*. London: Museum Press.

Watson, A. 2001. Composing Avebury: Archaeology and aesthetics, *World Archaeology* 33: 296–315.

Wheatley, D. 1993. Going over old ground: GIS, archaeological theory, and the act of perception, in Andresen, J., Madsen, T., and Scollar, I. (Eds.), *Computing the Past: Computer Applications and Quantitative Methods in Archaeology 1992*, pp. 133–38. Aarhus: Aarhus University Press.

________. 1995. Cumulative viewshed analysis: A GIS-based method for investigating intervisibility, and its archaeological application, in Lock, G., and Stančič, Z. (Eds.), *Archaeology and Geographical Information Systems*, pp. 171–86. London: Taylor & Francis.

________. 1996. The use of GIS to understand regional variation in earlier Neolithic Wessex, in Maschner, H. D. G. (Ed.), *New Methods, Old Problems: Geographic Information Systems in Modern Archaeological Research*, number 23 in Occasional Paper, pp. 75–103. Carbondale: Southern Illinois University Center for Archaeological Investigations.

Wheatley, D., and Gillings, M. 2000. Vision, perception, and GIS: Developing enriched approaches to the study of archaeological visibility, in Lock, G. (Ed.), *Beyond the Map: Archaeology and Spatial Technologies*, pp. 1–27. Amsterdam: IOS Press.

Williams, W. G. 1968. Stone age circles "form pattern of power," *Daily Telegraph*, September 23.

Woodman, P. E. 2000. Beyond significant patterning, towards past intentions: The location of Orcadian chambered tombs, in Buck, C., Cummings, V., Henley, C., Mills, S., and Trick, S. (Eds.), *Proceedings of the UK Chapter of Computer Applications and Quantitative Methods in Archaeology 1999*, pp. 91–105. Oxford: Archeopress, British Archaeological Reports 844.

Wright, D. J., Goodchild, M. F., and Proctor, J. D. 1997. Demystifying the persistent ambiguity of GIS as "tool" versus "science," *The Annals of the Association of American Geographers* 87: 346–62, http://dusk.geo.orst.edu/annals.html; accessed October 11, 2004.

Yokoyama, R., Shirasawa, M., and Pike, R. J. 2002. Visualizing topography by openness: A new application of image processing to digital elevation models, *Photogrammetric Engineering and Remote Sensing* 68: 257–65.

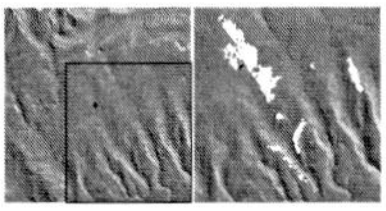

# 10 Reconsidering the Concept of Visualscape: Recent Advances in Three-Dimensional Visibility Analysis

*Eleftheria Paliou*

## The Concept of Visualscape

The development of current spatial analysis methods for exploring the visual structure of space has been determined by the constant interchange of concepts and ideas between studies in the fields of urban and landscape research. Within these works special emphasis is given to the notions of *viewshed, isovist, isovist field,* and *visibility graph* (Batty, 2001; Benedikt, 1979; Conolly and Lake, 2006: 225–33; Turner et al., 2001; Wheatley and Gillings, 2002: 201–16). Although these terms are associated with specific types of spatial representation and are used almost exclusively in relation to either built or landscape contexts, they nonetheless operate within a similar conceptual framework. This fact is reflected in their common references to environmental psychology works and concepts, especially Gibson's *optic array*—the nested complex of visual angles that have their apex at the eye of the observer and their base at the surfaces of the objects of the environment that surrounds her (Benedikt, 1979: 48; Gibson, 1979: 65–71; Llobera, 2003: 30; Wheatley and Gillings, 2000). Furthermore, conceptual interchange between applications of visual analysis in built and natural environments has been made manifest on numerous occasions; for example, the terms *isovist* (Benedikt, 1979) and *visibility graph* (Turner et al., 2001), which are now widely used in architectural and urban analysis, were first proposed by Tandy (1967) and De Floriani

*Computational Approaches to Archaeological Spaces* by Andrew Bevan and Mark Lake, Eds., 243–264 

and associates (1994), respectively, in landscape research. In contrast, Higuchi (1983) adopted a set of concepts to analyze the visual and spatial structure of landscapes that were originally suggested by Märtens (1890) and Hall (1966) for the analysis of built space.

In light of these developments, the adoption of unifying concepts that emphasize and take advantage of the theoretical and methodological affinities of current visual analyses could inspire new approaches to the study of human visual space and precipitate a fuller understanding of the interplay between humans and their surroundings. The notion of *visualscapes* was introduced with a similar intent. It was proposed by Llobera (2003) as an abstract and generic term to unify within a GIS the scope and methodology of studies of visual space found in various disciplines, such as geography, architecture, urbanism, and archaeology. According to Llobera, a visualscape is the spatial representation of any visual property associated with a spatial configuration that has an inherent visual structure that:

- *can be studied in its entirety or with respect to any relevant subset of the spatial configuration;*
- *can be spatially represented in various ways (e.g., scalar or vector fields);*
- *is essentially 3-dimensional . . . ;*
- *can be described by means of a multitude of parameters, as shown in some urban studies, and eventually mapped back into space for further research;*
- *occurs both in "natural" and urban landscapes.* (Llobera, 2003: 31–33)

Although the term was clearly intended to be employed in a GIS context (Llobera, 2003: 26; 2006: 149), products of analysis undertaken using architectural and urban computing software, such as isovists and isovist fields, have also been described as visualscapes (Llobera, 2003: 46), a fact hinting at a potentially more broad use of the term, which transcends the boundaries of GISc.

In practice, the use of concepts that seek to capture the visual properties of any spatial configuration, no matter how complex, depends on the development of methods that are able to consider and analyze spatial representations in a way that conforms well to the experience of the human observer in real life. Recent advances in 3-dimensional visibility analysis that have taken place in the fields of landscape and urban studies, and in archaeology in particular, have enabled new approaches to the analysis of human visual space. Despite aiming to serve a variety of different research needs, these new methods encourage us to rethink the pragmatic use of generic terms describing visual structure.

This chapter first reviews approaches to visibility analysis in 3D spaces that have been proposed in the fields of urban and landscape studies, and especially in archaeology. In the light of these works, the notion of visualscape as a unifying and functional concept within a GIS is reexamined.

## Approaches to Visibility Analysis in 3D Landscapes

Up to now, terrain visibility analysis has met great popularity within landscape studies, particularly in the field of GISc. At a very basic level it is a procedure that aims to identify those areas that can or cannot be seen from a single (or a set of) viewpoint(s) within a digital model of a terrain, either by calculating lines of sight between pairs of point locations or via viewshed generation. Analyses of this kind have been used to inform human action in modern landscapes (for example, mobile phone network design and path planning) and to explain human behavior and perception in contemporary and past environments (choice of settlement location, movement patterns, territory formation, and the like; see also Lake and Ortega, this volume). Over the years the toolkit of GIS practitioners has been enriched with advanced visual analysis techniques aiming to produce different types of mapped visual data—such as cumulative view-sheds (Wheatley, 1995), total viewsheds (Lake, Woodman, and Mithen, 1998 ; Llobera et al., 2010), as well as scalar and vector fields describing more abstract ideas of visual structure, that is, those related to "visual prominence," "exposure," "hierarchy," "genesis," and "change" (Llobera, 2003, 2006). Traditionally GIS-based visibility analysis has limited 3D functionality, because it is performed on models that do not incorporate three independent $x$, $y$, and $z$ axes.[1] Although this limitation does not pose difficulties for estimating visibility relationships within simple terrain models, it can be a problem when it comes to more complex landscape scenes that include vegetation or/and buildings.

It has often been pointed out that the presence of vegetation can greatly influence visibility patterns in the landscape (Llobera, 2007; Pyysalo, Oksanen, and Sarjakoski, 2009; Wheatley and Gillings, 2000). At least in the case of environments with high vegetation (such as forested areas), the results of visibility analysis could be flawed if the shape and distribution of tree coverage are not modeled with sufficient accuracy and detail. Obviously the acquisition of vegetation data for palaeo-environments is difficult, but even when the interest lies in modern landscapes, incorporating information on tree coverage into GIS viewshed models can be challenging. Recent studies have aimed to explore the effects of vegetation in landscape visibility by including simple tree models in the digital terrain and by proposing new computational algorithms (Dean, 1997; Liu et al., 2010; Llobera, 2007). The modeling of vegetation data can be implemented in many different ways and with various levels of success by merely adding canopy heights to the terrain elevation values with map algebra, incorporating 2,5D (TIN) (Dean, 1997) or 3D voxel models (Llobera, 2007) of trees/plants in the digital terrain, as well as including simple geometric shapes (for example, cones) describing the outer limits of tree crowns and trunks (Liu et al., 2010).

Nonetheless, there are usually restrictions to the number of vegetation types modeled and the complexity in the representation of tree species, so the success of the analysis depends on whether the vegetation model is sufficiently precise for its intended purpose. Lately, the use of airborne laser scanning technology has enabled the 3D recording of the exact form of contemporary landscapes, including tree coverage and buildings; however, the possibility of performing visibility (viewshed) analysis in these models has only started to be explored (Pyysalo, Oksanen, and Sarjakoski, 2009). When it comes to archaeological landscapes, where there are not enough data for an exact reconstruction of the palaeo-environment, algorithms that aim to estimate the probability of a location being visible in the presence of trees/plants of various densities and distributions are useful for defining the possible impact of vegetation on visibility patterns (Dean, 1997; Llobera, 2007).

Occasionally buildings in landscape scenes are also treated as obstacles in terrain visibility (Liu et al., 2010). Their incorporation in terrain representations is usually less problematic than the modeling of vegetation data, because built structures can occasionally be accurately described with simple geometric solid shapes or multipatch[2] geometry, which is a common feature of commercial GIS software. Still, when the aim is to analyze the visual properties of 3-dimensional buildings per se, standard GIS visibility tools prove insufficient to capture important aspects of the human experience of built space; humans tend to attribute special meanings and respond differently to certain architectural features—for example, the facade and entrance of a building are likely to trigger different movements and actions than its back spaces (Llobera, 2003, 2006); relief or painted decoration may attract a viewer's attention and play an essential role in shaping the perception of domestic spaces or monuments (Paliou, 2011). The importance of being able to perform visual analysis that can target features that may have had an added symbolic significance has been highlighted in the past (Llobera, 2003, 2006: 150; Paliou, 2013; Paliou, Wheatley, and Earl, 2011: 376). Most recent versions of commercial GIS software offer some basic tools for line-of-sight calculations (LOS) on building faces,[3] but, generally, to date the visual properties of 3-dimensional features in complex urban and "natural" landscape scenes have more efficiently been described and analyzed with the use of 3D modeling programs and ad hoc visibility algorithms.

A characteristic example of a fully 3D method that aims to explore the visibility of objects within their surrounding environment was first suggested by Groß (1991), who used computer graphic algorithms to analyze 3D landscape scenes. His approach was later enhanced by Bishop, Wherrett, and Miller (2000; also Bishop, 2003) as a completely automated method for performing visual impact assessments in a landscape context. The application of this methodology requires first the creation of a 3D model and the allocation of discrete colors to the objects or features of interest. At a second stage, the amount of visibility of the targeted features is estimated by calculating the counts of the color-coded pixels in a number of rendered camera views. Depth-information that indicates the distance (for instance, foreground, middle-ground, background) of visible objects from the camera can also be extracted from the images. Bishop's methodology not only determines whether a target object can be seen or not, but it can also establish how much of the surface of interest is visible or how much this surface cuts the horizon. Furthermore, it has been maintained that this methodology is capable of taking into account atmospheric conditions that could affect the perception of objects (Bishop, 2003: 685; Groß, 1991: 408), although no applications have yet been presented in this area. Nevertheless, in contrast to some GIS-based and urban analyses, Bishop does not attempt to calculate visual indices for "all" locations within the area of interest or map it back in space, because this would require prohibitive computing time (Bishop, 2003: 686).

More recently, Engel and Döllner (2009) have proposed an automatic visual analysis of 3D landscape scenes that seeks to produce mapped visual data for visual control and public safety assessments in modern environments. Their approach aims to create scalar fields showing changes in the visibility of a target surface as seen from a given observer path and is implemented with a virtual sensing device that is programmed to traverse the surface of a target object and calculate its visual contact with the path. Bishop's (2003) and Engel and Döllner's (2009) methods could allow for exact and detailed models of tree plans and buildings and are suitable for analyzing the visibility of objects in complex 3D scenes. Although these approaches have been applied in landscape analysis, they could also successfully be used in urban contexts.

## 3D Visual Analysis in Architecture and Urban Studies

Owing to the computation and time constraints associated with the analysis of large-scale 3D urban datasets, as well as the wide popularity of 2D Space Syntax methods (Hillier and Hanson, 1984), visibility analyses in the built environment have mainly been concerned with the 2-dimensional unoccupied space between buildings and walls. Axial[4] (Hillier and Hanson, 1984), isovist[5] (Benedikt, 1979), and visibility graph[6] analysis (Turner et al., 2001) primarily aim to explore the relationship between visibility and pedestrian movement and to identify patterns of copresence in built space that could encourage social encounters and interaction. These approaches have become widely used in architectural and urban computing, especially since the development of Depthmap, a program created at the Bartlett School of Architecture at University College London that incorporates a variety of built-in Space Syntax tools.

Nonetheless, in parallel with the increasing availability of software solutions intended for 2D urban analysis there have been attempts at the development of 3D analytical methods. Such approaches have been motivated by the idea that taking into account the third dimension is crucial for understanding human visual perception (Morello and Ratti, 2009; Yang, Putra, and Li, 2007) and that 3D methods coincide better with Gibson's definition of the optic array and Benedikt's original description of the isovist as a 3D entity (Benedikt, 1979).

Within the last few years a number of spherical analyses have been proposed, aiming to examine the visual properties of 3D open urban space. One of the first attempts toward this direction was made by Fisher-Gewirtzman and Wagner (2003), who introduced a metric tool intended to calculate the Spatial Openness Index (SOI), a concept that refers to the volume of open space potentially seen from a given point. Their method, which is implemented using JAVA language and VRML modeling, is able to measure the count of visible voxels of open space that fill up a sphere around an observation point. Fisher-Gewirtzman and Wagner attempted to demonstrate via a preliminary test that SOI is correlated to "perceived density" in urban space, which they suggest is an important indicator for environmental quality; nonetheless, their volumetric analysis was found to be so "extremely complex" that it could not be used for the comparison of several different spatial configurations (Fisher-Gewirtzman and Wagner, 2003: 581).

The Viewsphere, an ArcGIS extension programmed by Yang and colleagues (2007), was also created to calculate the visible volume from given observation points in urban space—namely, 3D isovists—and to produce indices of visual values that could closely relate to the sense of spatial enclosure. Yang and associates demonstrated that their tool can effectively perform volumetric measurements, while at the same time they acknowledged the limitations of their method, which is applicable only on 2,5D urban models (TIN or raster data) and cannot deal with real 3D objects "such as a cantilever structure, a bridge, a shopping mall, or a subway station" (Yang, Putra, and Li, 2007: 991).

More recently, Morello and Ratti (2009) have introduced a method of 3D isovist analysis that is based on the image processing of Digital Elevation Models (DEMs) using Matlab. Following rapid procedures, this approach succeeds in calculating the visible volume seen for every point in open space and creating a 3D matrix (*iso-visi-matrix*), which stores and sums all visible and hidden voxels from each viewpoint. The iso-visi-matrix is useful not only for exploring visual properties of open space but also for certain aspects of building visibility by simply slicing the voxel space through the facades of the built features of interest. It is interesting to note that Morello and Ratti have attempted to associate their analysis with Kevin Lynch's urban theory, mostly known by his influential book *The Image of the City* (Lynch, 1960). In this work Lynch identifies five visual elements that make cities more easily identifiable: paths, nodes, districts, edges, and landmarks. Morello and Ratti have showed that numeric values stored in iso-visi-matrices are able to capture important attributes of landmarks, such as their ability to be visible over near or long vistas. Thus their approach is a constructive and time-effective solution for urban analysis, but it nonetheless (as with Viewsphere applications) can be used only with relatively simple urban models.

The possibility of using 3D isovist and visibility graph analysis to explore 3-dimensional complex spaces has been explored by Derix, Gamlesæter, and Carranza (2008), who proposed a number of computational algorithms aiming to examine aspects of human experience in the World Trade Center (WTC) Memorial Museum at New York. Within the WTC Memorial Museum mobile visitors are expected to perceive the exhibited artifacts and remains from cantilevers, ramps, and balconies; 2-dimensional methods of visual analysis, which are restricted solely to the horizontal slice of space corresponding to the eye level of the viewer, would have been insufficient for investigating human experience in the building and making design assessments. Derix, Gamlesæter, and Carranza (2008) show that capturing and mapping numeric values of 3D isovists and visibility graphs in complex interior spaces is feasible, but the relation of these new types of data to human perception is not much discussed in their preliminary study. The mathematical generation of 3D isovists has also been explored by Van Bilsen (2008)[7] and Van Bilsen and Poelman (2009), who have attempted to demonstrate through the generation of a series of scalar fields of 3D isovist measures (area, perimeter, distance) how 3D analytical methods can be used for identifying areas of optimal visual control in built environments (for example, virtual drilling sites).

## Approaches to 3-Dimensional Visibility Analysis in Archaeology

### Methodology

In archaeology, as applies in the fields of geography and architectural and urban planning, the first attempts toward fully 3-dimensional visual analysis have been motivated by a need to overcome the pragmatic limitations of 2D approaches when it comes to facing problems such as how to estimate the effects of vegetation in landscape visibility patterns or explore the visual properties of intricate architectural structures. Moreover, they have been consistent with a more general trend in archaeological research toward approaches that seek to take fuller account of the complexity of human experience in landscapes and built spaces (Frieman and Gillings, 2007; Gillings and Goodrick, 1996; Llobera, 2003; Wheatley and Gillings, 2000). To date, the archaeological applications of visibility analysis in 3D spaces are few, and the focus has been mainly on built spaces. The work of Llobera (2003) points toward the use of 3D solid models in the context of GIS-based landscape visibility analysis; nonetheless, such an approach has yet to be implemented for research on past cultural landscapes. However, studies examining human visual experience in 3-dimensional historic and prehistoric built environments have been carried out since 2005 by members of the Archaeological Computing Research Group at the University of Southampton.

The methods developed at the University of Southampton are founded on the use of common functionalities of 3D modeling and GIS software[8]—namely, on affordable off-the-shelf programs that are frequently utilized by archaeological computing specialists to cover a variety of research needs. The methodology, which has been described in detail elsewhere (Earl, 2005; Earl et al., this volume; Paliou, 2009, 2011; Paliou and Wheatley, 2007; Paliou, Wheatley, and Earl, 2011), is based on the idea that a viewshed or isovist in a 3D digital environment corresponds to the area not in the shadow cast by a single point light source, which is placed at the viewer eye level. This conceptualization of the isovist is not new: in his seminal paper on isovist and isovist fields Benedikt (1979: 54) had proposed the dynamic generation and visualization of isovists perceived along a path by using a light source.

Drawing on this analogy, Earl (2005) and Paliou and Wheatley (2007) proposed a new method of analysis to investigate the visibility of objects of any form and shape modeled in 3D. The application of this method requires first the creation of a 3D model of the space and objects of interest (see color Figures I.7, I.8, I.9a). In this digital environment viewer locations are defined in equal intervals using a notional grid covering the areas potentially occupied by observers (Figures 10.2a, 10.4, I.9a). Grid spacing should be appropriate for the purpose of a particular analysis, but can be very fine[9] if necessary. The next step involves the execution of a simple script that animates the light source over each grid centroid (viewpoint) in succession at a specified observer's eye level. For each viewpoint location the texture (Figure 10.1) of the target object (that is, the whole environment or a subset of it) that incorporates patterns of light and shadow is extracted in a raster format using texture baking tools; such tools are built in to many 3D modeling programs and are routinely used by computer graphic designers to record various surface effects created by lighting on a 3D object. The textures of targeted objects can be imported into a GIS, and visual data can be summarized and mapped in a variety of different ways using map algebra operations, simple scripts, and batch processes.

Following this process various types of output can be produced that fall within Llobera's definition of visualscapes: times-seen (sometimes referred to as cumulative viewsheds or total viewsheds) maps—showing how many times the target object is visible from the specified observer space—can be created by summing up the textures (Figures 10.4b, I.9b); moreover, the count of visible pixels can be extracted and mapped on the observer locations, a procedure that produces scalar fields indicating changes in the visibility of the target object through the viewers' space (Figures 10.2b, 10.3, I.9c). Such maps could indicate the differences in visibility experienced by a mobile perceiver, if they are intersected with movement paths. Other types of information could also be mapped, for instance, those associated with illumination levels affecting visual clarity, the distance between the observer and the target object, and important viewing angles defining thresholds that express a change in viewing

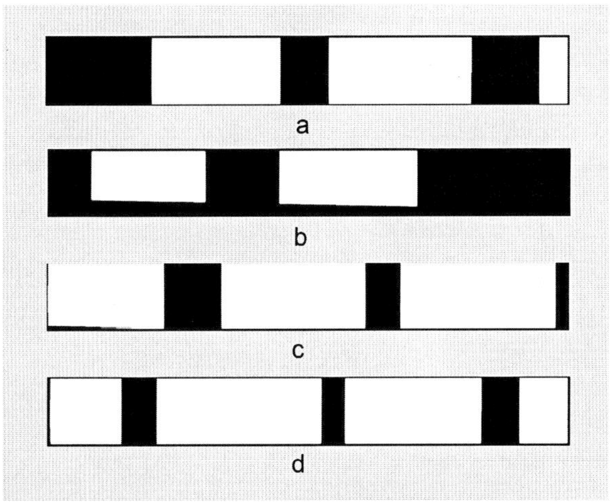


**Figure 10.1** Textures of the south frieze of the Miniature Fresco with information on the visibility of the painted wall surface from four different viewpoints (white/visible–black/nonvisible); (a–d) correspond to color Figure I.7(a–d), respectively (Paliou, 2011, Figure 4).

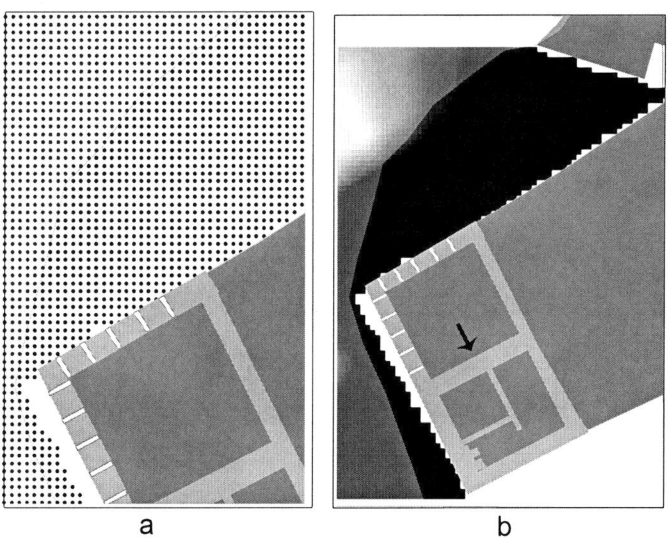


**Figure 10.2** (a) Grid of viewpoints on the public space outside Room 5, West House, Thera (Paliou, 2011, Figure 1); (b) map showing the changes in the visible area of the south frieze of the Miniature Fresco (white/visible–black/nonvisible, Room 5, West House; the arrow points to the position of the fresco) (Paliou, 2011, Figure 5).

experience (Earl, 2005; Paliou, 2009, 2011; Paliou, Wheatley, and Earl, 2011). Furthermore, a number of other operations known from viewshed analysis applications in landscape studies (cf. Fisher, 1994) could be performed, such as probability mapping and error modeling (Paliou, 2009: 77–81), processes that aim to consider possible errors in the reconstruction of features no longer preserved and visualize their spatial distribution.

## Applications

An area that could benefit from the application of these new methods is urban archaeology, and especially the social analysis of past architectural spaces and townscapes (Paliou, 2013). Built environments are redolent with visual cues that embody messages of power, status, and identity and that are consciously manipulated by human actors to create meaningful contexts of interaction and promote strategies of social recruitment and/or exclusion (Fisher, 2009: 455; Moore, 1996; Rapoport, 1990: 221). These cues are encoded in a 3D environment and on many occasions cannot be examined effectively or at all with 2D analytical approaches. The Late Bronze Age site of Akrotiri provides a good example of such a case (Paliou, 2011). The prehistoric settlement is well known for its exceptional level of preservation and the exquisite mural decoration that embellished many domestic buildings. It has been claimed (Doumas, 2005) that mural paintings decorating the first-floor rooms of Theran houses were intended to be visible to pedestrians outside the building through open windows, so as to reinforce the social status of the proprietor of the house (color Figure I.7). Nonetheless, the potential of the paintings to communicate messages to those traversing the street network cannot be assessed by observation at the site, because most of the decorated walls have collapsed, and murals cannot be appreciated within their original architectural context. Furthermore, visual access to first-floor rooms from outside the buildings would have changed dramatically for a person moving through the public street network; the likelihood of a pedestrian encountering the paintings would have depended, among other factors, on how visible the paintings were and from which parts of the street network they could be seen (Paliou, 2011). Visibility analysis in 3D space helped in tackling these issues. Following the procedures described earlier, scalar fields were created that recorded the visible area of wall decoration from public spaces outside Theran buildings (Figures 10.2b, 10.3; Paliou, 2009, 2011). These maps indicated that painted surfaces could have been visible from a great number of locations in the public street network. Other aspects of the visual experience of the paintings, associated with the ease with which the murals could be viewed (for example, vertical angle of view), were also taken into consideration. Results suggested that murals in the built environment of Akrotiri had considerable communicative potential, being able to address their messages to any pedestrian traversing the street

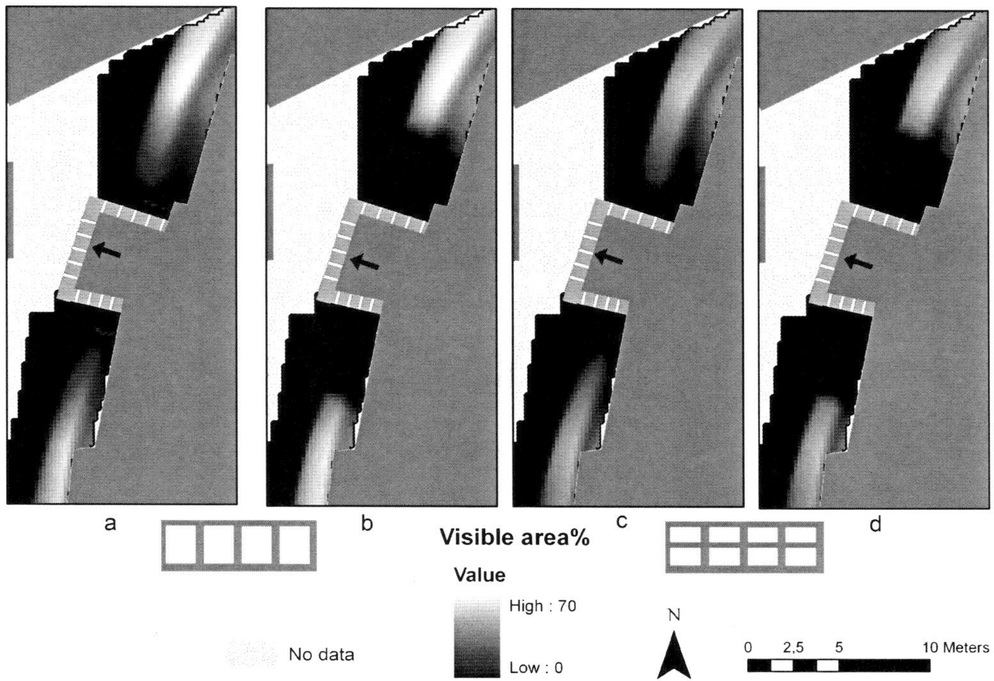


**Figure 10.3** Visible area of the west frieze of the Gate (Akrotiri, Thera) from the public spaces outside the building. Two alternative reconstructions of the window frame were considered for results in (a, b) and (c, d). In (b) and (d) only the amount of visible area up to 25 degrees above the eye level of the viewer was recorded—namely, those parts of the wall that could be seen with relative ease (Paliou, 2009, 2013); the arrow points to the position of the fresco (Paliou, 2011, Figure 13).

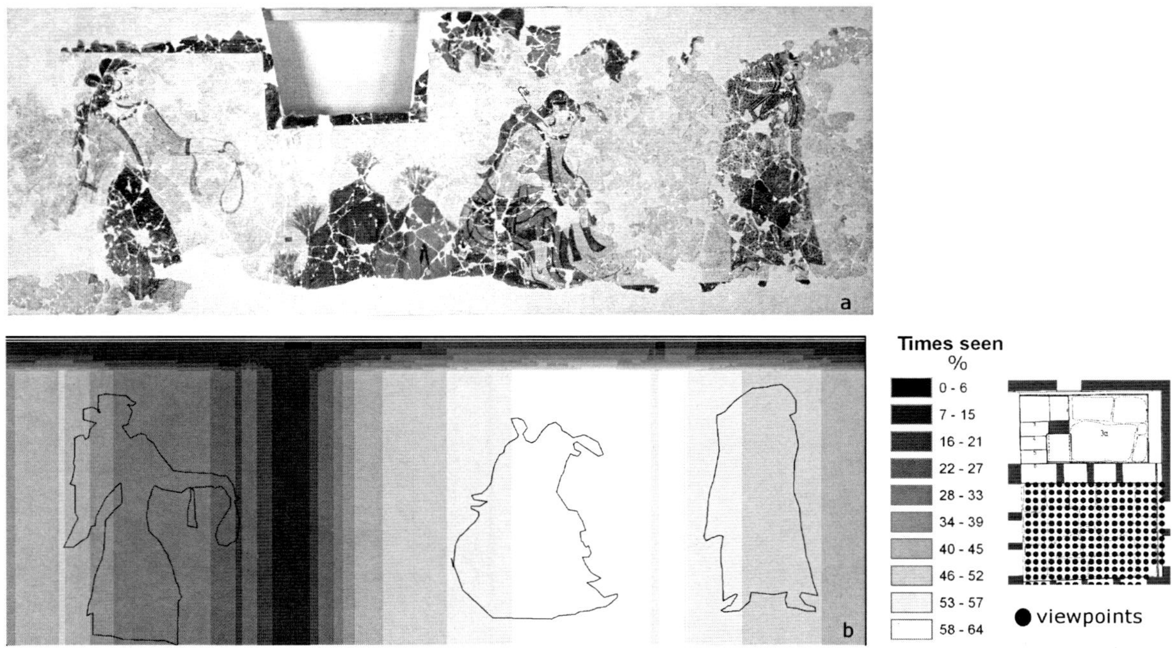


**Figure 10.4** (a) The wall painting of the adorants (courtesy of archaeological Society at Athens, Excavations at Akrotiri, Thera); (b) times-seen for the adorants from Room 3, expressing the percentage of the sampled viewer locations (213 in total, defined by a grid of 20-cm intervals) from which the painting is visible.

network in the course of daily life, as opposed to the few people allowed in the decorated rooms on formal occasions. Such interpretation may partly explain the wide distribution of mural paintings in the settlement, which has always puzzled students of Aegean painting (Paliou, 2011).

The same analytical methods were used to explore the visibility of mural decoration in fully 3D building interiors. Paliou, Wheatley, and Earl (2011; see also Paliou, 2009; Paliou and Wheatley, 2007) examined the relationship between visibility and iconography of Theran murals in Xeste 3, a public building in Akrotiri that is believed to have been used for public gatherings and rites of passage (Marinatos, 1984: 73–84; Marinatos and Hägg, 1986; Vlachopoulos, 2010). The viewing of the paintings during public ceremonies would have evoked personal and collective memories to the community members observing the rituals (Papageorgiou, 2000: 967), strengthening commonly accepted norms and promoting social cohesion. Nonetheless, visual access to the paintings is poorly understood from ground plans; murals decorating the east wing of Xeste 3 would have been observed through a row of door openings (*polythyra*), which significantly obstructed the viewing of pictorial features depicted on the walls, even when the doors were open (color Figure I.8). Times-seen analysis of painted compositions illustrating human representations showed that figures that appear to be distinguished because of their iconographic attributes—for example, their placement in composition, posture, dress, and hairstyle—have been depicted in the most visible parts of the decorated walls (Paliou, Wheatley, and Earl, 2011). The consistent correspondence between visual emphasis in pictorial space and visibility in actual space for three different compositions offered new evidence suggesting the intentions of the painter(s) and further supported some of the proposed interpretations of the paintings.

Patterns of visual communication during ritual performance are also indicative of the relations among interacting members of a religious congregation, who are senders, receivers, and observers (Clark, 2007; Moore, 1996). Clark (2007) examined such patterns in six different Byzantine churches in Jordan using isovist and visibility graph analysis. With these methods he managed to draw some interesting conclusions about changes in the spatial configuration of Jordanian churches through time and in the relationships between members of the congregation and the clergy. Nonetheless, the same methodology cannot successfully be applied to some popular architectural types of 6th-century Byzantine ecclesiastic space, where certain participants would have observed the liturgy from second-story balconies and galleries. Paliou and Knight (2013; Paliou, 2013) sought to examine visibility patterns in such spaces, using the still extant Basilica of San Vitale, in Ravenna, as a case study (color Figure I.9). At the time the building was consecrated (548 C.E.) male and female participants to the liturgy were likely to have occupied different locations in church space (Paliou and Knight, 2013). Female participants would have observed the liturgy from the second-story galleries (*matroneum*) (Figure I.9a) and perhaps the ambulatory side aisles, while males would have occupied the main nave area. The visual experience and patterns of copresence of those situated at the ground floor were investigated with 2-dimensional visibility analyses (isovist and visibility graph analysis), whereas 3D methods were used to examine visual access from the *matroneum* to the ground-floor chancel area, where ritual actions would have taken place (Figure I.9c). The results of the analysis (Figures I.9b, I.9c) highlighted differences in the ways the liturgy was perceived by female and male participants and in the interaction between these groups and the clergy. This work also demonstrated that mapped visual data can be fused with acoustic datasets to produce analytical products that have the potential to better represent the blending of the senses, as has been described by anthropological theorists (Frieman and Gillings, 2007; Ingold, 2000).

Three-dimensional visibility analysis could also be employed to examine visual aspects of the whole environment surrounding a perceiver, as opposed to a subset of spatial configuration as in the case studies presented earlier. This possibility has been explored by Earl (2005; Earl et al., this volume). Earl (2005) produced maps summarizing the properties (visibility, angularity, and distance) of all surfaces constituting the interior of Holyrood Church at Southampton for a 3D grid of viewpoints corresponding to the eye level of lying, sitting, and standing viewers. The significance of these data was assessed interactively by mapping on the model geometry. Such analysis was useful for investigating possible perceptual effects caused by the modifications in the church (for example, the removal of the rood screen and balconies) between the Georgian and Victorian periods (Earl, 2005). Furthermore, Earl proposed a number of ways in which such an analysis can be used for exploring the experience of a mobile perceiver in the church. In a more recent work, Papadopoulos and Earl (2013) applied texture viewshed analysis in combination with formal lighting analysis to various different types of Minoan built space (tombs, domestic spaces, and workshops), drawing attention to the possible relationships among the observed spatial patterns, social behavior, and use of space in the buildings under study. Earl and associates (this volume) offer additional examples of how these two techniques can be combined for the analysis of ecclesiastic space.

## Future Directions

Volumetric analyses similar to that proposed by Fisher-Gewirtzman, Burt, and Tzamir (2003), Yang and colleagues (2007), and Morello and Ratti (2009) have yet to be employed in the context of archaeology. Still, such approaches could be of relevance for archaeological investigations, because they conform to conceptual frameworks used in the past for the interpretation of historic and prehistoric built environments. Palyvou (2000), for example, has measured the amount of open urban space in Minoan buildings using 2D sections to estimate the proportionality of mass (built space) versus void (open space) and found that the latter exceeds the former in Neopalatial buildings. Such analysis could be linked to the aesthetic preferences of prehistoric people associated with desirable levels of natural light and air, as well as the wish for visual contact with the surrounding built and natural environment. Furthermore, Lynch's urban analysis, for which Morello and Ratti (2009) have sought to provide a quantifiable basis, has greatly influenced Favro's interpretation of the urban landscape of ancient Rome. In her book titled *The Urban Image of Augustan Rome* (1996) Favro discusses the changes in the effects of the physical city on a person's senses in the years between Julius Caesar's death and that of Augustus. To this end, she follows an experiential approach in which the changes that could have been experienced by a viewer in the central area of Rome are examined with illustrations and imaginative walks through the city and described via fictional narrative. Approaches similar to that proposed by Morello and Ratti (2009), as well as visibility analysis in fully 3D urban spaces (Earl, 2005; Paliou, 2011), could contribute to the formal investigation of Lynch's concepts in premodern contexts. Undoubtedly, the modeling of large-scale urban areas of historical cityscapes is an ambitious and difficult task, but initiatives such as the Rome Reborn project (Dylla et al., 2010) promise to produce a vast amount of 3D data that could be also used for the formal analysis of human experience in the reconstructed environments. Within this context the use of procedural modeling (Earl et al., this volume) as a mean for managing uncertainty and error in large-scale urban models could be a promising direction of research. It should also be noted that a similar approach would be worthwhile for landscape analysis; procedural modeling solutions have greatly facilitated the incorporation of detailed models of many different tree species in landscape scenes and the rapid representation of different vegetation densities; therefore, they could provide a suitable digital environment for probabilistic visibility calculations.

## Discussion and Conclusions

This chapter discussed recent advances in 3-dimensional visibility analysis by reviewing studies from the fields of geography, urbanism, architecture, and archaeology. An attempt was made to examine the conceptual contributions and benefits of the new methods for the study of visual space, which mainly reside in their capability of overcoming representational limitations of 2D and 2,5D approaches, offering new descriptions and analyses of the visual structure of space at various scales. The advantages of using these methods for the social and sensory analysis of premodern built environments were also mentioned here, although briefly, and some promising future directions for the use of these approaches in archaeological research were proposed. It is hoped that the preceding review and discussion have shown that these technological developments are of relevance to archaeological investigations and capable of enriching the interpretations of past environments.

An obvious conclusion that emerges from this study is that the new approaches to visibility analysis in 3D space can produce a great number of different outputs, which indicates that technical restrictions in the description of visual space have become much less of a concern over the last few years than they were in the past. It is equally apparent, however, that the conceptual implications of these methods, many of which have been presented merely in pilot studies, have not been fully explored; some of the products of analysis, especially those focusing on describing the properties of whole environments, are still hard to interpret within the context in which they have been employed. There is, therefore, the need for more empirical studies and experiments in modern environments to investigate how visual patterns in 3D space relate to human perceptions and behaviors.

The proliferation of 3-dimensional analyses also reflects the great variety of new lines of enquiry and methods (GIS, 3D graphics software, high-level programming languages, and so on) that are currently available for investigating the visual structure of space. An implication of this is that the scope and methodology of works concerned with the study of visual space can no longer be united and explored in a GIS, as perhaps was the case 10 years ago, given the existing GIS data structures and visibility analysis tools. In this respect the visualscape can no longer reach its full potential as an operational concept in a GIS context. It is also hard to imagine that the full range of approaches that are now available for the description of visual space could be implemented within a GIS environment in the near future. Still, the notion of visualscape, if it is not necessarily associated to a GIS application, succeeds in embodying the whole extent of ideas underlying current approaches to visual analysis in built environments and landscapes. This is corroborated by the fact that under these conditions almost all analytical products mentioned in this chapter could be described as visualscapes, despite being the outcome of different considerations and intentions; they occur both in natural and built spaces, they describe the visual structure of entire environments or subsets of space, and they have the potential to create a record of various different parameters of the spatial configurations under examination.

At the beginning of this chapter it was argued that the visualscape is a powerful notion that is prone to stimulate innovation, because by definition it aims to make explicit the theoretical and methodological affinities in the studies of visual space in built environments and landscapes. The developments in 3-dimensional visibility analysis suggest that it would be a more effective unifying and operational concept if it were used to describe the output of analysis independently from any particular implementation. A more broad definition and use of the term could encourage geographers, architects, urban planners, and archaeologists to realize and explore more fully the ever growing possibilities that currently exist for the study of visual space.

## Acknowledgments

I would like to thank Andy Bevan and Mark Lake, who invited me to contribute to this volume. Also many thanks go to David Wheatley and Graeme Earl for reading and commenting on earlier drafts of this chapter.

### Notes

1. See, however, note 3.
2. A 3D geometry used to represent the outer surface, or shell, of features that occupy a discrete area or volume in 3D space (from ArcGIS 10 guide).
3. For example, the latest version of 3D Analyst in ArcGIS 10 can perform simple LOS calculations (but not viewshed analysis) on the faces of buildings that are shaped by 3D multipatch geometry. Another application of 3D visibility analysis to relatively simple urban models, which consist of vector GIS data, has been proposed by Suleiman, Joliveau, and Favier (2012).
4. Axial analysis aims to identify most "accessible" and therefore most widely used street segments within an urban network by connecting the longest and fewest lines of sight that traverse each outdoor space in a continuous urban plan (Hillier and Hanson, 1984: 82–142).
5. An isovist is the area visible from a given point in built space (Benedikt, 1979). The 2D isovist usually corresponds to the 2D horizontal slice of space at the eye level of the viewer.
6. A visibility graph is a set of edge connections that connect locations in space that are mutually visible (Turner et al., 2001).
7. Gal and Doytsher (2012) have also proposed recently a 3D visibility algorithm that permits the rapid calculation of the visible faces of simple built structures, which is suitable for online and close to real-time applications.
8. For all examples described in this chapter various versions of 3Ds Max and ArcGIS were used. Nonetheless, the computational analyses could be also implemented with a variety of other 3D modeling and image-processing programs, as well as open source software.
9. In the examples presented in this chapter an analysis grid with intervals of 20 × 20 cm was used.

### References

Batty, M. 2001. Exploring isovist fields: Space and shape in architectural and urban morphology, *Environment and Planning B: Planning and Design* 28(1): 123–50.

Benedikt, M. L. 1979. To take hold of space: Isovists and isovist fields, *Environment and Planning B: Planning and Design* 6(1): 47–65.

Bishop, D. I. 2003. Assessment of visual qualities, impacts, and behaviours, in the landscape, by using measures of visibility, *Environment and Planning B: Planning and Design* 30(5): 677–88.

Bishop, D. I., Wherrett, J. R., and Miller, D. R. 2000. Using image depth variables as predictors of visual quality, *Environment and Planning B: Planning and Design* 27: 865–75.

Clark, D. L. C. 2007. Viewing the liturgy: A space syntax study of changing visibility and accessibility in the development of the Byzantine church in Jordan, *World Archaeology* 39: 84–104.

Conolly, J., and Lake, M. 2006. *Geographical Information Systems in Archaeology*. Cambridge: Cambridge University Press.

De Floriani, L., Marzanoa, P., Puppob, P. 1994. Line-of-sight communication on terrain models, *International Journal of Geographical Information Science* 8(4): 329–42.

Dean, D. J. 1997. Improving the accuracy of forest viewsheds using triangulated networks and the visual permeability method, *Canadian Journal of Forest Research*: 969–77.

Derix, C., Gamlesæter, Å., and Carranza, M. P. 2008. 3D isovists and spatial sensations: Two methods and a case study, in Haq, S., Hölscher, C., and Torgrude, S. (Eds.), SFB/TR 8 Report No. 015-05/2008 *Movement and Orientation in Built Environments: Evaluating Design Rationale and User Cognition*, pp. 67–72.

Doumas, C. 2005. La Répartition topographique des fresques dans les bâtiments d' Akrotiri à Théra, in Bradfer-Burdet, I., Detournay, B., and Laffineur, R. (Eds.), ΚΡΗΣ ΤΕΧΝΙΤΗΣ: *L'Artisan Crétois: Recueil d'articles en l' honneur de Jean-Claude Poursat, publié à l'occasion de 40 ans de la découverte du Quartier Mu. Université de Liège, Aegeum* 26: 73–81.

Dylla, K., Frischer, B., Mueller, P., Ulmer, A., and Haegler, S. 2010. Rome reborn 2.0: A case study of virtual city reconstruction using procedural modeling techniques, in Frischer, B., Crawford, J. W., and Koller, D. (Eds.), *Making History Interactive, 37th Proceedings of the CAA Conference March 22–26, 2009, Williamsburg, Virginia*, pp. 62–66. Oxford: Archaeopress.

Earl, G. P. 2005. Texture viewsheds: Spatial summaries of built archaeological spaces derived from global light mapping, *Proceedings of the 11th International Conference on Virtual Systems and Multimedia, Ghent, Belgium, October 2005*: 303–12.

Engel, J., and Döllner, J. 2009. Approaches toward 3D visual analysis for digital landscapes and its applications, in Buhmann, E., Kieferle, Pietsch, M., Paar, and Kretzler, E. (Eds.), *Digital Design in Landscape Architecture 2009: Proceedings of Presented Papers*, May 21–23, 33–41.

Favro, D. 1996. *The Urban Image of Augustan Rome*. Cambridge: Cambridge University Press.

Fisher, K. 2009. Placing social interaction: An integrative approach to analyzing past built environments, *Journal of Anthropological Archaeology* 28(2009): 439–57.

Fisher, P. 1994. Probable and fuzzy models of the viewshed operation, in Worboys, M. (Ed.), *Innovations in GIS: Selected Papers of the First International Conference on GIS Research*, pp. 161–75. London: Taylor & Francis.

Fisher-Gewirtzman, D., Burt, M., and Tzamir, Y. 2003. A 3D visual method for comparative evaluation of dense built-up environments, *Environment and Planning B: Planning and Design* 30(4): 575–87.

Fisher-Gewirtzman, D., and Wagner, A. I. 2003. Spatial openness as a practical metric for evaluating built-up environments, *Environment and Planning B: Planning and Design* 30: 37–49.

Frieman, C., and Gillings, M. 2007. Seeing is perceiving? *World Archaeology* 39(1): 4–16.

Gal, O., and Doytsher, Y. 2012. Fast visibility analysis in 3D procedural modeling environments, *Proceedings of the Third International Conference on Computing for Geospatial Research and Applications* (COM.Geo 2012), Article No. 27; http://dl.acm.org/citation.cfm?id=2345348; accessed October 20, 2012.

Gibson, J. J. 1979. *The Ecological Approach to Visual Perception.* Hillsdale, NJ: Lawrence Erlbaum Associates.

Gillings, M., and Goodrick, G. T. 1996. Sensuous and reflexive GIS: Exploring visualisation and VRML, *Internet Archaeology* 1; http://intarch.ac.uk/journal/issue1/gillings_toc.html: accessed January 20, 2010.

Groß, M. 1991. The analysis of visibility-environmental interactions between computer graphics, physics, and physiology, *Computers and Graphics* 15(3): 407–15.

Hall, T. E. 1966. *The Hidden Dimension.* New York: Doubleday.

Higuchi, T. 1983. *The Visual and Spatial Structure of Landscapes.* Cambridge, MA: The MIT Press.

Hillier, B., and Hanson, J. 1984. *The Social Logic of Space.* Cambridge: Cambridge University Press.

Ingold, T. 2000. Stop, look and listen! in *The Perception of the Environment: Essays on Livelihood, Dwelling, and Skill*, pp. 243–87. London: Routledge.

Lake, M. W., Woodman, P. E., and Mithen, S. J. 1998. Tailoring GIS software for archaeological applications: An example concerning viewshed analysis, *Journal of Archaeological Science* 25(1): 27–38.

Liu Liu, LiQiang Zhang, JingTao Ma, Liang Zhang, XingMing Zhang, ZhiQiang Xiao, and Ling Yang. 2010. An improved line-of-sight method for visibility analysis in 3D complex landscapes, *Science China Information Sciences* 53(11): 2185–94.

Llobera, M. 2003. Extending GIS-based visual analysis: The concept of visualscapes, *International Journal of Geographical Information Science* 17(1): 25–48.

________. 2006. What you see is what you get? Visualscapes, visual genesis a hierarchy, in Evans, T. L., and Daly, P. (Eds.), *Digital Archaeology: Bridging Method and Theory*, pp. 148–67. London: Routledge.

________. 2007. Modeling visibility through vegetation, *International Journal of Geographical Information Science* 21(7): 799–810.

Llobera, M., Wheatley, D., Steele, J., Cox, S., and Parchment, O. 2010. Calculating the inherent visual structure of a landscape ("total viewshed") using high-throughput computing, *International Conference Computer Applications and Quantitative Methods in Archaeology, Beyond the Artifact: Digital Interpretation of the Past, Prato, Italy, April 13–17, 2004*: 146–51.

Lynch, K. 1960. *The Image of the City.* Cambridge, MA: Harvard-MIT Joint Center for Urban Studies Series.

Marinatos, N. 1984. *Art and Religion in Thera: Reconstructing a Bronze Age Society.* Athens: D. & J. Mathioulakis.

Marinatos, N., and Hägg, R. 1986. On the ceremonial function of the Minoan polythyron, *Opuscula Atheniensia* XVI: 57–73.

Märtens, H. 1890. *Optisches Mass für den Städtebau.* Bonn: Max Cohen & Sohn.

Moore, D. J. 1996. *Architecture and Power in the Ancient Andes: The Archaeology of Public Buildings.* Cambridge: Cambridge University Press.

Morello, E., and Ratti, C. 2009. A digital image of the city: 3D isovists in Lynch's urban analysis, *Environment and Planning B: Planning and Design* 36(5): 837–53.

Paliou, E. 2009. *The Visual Consumption of Mural Painting in Late Bronze Age Akrotiri (Thera, Greece): A Computational Approach to Visibility Analysis in Three-Dimensional Built Environments.* Unpublished Ph.D. thesis, University of Southampton.

________. 2011. The communicative potential of Theran murals in Late Bronze Age Akotiri: Applying viewshed analysis in 3D townscapes, *Oxford Journal of Archaeology* 30(3): 247–72.

________. 2013. Visibility analysis in 3D spaces: A new dimension to the understanding of social space, in Paliou, E., Lieberwirth, U., and Polla, S. (Eds.), *Spatial Analysis and Social Spaces: Interdisciplinary Approaches to the Interpretation of Prehistoric and Historical Built Environments*, pp. 91–113. Berlin: Topoi, Berlin Studies of the Ancient World, De Gruyter.

Paliou, E., and Knight, D. J. 2013. Mapping the senses: Perceptual and social aspects of late antique liturgy in San Vitale, Ravenna, in Contreras, F., and Melero, J. (Eds.), *Proceedings of CAA 2010, Computer Applications and Quantitative Methods in Archaeology, International Conference, Granada, April, 6–9, 2010*: 229–36.

Paliou, E., Lieberwirth, U., and Polla, S. (Eds.). 2013. *Spatial Analysis and Social Spaces: Interdisciplinary Approaches to the Interpretation of Prehistoric and Historical Built Environments.* Berlin: Topoi, Berlin Studies of the Ancient World, De Gruyter.

Paliou, E., and Wheatley, D. 2007. Integrating spatial analysis and 3D modelling approaches to the study of visual space: Late Bronze Age Akrotiri, in Figueiredo, A., and Velho, G. L. (Eds.), *The World Is in Your Eyes, Proceedings of the XXXIII Computer Applications in Archaeology Conference*, March 21–24, 2005, Tomar, Portugal, 307–12.

Paliou, E., Wheatley, D., and Earl, G. P. 2011. Three-dimensional visibility analysis of architectural spaces: Iconography and visibility of the wall paintings of Xeste 3 (Late Bronze Age Akrotiri), *Journal of Archaeological Science* 38(2): 375–86.

Palyvou, C. 2000. Concepts of space in Aegean Bronze Age art and architecture, in Sherratt, S. (Ed.), *The Wall Paintings of Thera, Proceedings of the First International Symposium, Petros M. Nomikos Conference Centre*, Thera, Hellas, September 4–30, 1997, Athens: Thera Foundation.

Papadopoulos, K., and Earl, G. P. 2013. Formal three-dimensional computational analyses of archaeological spaces, in Paliou, E., Lieberwirth, U., and Polla, S. (Eds.), *Spatial Analysis and Social Spaces: Interdisciplinary Approaches to the Interpretation of Historic and Prehistoric Built Environments*, pp. 135–65. Berlin: Topoi, Berlin Studies of the Ancient World, De Gruyter.

Papageorgiou, I. 2000. On the rites de passage in Late Cycladic Akrotiri, Thera: A reconsideration of the frescoes of the "Priestess" and the "Fishermen" of the West House, in Sherratt, S. (Ed.), *The Wall Paintings of Thera, Proceedings of the First International Symposium*, Vol. 2, pp. 958–70. Petros M. Nomikos Conference Centre, Thera, Hellas, August 30–September 4, 1997. Athens: Thera Foundation.

Pyysalo, U., Oksanen, J., and Sarjakoski, T. 2009. Viewshed analysis and visualization of landscape voxel models, *Proceedings of the 24th International Cartographic*

*Conference, November 15–21, 2009, Santiago, Chile*, www.icaci.org/publications, March 9, 2010.

Rapoport, A. 1990. *The Meaning of the Built Environment: A Nonverbal Communication Approach, with a New Epilogue by the Author*. Tucson: The University of Arizona Press.

Suleiman, W., Joliveau, T., and Favier, E. 2012. 3D urban visibility analysis with vector GIS data, in *Proceedings of Geographical Information Systems Research UK (GISRUK), Portsmouth, 2011*, http://portail.univ-stetienne.fr/servlet/com.univ.collaboratif.utils.LectureFichiergw?CODE_FICHIER=1315906066024&ID_FICHE=185671, October 20, 2012.

Tandy, C. R. V. 1967. The isovist method of landscape survey, in Murray, H. C. (Ed.), *Methods of Landscape Analysis*, pp. 9–10. London: Landscape Research Group.

Turner, A., Doxa, M., O'Sullivan, D., and Penn, A. 2001. From isovists to visibility graphs: A methodology for the analysis of architectural space, *Environment and Planning B: Planning and Design* 28: 103–21.

Van Bilsen, A. 2008. *Mathematical Explorations of Urban and Regional Design*. Ph.D. dissertation, Delft University of Technology.

Van Bilsen, A., and Poelman, R. 2009. 3D visibility analysis in virtual worlds: The case of supervisor, in *Proceedings of Construction Applications of Virtual Reality (CONVR), 2009 Conference Sydney, Australia, November 5–6*: 267–78.

Vlachopoulos, A. 2010. L'Espace rituel revisite: Architecture et iconographie dans la Xeste 3 d'Akrotiri, Thera, in Boehm, I., and Müller, S. (Eds.), *Espace civil, espace religieux en Egee durant la periode mycenienne: Approches epigraphique, linguistique et archeologique*. Lyon: Maison de l'Orient et de la Méditerranée: 173–98, 223–29.

Wheatley, D. 1995. Cumulative viewshed analysis: A GIS-based method for investigating intervisibility, and its archaeological application, in Lock, G. R., and Stančič, Z. (Eds.), *Archaeology and Geographical Information Systems: A European Perspective*, pp. 171–85. London: Taylor & Francis.

Wheatley, D., and Gillings, M. 2000. Vision, perception, and GIS: Developing enriched approaches to the study of archaeological visibility, in Lock G. R. (Ed.), *Beyond the Map: Archaeology and Spatial Technologies*, pp. 1–27. Amsterdam: IOS Press.

———. 2002. *Spatial Technology and Archaeology: The Archaeological Applications of GIS*. London: Taylor & Francis.

Yang, P. P. J., Putra, S., and Li, W. 2007. Viewsphere: A GIS-based 3D visibility analysis for urban design evaluation, *Environment and Planning B: Planning and Design* 34(6): 971–92.

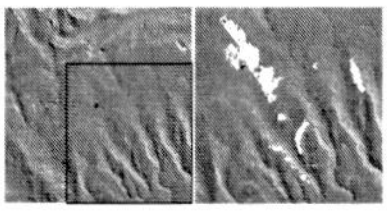

# 11 Formal and Informal Analysis of Rendered Space: The Basilica Portuense

*Graeme Earl, Vito Porcelli, Constantinos Papadopoulos, Gareth Beale, Matthew Harrison, Hembo Pagi, and Simon Keay*

Digital archaeological analyses of space have proceeded traditionally from two distinct perspectives. First, there has been a phenomenological appraisal of computer-graphic-modeled geometry and, second, a formal analysis of the surviving spaces and extrapolations from them. The former approach encompasses work on the visualization of raw geometric data captured through techniques such as laser scanning and photogrammetry. In addition, it has critiqued the processes and interfaces within which extrapolations from these data take place—namely, the construction of increasingly sophisticated graphical simulations in software such as 3ds Max, Maya, and Vue Infinite. Finally, it draws on the ability of such software and interactive visualization tools to stimulate new interpretations through interaction with the visualized geometry. This virtual phenomenology in turn has been seen to be significantly enhanced by an emphasis on physical realism (Chalmers, 2002), but it has not been without its critics (Lock, 2003). The use of physically realistic techniques facilitates the second, formal analytical approach by extending methodologies primarily stimulated by the architectural community to enable comparisons between built architecture (and their resulting spatial interactions), to evaluate their robusticity through stochastic methods, and to predict and analyze physical factors such as lighting in architectural designs. However, the novel research in physical (or "predictive") rendering is primarily constrained to the domain of computer graphics (Wilkie et al., 2009) and remains to be explored in depth in archaeological research. This chapter begins to explore their potential.

Archaeology has made use of both sets of methods, and with growing capture of 3-dimensional data and greater computational resources on the archaeologist's desktop they have both proliferated. In this chapter we introduce an integrated method drawing on all the techniques mentioned to address questions both of archaeological modeling practice and the analysis of their outputs. As a case study this chapter explores the Basilica Portuense at the site of Portus, Italy. It begins by identifying the interpretative context for the simulation and analysis of 3-dimensional digital forms. It considers the potentials of approaches focused on visibility and lighting within these environments and then outlines the case study and in particular focuses on areas of uncertainty explored in the analyses. Issues of uncertainty and the potentials of procedural modeling are outlined. Three methodological approaches are then explored—formal visibility analysis via Texture Viewsheds (Earl, 2005), lighting analysis, and finally a more phenomenological appraisal of the structure using physically accurate graphics. Finally, we identify conclusions and future work in these domains. As a whole, the chapter's contribution is in the development of a rigorous methodology and interpretative framework combining formal and informal spatial analyses, in the definition of a novel development of Texture Viewsheds, and in the development and exploration of a number of Roman archaeological and architectural questions for the Basilica Portuense.

## 3D Modeling for Spatial Interpretation

### Visibility

The creation of modeled landscapes and built environments has enabled a wide range of formal analyses of visibility. Paliou (this volume) provides a comprehensive overview of such formal methods, including both pseudo- and true 3-dimensional spatial analyses. In this chapter we develop the Texture Viewsheds (Earl, 2005) method outlined by Paliou and identify the benefits of comparing such visibility products with lighting analysis and more subjective appraisals of modeled space. The formal visibility methods correspond to a great extent with Llobera's (2003) concept of visualscape outlined in Paliou (this volume). In particular, our approach is designed to provide analytical detail for the total environment of the Basilica Portuense and subsections of it, is inherently 3-dimensional, and enables the analytical products to be remapped onto the modeled geometry on which they are based. It also builds on the 3-dimensional methods outlined by Paliou and Knight (Forthcoming) at the Basilica of San Vitale. Texture Viewsheds enable the archaeologist to generate numerical visibility indices that in turn can be considered alongside theoretical concepts of visibility. The analytical approach also acts alongside phenomenological appraisals of models to provide numerical correlates—offering a form of sensitivity analysis that might support or contradict experiential understanding of a given space.

Paliou (this volume) considers the potential of approaches such as isovists and VGA that summarize but are not based on true 3-dimensional representations of space. Our contribution complements these formal approaches, including those used in the religious spaces defined by Clark (2007a, 2007b) and Richardson (2003). It should also be considered in the light of Allison's (2001) critique of formal methods in a Roman context, such as access analysis, which are seen not to offer new ways of seeing Roman space but rather to confuse appreciation further and to enforce analogic simplifications of concepts such as public and private, and indeed "access" itself. Giles's (2007) detailed critique of visual analyses is also relevant here. Giles identifies the need to consider premodern visuality, multimodal understandings of space and of the bias in current methods, including those outlined here, toward modern visual sense. We return to these issues in the conclusions to this chapter, but we propose that what follows should first be read in the light of these compelling critiques.

## Lighting and Physical Accuracy

### Simulation

Lighting is increasingly seen as a method both of formal appraisal of archaeological environments and of accurate simulation of their appearance. At the core of this approach lies an attempt to simulate the physical transport of light within a scene to a degree that quantitative appraisals of a modeled environment such as the amount of available light and the color of a perceived object closely parallel those obtained experimentally in an equivalent real space. Such quantitatively robust outputs are in turn used to justify the direct use of simulated environments in the analysis of spatial experience.

Research in physical accuracy of lighting in archaeological environments is exemplified in the work of Devlin and Chalmers (2001) and Devlin, Chalmers, and Brown (2002). The work of Chalmers has focused on increasingly physically accurate simulation of light transport and of all components of the visual scene. A physically accurate scene in this sense should provide means to introduce physically accurate artificial and natural lighting (Kider et al., 2009; Masuda and Yamada, 2004; Sundstedt, Chalmers, and Martinez, 2004; Zányi, 2007; cf. Dawson et al., 2007). In addition the light created by such luminaires should react appropriately with the scene. Thus, smoke and dust particles should influence them (Gutierrez et al., 2008; Hawkins, Einarsson, and Debevec, 2005) and the illumination of the surrounding environment such as the sky (Happa et al., 2009a), and environmental factors such as rain should behave accurately in the scene (Garg and Nayar, 2006;

Wang et al., 2006). The light should react with surfaces in ways appropriate to their physical form—for example, the color expressed as the dynamic range of the surface materials (Happa et al., 2009a; Happa et al., 2010), the surface reflectance properties (Callet et al., 2010; Müller, Bendels, and Klein, 2005), the scattering of light within materials such as marble and skin (Hašan et al., 2010), and the perception of these surfaces as a function of motion, depth of focus, and accommodation to brightness (Chalmers et al., 2007; Ledda, Santos, and Chalmers, 2004). Those most interested in near-perfect digital replication of complex surface/light interactions (for example, gemstone workers) are increasingly adopting spectral rendering approaches that represent the complex interactions that are also of significance in archaeological contexts—for example, dispersion in glass and lithic objects, absorption, and metemerism (Weidlich and Wilkie, 2008). More recently, work in physically real simulation has also incorporated the potentials of multimodal environments that include touch, sound, and smell alongside visual stimuli (Chalmers and Zányi, 2009).

It is not, however, the case that all problems in creating physically accurate scenes have been overcome. First, there remains only limited work in the formal validation of digital constructed cultural heritage. Although some studies have demonstrated, for example, the effects of motion or high dynamic range imagery on perception of modeled objects (Hasic, Chalmers, and Sikudova, 2010), there remains very little coherent study. In particular, studies of the perception of individual surface types (such as marble) remain isolated from the cultural heritage questions to which they pertain. For example, how does the marble behave under particular illumination in a given cultural heritage environment? The work of Kider and associates (2009) discussed earlier goes some way in this but does not provide a quantitative, perceptually driven summary and, although technically compelling, may be critiqued on archaeological grounds. Similarly, innovations in physical simulation of measured marble BRDF (the Bidirectional Reflectance Distribution Function, which describes reflection of light at a surface) show promise in enabling ground-truthed testing of measured and analytical BRDF models (Dong et al., 2010), but they are only at an early stage. For example, one might produce a physical replica object of one of the components of the Basilica Portuense scene that exhibits subsurface scattering and compare its interaction with real light sources to the predicted CG model (Hašan et al., 2010). Although both are based on the same BRDF model, the interaction with the physical luminaire would be of considerable value if the technique were to be more widely implemented.

The second problem in creating physically accurate cultural heritage environments is algorithmic. In many cases, cultural heritage research questions have aligned very well with computer graphic innovations in modeling and rendering. However, because of the expense of such research, where adjacencies do not always exist, innovation can be slow. Indeed, even where a need beyond cultural heritage exists it may be that research has not been undertaken for other reasons. For example, as Happa and associates (2009) note, there is limited research in rendering environments with very low lighting levels, including environments at night. Many of the luminaires associated with archaeological environments function at such low energy levels that the algorithms available, and indeed the hardware to consume and interact with them, are less well developed than in environments lit by modern artificial luminaires.

The final problem remains access to data of sufficient quality to merit the use of physically accurate simulations. Research at the forefront of computer graphics commonly employs nonstandard capture devices and makes use of file formats that are particular to only limited workflows. Thinking only of the research identified here in terms of physically accurate rendering, one would ideally need to obtain a spectrophotometer, color charts and gray cards, High Dynamic Range Imaging (HDRI) image probes, high dynamic video, an HDRI monitor, and a gonio-reflectometer in order to obtain the requisite data. These are only a small subset of the tools required to produce physically accurate simulations. One possibility for archaeological science does lie in recent work using the physical properties, including the chemical composition and other characteristics of an object directly in rendering software. Rather than using a surface reflectance model as input to a simulation, work such as the modeling of a bronze writing tablet using spectral rendering has come as close as possible to a wholly physical visualization process (Robin et al., 2010). Since chemical analysis of archaeological material, and access to libraries of such data, are both common, such an approach could place physically accurate rendering in the hands of the archaeologist. In turn, the reliance on physical properties over simulation could minimize the computer graphic expertise required to produce an accurate simulation. Unfortunately, although the software tools commonly available to archaeologists have become extremely powerful, for reasons of necessary ease of use and robustness it is difficult to modify them to accommodate the latest advances in lighting and rendering. So, even where computer graphics work has produced freely available metrics corresponding to particular common cultural heritage materials—for example, the MERL (Matusik et al., 2002) and CUReT (CURET) measured BRDF databases—the means to apply these within a cultural heritage domain are limited. We are therefore constrained in most of our work to physical analytical BRDF (and in the case of the subsurface, BSSRDF) models.

Despite these problems it is our belief that the value of such physically accurate studies lies in the sense that the viewer has a more direct appraisal of the scene than one where the simulation lacks this physical rigor. Appearance of photorealism has little value if that appearance has no quantifiable validity, unless the research process is explicitly to be informed by the production of archaeologically informed works of digital art. Furthermore, Chalmers has defined "authentic illumination" (Chalmers, Roussos, and Ledda, 2006) as a composite of this accurate physical simulation of light transport with archaeological data that have been well researched and that are therefore believed to provide a good representation of the physical form and components of the space when constructed or used. By valuing such physical characteristics one is inevitably drawn to an interpretation of perception that persists through time, rather than one that is entirely temporally and culturally contingent. However, it is rarely the case that such studies propose to recreate the past as it was but rather to create representations that can at least be validated archaeologically and physically be faithful to the evidence.

The directions such representations can be taken in depend on the epistemological direction of the research of which they form a part. For example, Dobbins and Gruber (Forthcoming) present simulations of the "Drinking Contest at Antioch" and the House of the Faun at Pompeii. These studies concentrate on comparative analyses of shadows and their effect on a contextual appreciation and appraisal of mosaic decorations. Kider and colleagues (2009) identified the accurate simulation of pools of light as being of crucial significance in archaeological interpretations of spatial function and of social interaction as mediated by environment. Finally, Callet and associates (2010) explored the possibility of distinct light design in medieval contexts. In these we see archaeological narratives in turn of phenomenology, of functional interpretation of space, and of aesthetics, all based on an explicit approach to accurately simulate environments. The archaeological knowns and unknowns are similar: the precise architectural finish, the relationship between form and intention, and the inherent spatiality of cultural practice.

## Analysis

The majority of work in the area of archaeological lighting relates to the production of attractive images of the type outlined previously. However, the same technologies used to define patterns of color, light, and shade on a rendered scene can be used to provide metrics indicating the motion and effects of light energy within the same scene. Indeed, it is a necessary component of the physically accurate rendering process that such simulations are undertaken. Formal lighting analysis halts the rendering process once light has been propagated within a scene, and instead of producing a simulation of the resultant perceived color for each point in space it provides a numerical summary of the energy distributed. This ability is vital to lighting designers, whether for the production of a lighting scheme for a new building or a theatrical production. The cost of construction and the very limited margins for error and modifications mean that lighting studies are a habitual part of the design process. As a consequence their physical accuracy is essential.

A small number of research projects have attempted to employ these formal analysis tools in an archaeological context, in order to quantify the quality and amount of light available in particular areas and the potential impact of such lighting on the behavior in them. The work by Dawson and associates (2007) and Dawson and Levy (2005) explored taskscapes in arctic houses through the simulation of structures and of their illumination. We employed similar techniques in work at the Minoan cemetery site at Phourni in Crete (Papadopoulos, 2010; Papadopoulos and Earl, 2009), creating multiple alternatives to explore and analyze. More recently our work at Zominthos's ceramics workshop (Papadopoulos and Sakellarakis, Forthcoming) used lighting analysis to explore the relationship between tasks and illumination. Critiques of such task-focused approaches center on the relevance of modern, Western lighting statistics to simulated past environments and behaviors. We agree that this use of modern lighting statistics poses questions for interpretation but do not agree that they cannot provide useful, perceptually based starting points for critique and development, particularly when positioned in broader ethnographic frameworks (as in Papadopoulos and Earl, Forthcoming).

## The Basilica Portuense

The imperial town of Portus was established to support the increasing flow of goods from across the Mediterranean through Rome and as a redistribution hub for the empire as a whole. The archaeological remains of the port are situated northwest of the Republican river port of Ostia on the Tyrrhenian sea and about 25 km from Rome. The port included warehouses, baths, houses and apartments, and the structures needed for the loading, unloading, and administration of goods. Among these, an early-Christian basilica has been identified to the south of the entrance to the hexagonal Trajanic basin, adjacent to the transverse canal linking the port, via a further canal, to the River Tiber. The basilica stood in an area that appears to have been dominated by commercial, administrative, and storage buildings, some of which were converted to create the basilica (Paroli, 2005: 258) (Figure 11.1). Computer simulation work on the basilica forms one part of the wider Portus and Portus in the Roman Mediterranean AHRC projects that incorporate computer applications throughout (Earl et al., In Press).

Throughout its early history the building underwent a series of transformations. The focus of the research described in this chapter has been on examining the effect of these transformations on the appearance of and visibility within the space, and on the use of such metrics to inform decisions about reconstructing the missing detail. These two interconnected approaches define an interpretative process that is based equally on the expression and the qualification of architectural uncertainties and an attempt to explore the implications of different ways of envisioning the past life of the building.

The final shape of the Basilica Portuense was characterized by three naves, with an apse at the end of the central nave, which was divided from the

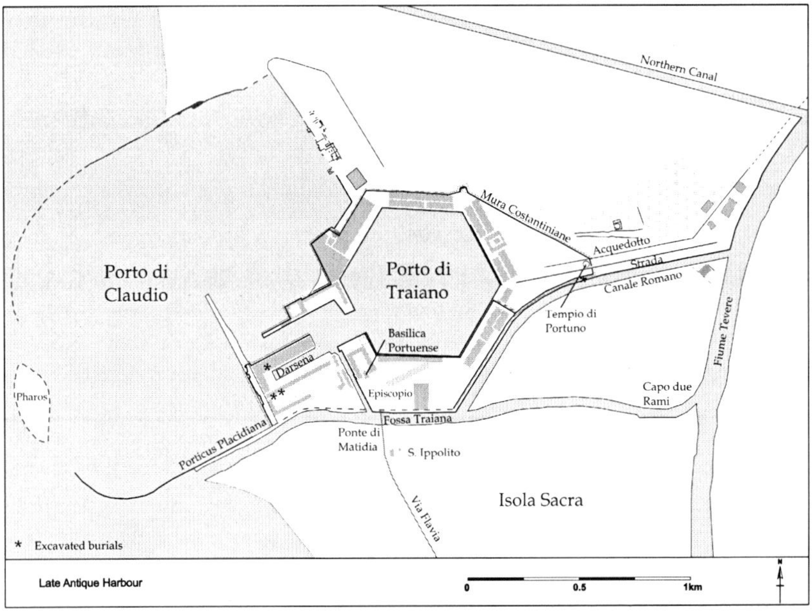


**Figure 11.1** Map showing Portus and the Basilica Portuense (after Keay et al., 2005)

lateral aisles by means of 10 pairs of columns, while the façade presented a triforium entrance. The Basilica remained in use during the Middle Ages, in part as a consequence of restoration work in the 11th and 12th centuries. It began to be abandoned in the 13th century (Borrello et al., 2002: 1267). In recent years the site was excavated and interpreted by a group of archaeologists under the direction of Dr. Lidia Paroli of the Soprintendenza dei Beni Culturali di Ostia. The hypothesized architectural schemas explored here are based on her work and on the interpretations of the architects Dr. Roberta Loreti and Dr. Giuliana Irace. Such transformation is not unusual in the development of early-Christian places of worship. The first Christians living in the Roman provinces could not profess their religion freely, and secret meetings were regularly attended. For three centuries they suffered persecution, most notably under Diocletian. Eventually Christians gained their freedom of religion with the Edict of Milan, in 313 C.E. This edict led to the adaptation of many ancient buildings around the empire for the Christians' liturgies, particularly domuses, which were donated by the Roman Aristocracy as sign of devotion to the God (Nicolai, 2001: 93). These "Titularis" basilicas took their names from aristocratic citizens who, after converting to Christianity, supplied properties and furniture and supported the construction of the churches. Many examples are still present, including several in Rome, such as Titulus Sabinae (Basilica of Santa Sabina), Titulus Lucinae (Basilica of San Lorenzo in Lucina), Titulus Vestinae (Basilica of San Vitale), and Titulus Clementi (Basilica of San Clemente).

A similar history is proposed for the early-Christian Basilica Portuense, which may have been part of the Tituli Pammachii. In fact, the building was identified as an ancient guesthouse for pilgrims in the "Xenodochium of Pammachius" by Rodolfo Amedeo Lanciani (1866: 100–03), with a dedication to Saint Peter and Saint Paul as proposed by Fevrier (1958: 316). Giovanni Battista De Rossi (1866: 99–100) suggested the dedication was actually to the Virgin Mary and that the construction of the Basilica was against a hospice. The later excavations have endorsed this thesis, although there remains uncertainty.

In the earliest phases of the "basilica," a central and smaller building occupied the actual central area of the church. This building survived a variety of modifications in the first and second centuries (see Paroli, 2005, Figures 7.19–7.20) (Figure 11.2). Later, many sections of the building described were

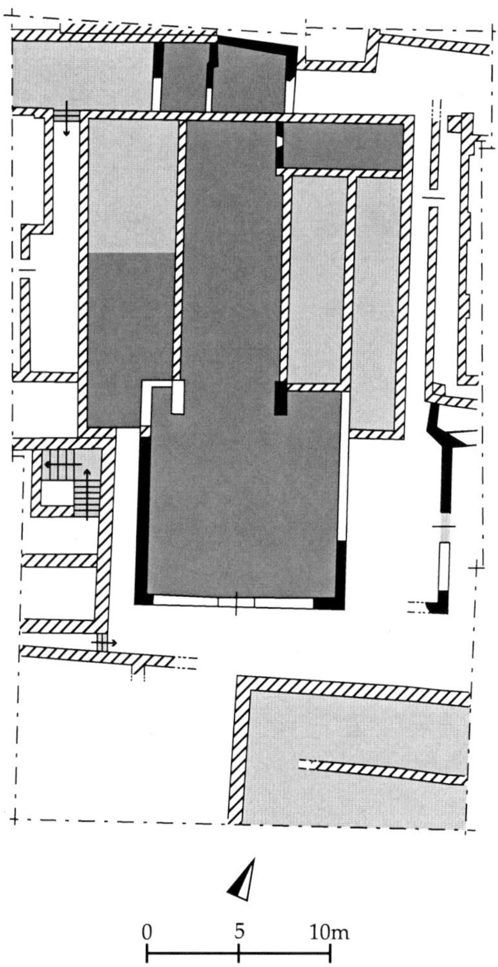


**Figure 11.2** Plan of building phases 1–2 (after Paroli, 2005)

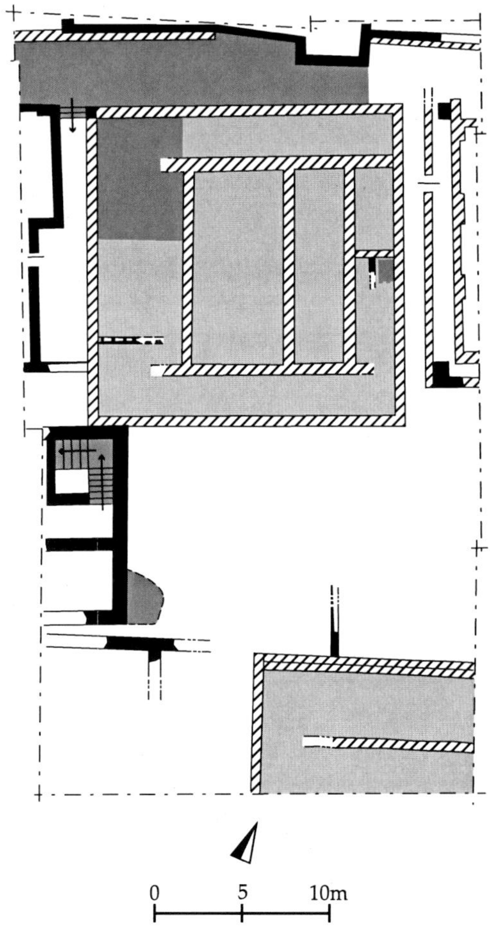


**Figure 11.3** Plan of building phase 3 (after Paroli, 2005)

reused and adapted to construct a newer and longer building, which, given the shape, was probably still utilized for commercial or administrative purposes (Figure 11.3). Transformation into a basilica is attributed to the end the second half of the 4th century (Paroli, 2005: 261). This was composed of eight pairs of columns with unfluted shafts and composite capitals with curled acanthus leaves. These divided a central and wider aisle from two narrower side aisles, with a narthex and a number of side doors (Figure 11.4). At the beginning of the 5th century two lateral rooms were added, one at the end of each lateral nave, and an apse was constructed. A marble schola cantorum with solea, existing also in the previous phases, was restored and improved. It was almost

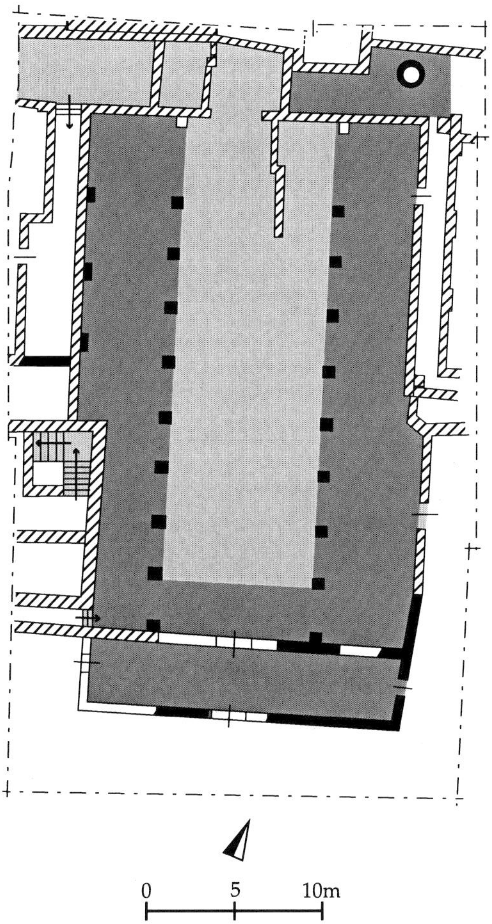


**Figure 11.4** Phase 1 (after Paroli, 2005)

certainly richly decorated. At this point a doorway was also closed off (Figure 11.5). At the end of the 6th century the Basilica was extended to the south. Two additional pairs of columns were constructed, with Ionic capitals and a new façade, composed of an entrance decorated with a monumental triforium, two lateral entrances, and an external narthex (portico) regulating the access to the main building (Figure 11.6). The portico was not entirely excavated but was likely to have been composed of shorter columns and to have covered the width of the façade. It was previously considered as a "quadriportico" by Lanciani (1865), but the recent excavation rejected this. During the 8th century a hexagonal baptistery was added at the end of the western nave (Borrello et al., 2002: 1282).

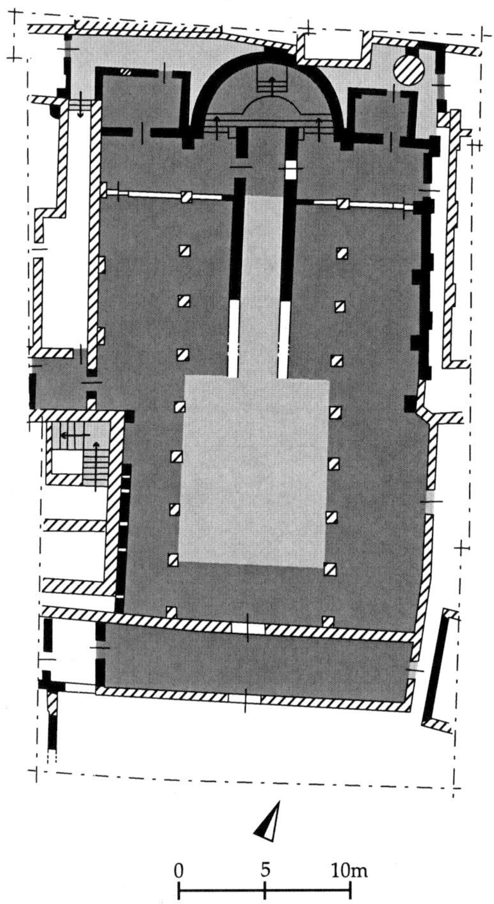


**Figure 11.5** Phase 2 (after Paroli, 2005)

## Modeling and Rendering Methods

For the analyses described shortly we chose to use 3ds Max Design 2011, which has a number of tools tailored to architectural spaces. This and all other analyses described in this chapter were undertaken in a Windows7 environment—other than the supercomputer, which runs Windows HPC Server 2008. The lighting analyses make use of a simulation of the light energy within the scene, taking into account the luminaires, environmental effects, and the materials with which the light energy interacts. Two sets of statistics are commonly produced by such lighting analyses: first, the illuminance or light arriving at a surface, commonly divided into the direct illuminance (that is, the light that

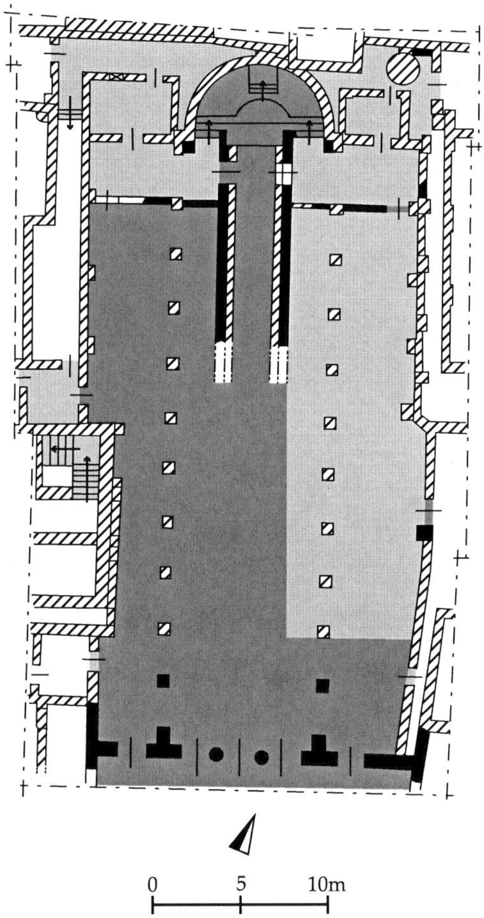


**Figure 11.6** Phase 3 (after Paroli, 2005)

travels from the luminaire to the surface without any intervening reflections) and the indirect illuminance (which summarizes the light reflected from surfaces but not received direct from the light source); second, the luminance—the light that reflects from an illuminated surface. The accuracy of lighting analyses depends on a number of factors:

1. Environmental effects such as participating media and the atmosphere
2. Fit of simulated luminaire to its real-world equivalent, in terms of energy and light distribution (particularly difficult when employing ancient light sources that may well have no experimentally derived CIE or other metrics)

3. Quality of geometry defining the space analyzed
4. Quality of the material approximations employed
5. Quality of the (necessarily approximate) solution calculating the interactions between these variables

For this chapter and previous papers we undertook detailed comparisons of the impact of each of these factors on the lighting metrics. The relationship between the factors is extremely complex, particularly given the many parameters associated with each. For example, one of the software rendering engines used (Mental Ray) makes use of many parameterized approximations, including factors effecting caustic (focused) and global illumination (bounced) photons. (Results are also different when using the iRay path tracing rendering mode, which offers no parameters other than the time or number of iterations taken to execute the rendering. However, iRay is not currently able to generate lighting metrics and was therefore used only in the visual simulations.)

The lighting analysis is influenced to a great extent by the materials present within the scene. Since it assumes that light energy moves around in a physically accurate way only materials accurately simulating energy absorption, opacity, subsurface scattering, refraction, and reflection should be used in an environment where lighting analysis is undertaken. Solutions such as BTF (Dana et al., 1999) and BRDF implementations in physically accurate Mental Ray (.mi) shaders enable more accurate approximations to the complexities of these interactions, but they remain approximations. Similarly, although increasing possibilities are emerging for spectral rendering (including a Mental Ray color space and systems such as Maxwell), the availability of the necessary spectral surface reflectance and luminaire data limits their application where novel data collection is not possible. Each one of these factors will directly affect the energy simulation within the solution. For the sake of brevity we do not outline our evaluation of these here.

## Modeling the Basilica

Our study is limited only to the three phases in which the architecture and dates correspond to the notion of an early-Christian basilica. In addition, our study has been limited to the evidence present in Paroli (2005). Additional interpretations are presented in Borrello and associates (2002) relating to, for example, the short and small triforiums in phase 2, the postulated changing height of the schola in phase 3, and the presence of polibolatis mosaic in the western nave. It was not possible to examine all of these for this chapter, but our work continues.

Roman building styles were adopted for the foundations and perimeter walls of the basilica, with opus latericium, opus vittatum, and reticulatum still appreciable in situ. The church was constructed using the ancient Roman Foot and in its final form covered an area approximately of 20 m width and 45 m length, similar to the size of the plan of Saint Peter in Vincoli, Rome (Pensabene et al., 2007: 461). The perimeter reused walls of earlier buildings or, as on the western side, common walls shared by other existing buildings. These have been interpreted as potential residential buildings, composed of two or more floors and probably used by the clergy. The ground plan of the Basilica Portuense is therefore unusual in the specifics but conforms to the general structure of one larger and two smaller naves (Borrello et al., 2002: 1264). The two rows of columns are assumed on the basis of parallels to have supported arcades, hence raising the central ceiling above those of the lateral aisles. The pavement discovered shows different styles in use during the three phases. In the latest, the Basilica was principally decorated with opus sectile all over the aisles. However, the building certainly incorporated archaic mosaics in the central nave, including one probably preserved by a postulated early-Christian Ambon, and a pavement of bipedalis terracotta brick tiles decorated the northern half of the eastern aisle. In the clerestory the remains of some furniture in marble were discovered, suggesting a central altar supported by four small columns and a cathedra surrounded by two semicircular benches. These matched the shape of the apse and were closed elegantly by two opposing dolphin armrests.

The foundations and flooring of the Basilica are relatively well preserved. Therefore, the main complexity in modeling the Basilica arose from the elevations. Structural simulation using finite and discrete element analysis and other approaches was considered, but for the purposes of this chapter we have relied on extensive discussions with the excavator, Lidia Paroli, the architects, and on a wide range of parallel churches—in particular Sant'Agata dei Goti, Santa Sabina, and San Vitale.

We considered a wide range of factors in the construction and evaluation of the various models explored in this chapter, but here we concentrate on those with a likely impact on visibility and lighting. The height and morphology of the roof were based on parallels but also on a detailed analysis of its potential construction. Modeling of multiple options was particularly important for the lateral naves, where the wall line is irregular. The roof line in turn influenced the presence and shape of clerestory windows, which have been informed by parallels and modified by the consequences of the lighting analyses described shortly. Similarly, the windows in the apse are significant. Evaluation of Sant'Agata dei Goti and some structural details suggests that the Basilica Portuense may have had an unusual configuration of two large windows in the apse rather than the common three. It may also be possible that there were smaller windows along the eastern aisle, similar to those still viewable in the same Sant'Agata dei Goti or in Santa Sabina on the Aventine Hill in Rome.

The buildings on the west make it unlikely that there were windows here. These two assumptions are also supported by the later construction of the triforium gate. Effectively, without the gate and without the windows described, the archaic lower part of the building was even darker during the day hours. Similarly, the portico affects the light entering the building, and the model of this portico was based in part on evidence at San Vitale in Rome.

Digital materials were generated to provide the necessary physical and aesthetic effects on the analyses described shortly. The marble of the capitals, the furniture in the apse, the schola cantorum, and the ambon proved particularly challenging. Physically accurate materials were used as far as possible to represent some common Roman marble types, including Lunense, Cipollino, and Proconnesio. Where possible the choice of marble was based on surviving materials but inevitably included a great deal of extrapolation from parallels. We were not able to undertake detailed microscopic analysis or gonio-reflectometry on fragments from the site. However, in part we have been able to use the knowledge of other marble fragments found at Portus and recorded using Reflectance Transformation Imaging. In addition to providing a color calibrated dataset this technique also offers possibilities for representing the surface structure (via calculated normal) and hence the interaction with incident light (Earl et al., 2010; Earl, Martinez, and Malzbender, 2010).

Some material elements in the final model had to be based on photographs of the surviving and parallel datasets. It was not always possible to color calibrate these, and therefore the analyses for lighting sometimes used simplified but physically accurate materials as a replacement. This is the case, for example, with a maritime mosaic and another mosaic representing black plaits excavated on the northeast side of the central nave. Similarly, early-Christian mosaics preserved in the Basilica of San Giovanni in Laterano were placed in the upper walls of the central nave of the Basilica Portuense, whereas the painting on the half-dome derives from an original early-Christian mosaic that existed on the half dome of Sant'Agata dei Goti. The original mosaic was destroyed during a fire in the Middle Ages but had been represented in earlier drawings (Huelsen et al., 1924). As far as possible such imagery was used as a mask to enable physically accurate components to be used to calculate reflectivity. Indeed, an attempt was made to synthesize the broad spectrum of pigments contained in such paintings via a simple geometric material that matched their proportions but not their distribution. Thus in the final renderings an attempt was made to use the locations of these paintings and their broad colors to add appropriate reflected color to the scene rather than as particular points of interest. Final dressing of the model included addition of curtains and artificial luminaires. These were generally kept very simple, with the intention of adding an indication and having an impact on lighting. However, the large chandeliers are based on discussions with the excavator and other experts, and on studies of similar early-Christian luminaires.

### Uncertainties in the Morphology of the Basilica as Modeled

Although it is important to identify uncertainties in computer simulation, the definition of a precise series of options can lead to a reduced awareness of the myriad of additional uncertainties. Computer modeling is inevitably a hybrid of art and science; it is a process of informed extrapolation, in a social, cultural, and methodological context. However, since this chapter considers formal analyses that are themselves solely based on these simulations, we have attempted to systematize the uncertainties, at least as they are perceived by us and the others involved in the research. The previous sections have outlined a range of uncertainties in terms of morphology, material, and illumination. In addition to a host of others we considered the following uncertainties in some detail in the manual and procedural modeling approaches employed:

- Façade from phase 2
- Location of windows in the nave, apse, and eastern aisle in phases 1, 2, and 3
- Morphology of the portico in phase 3
- Morphology of the postulated ambon
- Exact height of the schola cantorum
- Uncertainties about the original form of the excavated remains of benches and altar
- Structure of the roof trusses
- Uncertain longitudinal wall in phase 1
- Poor preservation of opus sectile and bipedalis indicated by the surviving impressions (opus sectile reuses polychromatic pieces, and so the specific types will have an impact on the reflected light within the scene)
- Several possible arrangements for the pavement in the central nave
- Decoration of the walls; only limited surviving fragments of painted plaster

## Formal Analyses of the Basilica Portuense

### Uncertainty and Procedural Modeling

The uncertainties inherent in the simulation of past environments from data of variable quality is at the core of critiques of computer graphic modeling in cultural heritage. The issues surrounding authenticity and potential means to demonstrate uncertainty have been covered in considerable depth elsewhere. For this study we concentrate on means to introduce uncertainty into formal analyses. In the context of visibility-based studies we have considered the possibilities of manual variation in model parameters in order to conduct sensitivity analyses of a form associated with landscape viewshed analysis (Paliou et al., 2010; see also Paliou and Wheatley, 2007). The potential of procedural modeling techniques for demonstrating ambiguity is beginning to be realized, particularly as technologies such as CityEngine have emerged (Haegler, Müller, and Van Gool, 2009; Müller et al., 2007). CityEngine uses the CGA shape grammar (Müller et al., 2006; Wonka et al., 2003) to define partitioned geometry automatically according to rules such as the range of spacing of architectural components. Other approaches such as that defined by Merrell (2007) provide alternative methods for parameterizing model construction.

In this chapter we advance the use of the same procedural approaches to produce variation in geometry before formally calculating visibility or lighting. By scripting stochastic parameters the interpretation necessarily requires a clear understanding of the uncertainties in the data and their interpretation, but in return enables very rapid numbers of variations to be produced. In turn the formal analyses undertaken can be based on a larger, more statistically significant population. Such techniques are not seen as enhancing the quality of given modeled outputs but rather as a means to characterize the certainty surrounding them. In lighting and rendering, this stochastic component is essential. As we have identified, we must consider not only structurally constrained parameters such as the location and size of windows but also the possibilities of furniture such as screens and benches, and the properties of artificial luminaires. For the purposes of this chapter we used the procedural modeling capabilities to define variation in the size and shape of the clerestory windows, within parameters specified on the basis of surviving basilicas. First a sample of 10 permutations for the clerestory was generated using CityEngine. Figure 11.7

**Figure 11.7** Visualization of the clerestory geometry generated procedurally in CityEngine (10 overlapping permutations, each with 10% opacity)

shows the variation in the stochastic meshes produced. These were then compared as part of the lighting analysis described shortly. Many more permutations were used to examine the visibility of the areas in the clerestory where decorated panels are postulated.

## Texture Viewsheds

In 2005 we introduced the Texture Viewsheds method (Earl, 2005). This used a moving light source to produce patterns of light on surfaces within 3-dimensional models, corresponding to lines of sight. The technique allowed the summarizing of spaces according to relative visibility, angle of visibility, and distance of viewer from the point viewed. In turn, visibility maps for the spaces summarized were used to define automated pathways through those spaces, optimizing visibility of particular areas. In this way it was possible to produce simulated pathways potentially corresponding to deliberate design. The obverse technique summarized the visual experience given a particular path through space. The numeric summaries produced through this method included the cumulative visibility of surfaces from all viewpoints and an indication of the area of a given space or subspace visible from particular observer locations. Through this technique it was possible to produce summaries that were hypothesized to parallel subjective appraisals of concepts such as enclosure.

In our revised implementation of Texture Viewsheds for this chapter we employ the raster-based methodology already outlined only as a means to produce targeted fine-grained coverages. We use the VIPS framework (Cupitt and Martinez, 1996; Martinez and Cupitt, 2005) to produce the graphical summaries. The sample resolution and hence the spatial precision of the raster approach are limited only by the chosen output raster and the size of the unwrapped geometric component. Figures 11.8 and 11.9 show some sample Texture Viewsheds outputs from this technique.

Given the computational overheads required for a full Texture Viewshed coverage at high resolution, this chapter introduces an additional approach that uses only the modeled geometry. This latest method is undertaken entirely in 3ds Max, using MaxScript to obtain a summary of visible vertices. Post-processing of these data in a database enables outputs of summary statistics such as angle of view, surface normal of viewed location, distance between observer and viewed location, angle of view in XY and XZ planes, visibility by surface areas, and cumulative and probabilistic visibility. These outputs are then in turn reloaded to the scene in order to produce a visual representation of patterns of visibility in the modeled scene. In terms of previous work this method is closest to that proposed by Engel and Döllner (2009). Our contributions are implementation of the system in a parallel environment to other analyses, reuse of the model geometry (that is, 3ds Max Design 2011), and

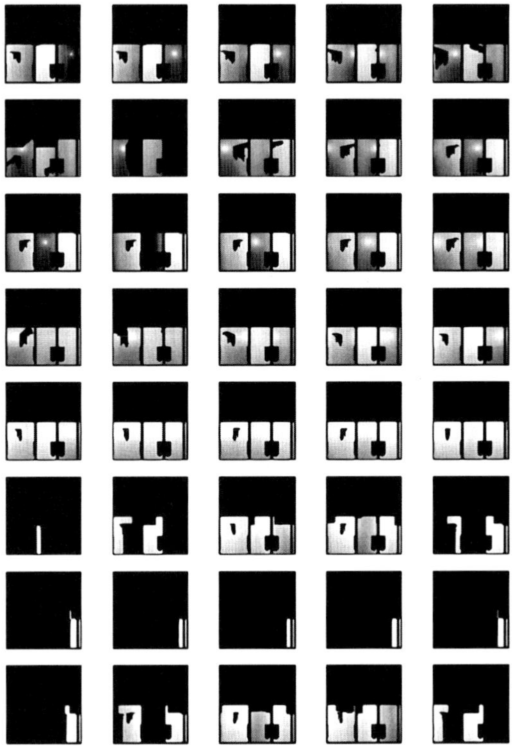

**Figure 11.8** Sample of Texture Viewsheds produced

**Figure 11.9** Example of a cumulative viewshed summary

the rapid computation of summary statistics conforming to extant archaeological and more general spatial interpretation. Also, in combination with the approach advanced in 2005 we are able to produce extremely high-resolution surface visibility metrics that are impractical using a geometric approach. Finally, we produce summaries of surface visibility values in addition to the identification of viewer statistics mapped onto their locations.

The modeled scene is prepared as for the standard Texture Viewsheds approach. A watertight mesh is generated with no overlapping surfaces. Each surface is then subdivided to produce an approximate equal mesh density across the model. This approach allows for general comparisons to be made in terms of relative visible areas. Such subdivision can be done relatively easily at the end of the modeling stage. In addition, a maximum size of face can be generated by projecting a vector shape grid onto each modeled object. As a check MaxScript allows any faces outside a given area threshold to be identified. In some cases it may be preferable to generate high-density sample regions in order to examine particular areas in detail. Generating high-density sample regions invalidates any visibility density summaries, however. Alternatively, the area of the surface can be compared to the proportion of the total number of vertices that is visible.

A voxel data representing the open area in a space within which observers can move is then approximated within the scene by stacked planar geometry where the separation of the geometry matches that of the gridded planes (color Figure I.10). These slices in turn are used to iterate through viewer positions by moving a light source to each vertex in turn. Having generated the visibility statistics (Table 11.1) the geometry is remapped via vertex colors to represent the chosen pattern of visibility (Figure 11.10). The remapped model can then be explored interactively.

Undertaking an analysis of this sort is computationally expensive. The simplified Basilica Portuense model analyzed contains approximately 9,500 model vertices. The visibility of each vertex was tested against one of 1,656 observer locations. The observers were defined by a 1 m × 0.5 m voxel representation three layers thick—sampling visibility throughout the building at observer heights of 1 m, 1.5 m, and 2 m. Visibility data were then remapped to the model surfaces. The analysis time on a standard specification 64 bit, 2.67 GHz dual core system with 4 GB RAM was approximately 0.14 seconds per vertex per observer. As a consequence, we used the University of Southampton Iridis3 supercomputer to enable the analysis to be completed on up to 30 dual quad core 2.27 GHz nodes in less than a day. This relatively short time was essential—had we been restricted to undertaking the analyses in series we would have had no opportunity to iterate experimental designs. We believe that the length of time and the amount of effort required to produce photorealistic archaeological models has in turn limited experimentation and robust critique of such simulations. Thus the value of our approach is predicated on

**Table 11.1** Sample Observer Visibility Data

| *ObserverID* | *Vertex* | *ObsX* | *ObsY* | *ObsZ* | *VerX* | *VerY* | *VerZ* | *Color* | *XYAngle* | *XZAngle* | *Distance* |
|---|---|---|---|---|---|---|---|---|---|---|---|
| 1 | 1 | 22.3832 | 9.81843 | 2 | 27.7309 | 38.7952 | 5.97076 | 24.8358 | 79.5435 | 7.67477 | 29.7325 |
| 100 | 1 | 18.3832 | 16.5684 | 2 | 27.7309 | 38.7952 | 5.97076 | 44.5642 | 67.1902 | 9.35136 | 24.4372 |
| 1000 | 1 | 15.3832 | 39.0684 | 1.5 | 27.7309 | 38.7952 | 5.97076 | 83.3449 | 1.26758 | 19.8993 | 13.1351 |
| 1001 | 1 | 16.3832 | 39.0684 | 1.5 | 27.7309 | 38.7952 | 5.97076 | 82.8049 | 1.37924 | 21.4977 | 12.1998 |
| 1002 | 1 | 17.3832 | 39.0684 | 1.5 | 27.7309 | 38.7952 | 5.97076 | 82.1208 | 1.51247 | 23.3595 | 11.2756 |
| 1003 | 1 | 18.3832 | 39.0684 | 1.5 | 27.7309 | 38.7952 | 5.97076 | 81.2392 | 1.67419 | 25.5509 | 10.3655 |
| 1004 | 1 | 19.3832 | 39.0684 | 1.5 | 27.7309 | 38.7952 | 5.97076 | 80.0805 | 1.87461 | 28.1591 | 9.47352 |
| 1005 | 1 | 20.3832 | 39.0684 | 1.5 | 27.7309 | 38.7952 | 5.97076 | 78.5239 | 2.12951 | 31.3009 | 8.60535 |
| 1006 | 1 | 21.3832 | 39.0684 | 1.5 | 27.7309 | 38.7952 | 5.97076 | 76.3822 | 2.4646 | 35.1323 | 7.76894 |
| 1007 | 1 | 22.3832 | 39.0684 | 1.5 | 27.7309 | 38.7952 | 5.97076 | 73.3584 | 2.92473 | 39.8591 | 6.97574 |
| 1008 | 1 | 23.3832 | 39.0684 | 1.5 | 27.7309 | 38.7952 | 5.97076 | 68.9765 | 3.59582 | 45.7426 | 6.24223 |
| 1009 | 1 | 24.3832 | 39.0684 | 1.5 | 27.7309 | 38.7952 | 5.97076 | 62.477 | 4.66572 | 53.0823 | 5.59195 |
| 101 | 1 | 19.3832 | 16.5684 | 2 | 27.7309 | 38.7952 | 5.97076 | 41.4411 | 69.4152 | 9.49436 | 24.0724 |
| 1010 | 1 | 25.3832 | 39.0684 | 1.5 | 27.7309 | 38.7952 | 5.97076 | 52.6842 | 6.63791 | 62.1354 | 5.05711 |
| 1011 | 1 | 26.3832 | 39.0684 | 1.5 | 27.7309 | 38.7952 | 5.97076 | 37.8508 | 11.4597 | 72.9024 | 4.67748 |
| 1012 | 1 | 27.3832 | 39.0684 | 1.5 | 27.7309 | 38.7952 | 5.97076 | 14.6492 | 38.1546 | 84.3506 | 4.49258 |
| 1018 | 1 | 14.3832 | 40.1934 | 1.5 | 27.7309 | 38.7952 | 5.97076 | 83.6207 | 5.9801 | 18.424 | 14.1459 |
| 1019 | 1 | 15.3832 | 40.1934 | 1.5 | 27.7309 | 38.7952 | 5.97076 | 83.1594 | 6.46046 | 19.7873 | 13.2064 |
| 102 | 1 | 20.3832 | 16.5684 | 2 | 27.7309 | 38.7952 | 5.97076 | 38.045 | 71.7071 | 9.62686 | 23.7442 |
| 1020 | 1 | 16.3832 | 40.1934 | 1.5 | 27.7309 | 38.7952 | 5.97076 | 82.5846 | 7.02432 | 21.3565 | 12.2766 |
| 1021 | 1 | 17.3832 | 40.1934 | 1.5 | 27.7309 | 38.7952 | 5.97076 | 81.8577 | 7.69536 | 23.1787 | 11.3587 |

**Figure 11.10** Geometry remapped with Texture Viewsheds visibility data

being able to develop, refine, and reject simulation parameters without worrying about implications of time. Where facilities restrict the complexity of the analysis we propose that the stochastic elements be retained while the number of observers is decreased. Without sensitivity analysis the visibility data for archaeological models with any underlying uncertainty are in essence meaningless.

## Lighting Analysis

The lighting analyses conducted in 3ds Max Design 2011 of the Basilica Portuense started with the creation of a daylight system and the definition of the exact geographic coordinates for the building. The scene was checked for invalid lights and materials, according to the requirements identified, and the frame buffer set to floating point. Similarly, optimal settings were chosen for the simulation of light energy within the scene. The mr Physical Sky was used as an environment shader, with CIE Clear Sky used as a sky model to define the typical brightness distribution of the skylight. The tested date was March 20, 2011, with samples at 9 A.M., 1 P.M., and 5 P.M. Although precise ephemeris data exist for the Roman period and have been employed elsewhere to guarantee sun angles (for example, Papadopoulos and Earl, 2009) we did not judge them to have had a sufficient influence in this case, particularly given the broad time samples examined.

The results are expressed in three ways: rendered image overlay, pseudocolor exposure control, and lighting meters. These three outputs were produced for all the structural models tested as well as for the different times during the day. The pseudocolor and the rendered image overlay are view dependent, which means that the values obtained illustrate the amount of energy returning to the virtual camera. For both pseudocolor and image overlay, the results were delivered in Luminance (cd/m$^2$), indicating how much luminous power will be perceived by an eye looking at the surface from a particular angle of view. Luminence is thus an indicator of how bright the surface will appear. The light meters, however, which measure only Illuminance (lux) are view independent and thus depict the intensity to which something is illuminated, taking account of the physical qualities of the objects.

A range of modifications were made to the master model of the Basilica Portuense in order to examine the significance that the contribution of light might have at the interior. The primary model of the Basilica was modeled with several doors, for which the height is uncertain, whereas the triforium has been simulated with three curtains, the middle of which is semi-open. In order to test all the possible options for the amount of light that entered to the interior, we decided to render additional structural models with less or no curtains and some or all doors open. In addition, lighting analysis was undertaken on a further variation, with four windows at the east side of the building (Figures 11.11 and 11.12). Because there are several reasons why windows at the other sides of the buildings are not structurally possible, and are mainly related to the existence of other structures in contact to the Basilica, we decided not to produce and test more structural models with windows at the other sides as well. The other buildings attached on the western side existed at least since the beginning of the 3rd century and were at least two stories high, covering the entire height of the western walls of the Basilica.

In addition, the 10 variants of the clerestory windows were tested to see if they might have had an effect on the light in the interior. The alternative models of the clerestory windows deliberately did not include glass or frames

**Figure 11.11** Rendering of Basilica without additional windows to the east

**Figure 11.12** Rendering of Basilica with additional windows to the east

in order that the influence of the geometry alone could be isolated. Therefore, the specific energy values are not discussed (since these are related to the transmission and other properties of the glass) but rather the relative impact of the morphological changes on the light entering the space. In a future study we will undertake a more detailed sensitivity analysis, using physically accurate glass parameters from the period. Still, taking into account the clerestory modifications, a total of 160 individual model permutations were considered in this lighting study.

As described, the lighting analyses focused on three broad questions:

1. What is the pattern and degree of light and shade in the building, and does it change throughout the day?
2. What are the effects of features such as drapery and windows, and how robust are the lighting results given the uncertainty of such variations in the model's underlying assumptions?
3. What are the potential effects of these results on hypothesized design choices and use of the space?

The results of these analyses demonstrate that the contribution of light in the interior during the whole day is quite significant, whereas the amount of light linearly decreases as it reaches the floor of the Basilica (color Figure I.11). A great amount of light is accumulated at the upper part of the structure where the clerestory windows exist (Figure 11.13). Light also enters through the open side doors, and the windows of the sanctuary at the lower part. To be more specific, the amount of light reaching the floor does not exceed 5–10 $cd/m^2$ during the day, whereas at the upper parts of the nave the values are dramatically increased because of the windows, in several cases exceeding

**Figure 11.13** Rendering that demonstrates low light levels and the likely effect of luminance accommodation as an observer inside looks from the interior to the exterior

1,000 cd/m$^2$—although in several areas, usually near closed doors or where no openings exist, the values are considerably low. However, at the areas where side openings have been modeled, the values increase, reaching 40–60 cd/m$^2$ and then gradually decrease (color Figure I.12), not having a great effect on the nave of the Basilica. In general, it can be argued that the amount of light up to the human height is noticeably low during the whole day, whereas it slightly increases when light shafts directly affect the interior. In contrast, the upper part is well illuminated during the day, and any existing paintings should have been equally illuminated as well. The narthex is always over-lit, since it belongs to the exterior of the Basilica and constitutes a transitional area to the interior.

Certain areas of the Basilica remain under-lit or over-lit during the day, which may have been related to ceremonial practice. For example, the sanctuary seems to be lit only by the two windows modeled at that area, whereas any light coming from the clerestory, any side openings, and the triforium seem not to significantly affect its illumination (Figure 11.13). In any case, there is a variety in the obtained values, with higher levels in the morning and early afternoon, and lower values in the evening. The triforium is less illuminated in contrast to other areas at the nave of the building, apart from the time of the day when light beams enter through the windows of the apse. In considering this fact we cannot wholly reject the possibility of windows covered with curtains, although the function of the building, the location of the building relative to surrounding buildings, and their arched form all suggest that curtains were unlikely. The ambon seems to be well illuminated during the whole day, mainly because of the light coming from the clerestory windows (color Figure I.12).

The fact that a small amount of light reaches the floor suggests that mosaics and the opus sectile floor were not well illuminated, with the exception of any light rays coming from the clerestory and the triforium. The physical properties of these floors have been investigated in detail, and such isolated pools of light would have created colorful and potentially sparkling variations (color Figure I.14). It was also observed that the greater significance of light in the interior is attested when the sun is at specific angle—that is, in the morning hours (9–11 A.M.) and late afternoon (2–5 P.M.), significantly affecting the illumination of the building and also producing a series of effects (for example, reflections) as the light interacts with the materials used for the columns and the other marble elements (Figure 11.14).

The tested variations—namely, no curtains at the narthex, all the curtains closed, and all the doors open—resulted in great variations in the lighting values obtained. For example, a model with the curtains closed produced lower lighting values at the nave and even lower at the aisles (color Figure I.15). These values range from 10–50 $cd/m^2$ in morning hours, whereas the areas where rays of light come through the side doors or other openings produced values ranging from 50–120 $cd/m^2$, exceeding that amount toward the midday. The values at the upper part of the nave walls (clerestory) remained stable, since the lighting at that part is dependent on the clerestory windows. In phase 3, when the curtains are open at the triforium, the light entering from the exterior passing through the exonarthex is significantly increased, whereas the values are even higher during the afternoon. However, when the model is rendered with the curtains and the side doors open, the values obtained considerably increase (Figure I.16). The greatest part of the nave is well illuminated, whereas the values linearly decrease as the rays of light reach the sanctuary. Similarly, the aisles are better illuminated, with values exceeding 1,000 $cd/m^2$

**Figure 11.14** Rendering of opus sectile floor

at the front part of the Basilica. The great amount of light coming from the various openings significantly affects the illumination of the whole building, producing values atypical for such environments. The aisles of the Basilica, especially toward the sanctuary, are significantly dim. However, the alternative model with the four windows at the east side produced higher values at that area, also partly affecting the illumination of the nave (Figure I.17). The highest values were observed in the morning hours, whereas the light gradually decreases toward the evening—although it cannot be argued that the arrangement of the windows has a significant effect on the overall illumination of the Basilica.

The procedural clerestory variations were tested with light meters to observe the light differences, concentrating in the upper parts of the Basilica, where there is likely to be greatest impact on the distribution of light. As shown in Table 11.2, each structural variation resulted in higher or lower amounts of light entering the interior. The difference between model iterations 2 and 3 as well as 7 and 8 is negligible, whereas the difference between 3 and 4 or 6 and 7 is more than 150%. The latter disparity is mainly due to the greatest difference in the dimensions of the windows. Overall, the minimum 40.8 lux and maximum 212 lux values represent a maximum variation of more than 500%. Per pixel comparisons of the type outlined may inevitably be biased by discrete factors, however; for example, a slight change in window size may result in a sufficient angular variation to allow in direct light from the luminaire. We therefore need to undertake additional work to validate and cross-check the results with other lighting analysis methods and to assess variations across wider areas. Still, this comparative study proves that the way that light interacts in an environment as well as the visual effects produced highly depend on the architecture of that environment, and even slight changes can result in different outcomes (see also Papadopoulos, 2010; Papadopoulos and Earl, Forthcoming).

**Table 11.2** Impact of Structural Variations on Light Entering Interior

| *Model Variant* | *Illuminance Value (Lux)* |
|---|---|
| 1 | 113 |
| 2 | 65.4 |
| 3 | 62.4 |
| 4 | 212 |
| 5 | 152 |
| 6 | 40.8 |
| 7 | 105 |
| 8 | 99.1 |
| 9 | 87.3 |
| 10 | 43.0 |

## Informal Appraisal of the Constructed Basilica Portuense

The final means of engagement with the modeled Basilica Portuense was through the many alternative rendered outputs generated and through the collective effort of modeling and rendering the building. This chapter identifies only a tiny fraction of the many hundreds of rendered scenes and objects produced in its writing and throughout the modeling of the Basilica. In Cripps and associates (2006) we argued for the significance of modeling as a situated, interpretative practice. Indeed, it may be argued that one of the main effects of the amount of time required to undertake a detailed study such as this is that the modeler is required to spend a great deal of time in the virtual space making decisions and exploring the evidence. Creation of our rendered outputs frequently led to a rethinking of the formal approaches and their results. The process of researching, creating, and visually experiencing these modeled environments was profound, and it provided an extremely detailed appraisal both of the site as known and of the great extent both to which the site can be reconstructed and will remain unobtainable. This informal influence of the modeling process is evidenced throughout the conclusions outlined in the next section, and it pervaded the research process, but it cannot necessarily be linked explicitly to given arguments. In this sense, the "paradata" for much of our research remains intangible (Beacham, Denard, and Niccolucci, 2006).

## Conclusions

The roles of changing visibility and the interaction of light with surfaces within religious structures is clearly complex, variable, and frequently unobtainable in archaeological contexts. Contemporary and recent ethnographies highlight the significance of light in defining spatial behavior and practice in religious spaces, and alignments and statistically significant variations in the pattern of light and shade in older religious spaces imply a significance of illumination in devotion. In a Roman setting, Vitruvius spoke specifically about the effects of light on buildings. Ecclesiastical texts and philosophy and also architecture suggest but do not prove builders' consideration either of the functional or the mental role of light. Furthermore, the relationship of building morphology, orientation, lighting, decoration, and practice is sufficiently complex and indeterminately interdependent that categorical statements elude us. Nonetheless, a significant amount of work has been undertaken regarding the discussion of illumination and especially of its visual effects in religious architecture (for example, see Krautheimer, 1965: 153–59; Whittemore, 1933: 12–28).

Comparatively fewer projects have considered the intentional manipulation of light in such contexts. Potamianos (1996, 2002) discussed the close relation of light to architecture by observing a series of Byzantine churches and comparing their orientation and size as well as number and dimension of openings, trying to talk about the importance of light from a historical and empirical point of view. Antonakaki (2007) attempted a comparative study of a church and an Ottoman mosque in which the spatial relation between the believers and the buildings was explored. An earlier attempt was made by Triantafyllides (1964), who estimated the amount of light in churches by using a light meter at specific locations, although the results were not related to any liturgical events. Unver and Enarun (1999) provided a set of proposed illumination quantity and quality for specific areas in churches, and Liritzis and Vassiliou (1997) undertook an investigation of the orientation of Byzantine churches in relation to the solstices and equinoxes as well as the name day of the saint. Last, there are a few case studies that examine the illumination in buildings with special architectural characteristics, sculpture, frescoes, and monumental art (see, for example, Iliadis, 2001, 2005).

In our own case study we deliberately explored the full range of archaeological uncertainties in order to problematize and evaluate formal and informal modes of analysis. Given the range of variables, and the numerous lighting and structural models that can be produced on the basis of their ambiguities, we propose that however sophisticated the methodology the resulting conclusions remain vague and must be clearly qualified. This analysis can only ever provide one input to the various approaches employed in archaeological interpretation. Still, allowing for the considerable flexibility in the models produced, our findings do propose a number of points for consideration.

The distribution of light to the interior is not uniform in any of the model permutations, and the arrangement of the Basilica in zones (exonarthex, nave, aisles, and sanctuary) parallels the distribution of the illumination and visibility, creating a blending space (Antonakaki, 2007: 11–57). The daylight levels observed are relatively low at ground level, whereas the upper area is well lit because of the clerestory windows, creating a clear visible demarcation. The visibility analyses clearly demonstrate that the main axis of the building forms the visual focus, whereas the most visually integrated area is the center of the church, at the point where the Ambon is located. The aesthetic impression of the church is defined by movement from the entrance to the sanctuary (Kalligas, 1946), and the light is concentrated on specific areas, potentially to direct the eye movement and indicate where the interest in space should be focused (Potamianos, 1996: 154). The aisles are less integrated and suggest a secondary role (Davies, 1982: 95). The Texture Viewsheds analysis demonstrates a significant reduction in visibility of the aisles, particularly at human height. By comparison, the hypothesized area of paintings in the clerestory dominates the space (Figure 11.15), with the sections at either end particularly prominent in terms of angle of view (Figure 11.16).

When one moves along the main axis of the church, a series of transitions becomes apparent mainly occurring with a decrease in light, supported by the visibility results (color Figure I.17). The initial transition from the exterior to

**Figure 11.15** Visibility of hypothesized painting, with 2 m voxel layer reclassified according to angular prominence of the painting

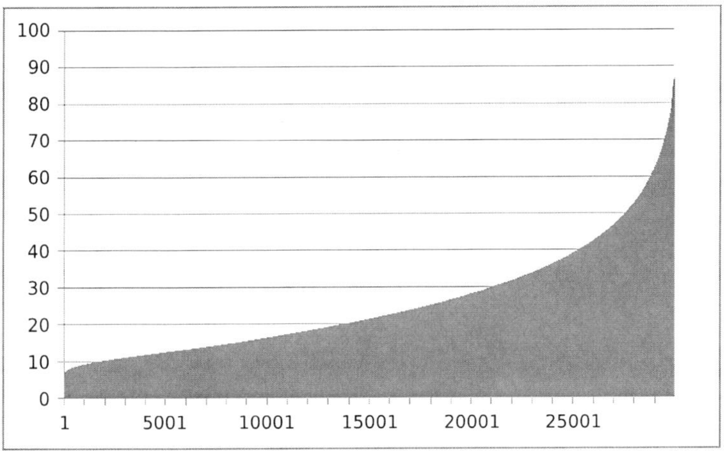


**Figure 11.16** Graph showing angular visibility of one of the hypothesized clerestory paintings (Y axis shows observer angle of view; X axis shows observer)

the exonarthex is evident with a slight decrease in light, followed by the entry into the dark interior. Through this transition, the viewer gradually adapts to the interior and is reoriented via specific patterns of light. Such a pronounced visual transition might relate to progression toward areas of higher sacredness (Humphrey and Vitebsky, 1997). The light coming from different directions minimizes strong directional shadows, assigning a symmetrical balance to the space and delineating 3-dimensional forms (Davies, 1982: 95), potentially also evoking mystical feelings and devoutness (Antonakaki, 2007: 57–66).

**Figure 11.17** Rendering of atmospheric effects

The Basilica is dominated by light from the clerestory windows, which might have been intensified by the use of translucent glass, and is reflected on the various specular surfaces, creating clouds of light and other atmospheric effects (Figure 11.17). In addition, the various polychrome surfaces and mainly the marbles would have created a bleeding effect, producing random or even intentional colored animations of the interior space. The use of flame illumination cannot be excluded (Figure 11.18) and actually should be emphasized, because it might have provided more light both for the common practices and the various rituals, therefore serving a functional and symbolic role, contributing to the enhancement of the religious atmosphere and adding a sense of mystery and devotion. Although we have employed such artificial luminaires in our rendering work we plan to include it in formal lighting analysis at a later stage of the project. Presumably, also the overall aesthetic would have been enhanced not only by the vision and the interplay of light and shadows but also through the rest of the senses—when listening to the priests, hearing the echoes and the resonance of steps, and smelling the burning of incense and candles.

This integrated analysis can provide only an indication of the various methodological tools that can be employed in the reasoning process of attempting an enhanced archaeological interpretation. The various constraints that the dataset poses and the numerous lighting and structural models that can be produced occasioned by the ambiguities of the records cannot directly provide

**Figure 11.18** Illustrative rendering of Basilica Portuense under artificial illumination (Note: we do not currently have sufficient data to render physically accurate simulations of artificial illumination in this model.)

all the answers to the research questions we have identified in this chapter. It is clear that there are significant structures to the illumination within the building and that patterns of visibility are changing and complex. Our sensitivity analyses suggest that within the architectural and other parameters explored these patterns are robust. What remains at issue is intentionality. Within the range of architectural options explored the axial focus and division into discrete patterns of light and shade are duplicated, albeit with varying details. It is certainly the case, however, that were one to deviate from the available architectural correlates it would be possible to produce very different visual properties—for example, by varying the form of the clerestory further. The visual structure we have identified might therefore be a consequence of the generalized basilica form adopted. This form may in turn have been influenced by the lighting of the space but equally by other factors, such as structural integrity and antecedent architectural designs.

The Basilica Portuense remains a complex visual space, and our engagement with it has been a self-consciously ocularcentric one. In this paper we attempted to identify vision-specific methods for exploring the Basilica in a way that as far as possible relates to the physical possibilities of the imagined original place. We have not ignored the possibilities of alternative sensoriums, however; indeed, as our work continues we look forward to an increasingly multisensory exploration of the Basilica. Already the possibilities of technologies such as acoustic rendering are clear. Whatever we can add to the vitality of the engagement with the modeled past must surely assist in our living in the modeled world—in gaining a sense of presence (Ch'ng, 2009)—as a means to try to understand it in new ways.

In the context of much more recent religious spaces Giles has made clear the extent to which they were places of memory and thus potentially places as much about what could no longer be seen, what had been covered over or destroyed, as places to be seen (2007: 115). She also outlines the importance of movement, supporting our own emphasis on unfolding visual narratives in the visualized and analyzed spaces of the Basilica.

> From the late fourteenth century onwards, changes in liturgy and architecture made the rite of the Mass less physically accessible to the laity. In response, an increasing emphasis was therefore placed on medieval seeing as a means of participating, visually and bodily, in the divine. . . . Traditionally, archaeologists have modelled these lines of sight as straight lines (Graves, 2000; Roffey, 2004). Yet perhaps we need to take more account of the circuitous journeys and movements that this seeing involved: negotiating or pushing one's way through the crowd, peering around tombs or other monuments, jostling for a glimpse through the squint or hagioscope and kneeling, stooping, stretching, craning and straining to see and hear the moment of the elevation of the Host, signalled by the ringing of the sacring bell. (Giles, 2007: 115)

We believe that in creating the Basilica Portuense to explore we are enabling new versions of such circuitous journeys to be enacted and quantified, to generate serendipitous discoveries of places and objects and to come to some sense of the religious through the lens of crumbled walls, fragmentary mosaics, and computer wireframes. In terms of Allison's (2001) critique of formal analyses we would accept that many of the formal lighting and visibility conclusions could in part be derived from the extant ground plans, and also that analogous associations of postulated architectural forms with religious practice are at best optimistic. However, it is our contention that our practice of these methods (if not always their results) significantly increases awareness of the shape both of the available archaeological evidence and of its potential original form and as such creates a new location to build interpretations. It is through this self-conscious process also that we have peopled this intriguing digital place.

## References

Allison, P. M. 2001. Using the material and written sources: Turn of the millennium approaches to Roman domestic space, *American Journal of Archaeology* 105(2): 181–208.

Antonakaki, T. 2007. Lighting and spatial structure in religious architecture: A comparative study of a Byzantine Church and an early Ottoman mosque in the city of Thessaloniki, *Proceedings of the 6th International Space Syntax Symposium*, June 12–15, 2007, Istanbul: 1–14.

Beacham, R., Denard, H., and Niccolucci, F. 2006, An Introduction to the London Charter, in Ioannides, M., Arnold, D., Niccolucci, F., and Mania, K. (Eds.). *The evolution of Information Communication Technology in Cultural Heritage: Where Hi-Tech Touches the Past: Risks and Challenges for the 21st Century*, Short papers from the joint event CIPA/VAST/EG/EuroMed, pp. 263-69.

Beale, G., Earl, G. P., Chalmers, A., Happa, J., and Williams, M. 2010. Polychromy, digital simulation and an Amazon from Herculaneum, in Moore, A., Taylor, G., Harris, E., Girdwood, P., and Shipley, L. (Eds.), *Conference Proceedings, TRAC 2009.*

Benedikt, M. L. 1979. To take hold of space: Isovist fields, in *Environment and Planning B: Planning and Design* 6: 47–65.

Blomberg, M., and Henriksson, G. 2002. Literary and archaeoastronomical evidence for the origins of the Hellenic calendar in the Aegean Bronze Age, in Maravelia, A. (Ed.), *Ad Astra per Aspera et per Ludum European Archaeoastronomy and the Orientation of the Monuments in the Mediterranean Basin, Thessaloniki.* Oxford: Archaeopress.

Borrello, L., Maiorano, M., Paroli, L., Serlorenzi, M. 2002. La basilica urbana di Porto, in Guidobaldi, F., and Guiglia Guidobaldi, A. (Eds.), *Ecclesiae Urbis, Atti del Congresso Internazionale di studi sulle chiese di Roma IV–X secolo, Roma, 4–10 settembre 2000, Roma 2002*, pp. 1263–85.

Boyce, R. P. 2003. *Human Factors in Lighting*. Oxford: Taylor & Francis.

Callet, P., Dumazet, S., Leclercq, C., and Politi, C. 2010. Natural lighting, gilts, and polychromy of Notre-Dame de Paris Cathedral, in Artusi, A., Joly-Parvex, M., Lucet, G., Ribes, A., and Pitzalis, D. (Eds.), *Conference Proceedings, 11th VAST International Symposium on Virtual Reality, Archaeology, and Cultural Heritage, September 21–September 24, 2010, Paris, France*, pp. 63–70.

Chalmers, A. 2002. Very realistic graphics for visualising archaeological site reconstructions, in *Conference Proceedings, 18th Spring Conference on Computer Graphics*, pp. 7–12. New York: Association for Computing Machinery Press.

Chalmers, A., Debattista, K., Mastoropoulou, G., and dos Santos, L. P. 2007. There-reality: Selective rendering in high fidelity virtual environments, *International Journal of Virtual Reality* 6(1): 1-10.

Chalmers, A., Roussos, I., and Ledda, P. 2006. Authentic illumination of archaeological site reconstructions, in *CGIV: IS&T Third European Conference on Color in Graphics, Imaging, and Vision, Poitiers, France.*

Chalmers, A., and Zányi, E. 2009. *Real Virtuality: Emerging Technology for Virtually Recreating Reality*, British Educational Communications and Technology Agency (BECTA), http://dera.ioe.ac.uk/1513/.

Ch'ng, E. 2009. Experiential archaeology: Is virtual time travel possible? *Journal of Cultural Heritage* 10(4): 458–70.

Clark, D. L. 2007a. Public and private space at Deir 'Ain 'Abata: An application of space syntax analysis to the monastery, in Politis, K. D. (Ed.), *Excavations at the Sanctuary of St. Lot, Deir 'Ain 'Abata, Jordan 1998–2003*. London: British Museum Press.

———. 2007b. Viewing the liturgy: A space syntax study of changing visibility and accessibility in the development of the Byzantine church in Jordan, *World Archaeology* 39(1): 84–104.

Classen, C. 1993. *Worlds of Sense: Exploring the Senses in History and Across Cultures.* London: Routledge.
Cripps, P., Earl, G., and Wheatley, D. 2006. A dwelling place in bits. *Journal of Iberian Archaeology* 8: 25–39.
Cupitt, J., and Martinez, K. 1996. VIPS: An image-processing system for large images, in *Conference Proceedings, IS&T/SPIE Symposium, Electronic Imaging: Science and Technology, 2663, Very High Resolution and Quality Imaging, 1996*, 19–28.
CURET. Columbia-Utrech Reflectance and Texture Database, www.cs.columbia.edu/CAVE/software/curet/.
Dana, K. J., Van Ginneken, B., Nayar, S. K., and Koenderink, J. J. 1999. Reflectance and texture of real world surfaces, *ACM Transactions on Graphics* 1(18): 1–34.
Davies, J. G. 1982. *Temples, Churches, and Mosques: A Guide to the Appreciation of Religious Architecture*. Cleveland, OH: The Pilgrim Press.
Dawson, P., and Levy, R. 2005. Using computer modeling and virtual reality to explore the ideological dimensions of Thule whalebone architecture in Arctic Canada, in *Internet Archaeology* 18(1); http://intarch.ac.uk/journal/issue18/.
Dawson, P., Levy, R., Gardner, D., and Walls, M. 2007. Simulating the behaviour of light inside Arctic dwellings: Implications for assessing the role of vision in task performance, *World Archaeology* 39(1): 17–35.
De Rossi, G. B. 1866. *Bullettino d'archeologia cristiana*, IV(6): 99–100.
Devlin, K., and Chalmers, A. 2001. Realistic visualisation of the Pompeii frescoes, in Chalmers, A., and Lalioti, V. (Eds.), *Conference Proceedings, First International Conference on Computer Graphics, Virtual Reality, and Visualisation, AFRIGRAPH 2001*, pp. 43–47. New York: ACM Press.
Devlin, K., Chalmers, A., and Brown, D. 2002. Predictive lighting and perception in archaeological representations, in *Conference Proceedings, UNESCO World Heritage in the Digital Age, 30th Anniversary Digital Congress.*
Dobbins, J., and Gruber, E. Forthcoming(a). Illuminating historical architecture: The House of the Drinking Contest at Antioch, in Contreras, F., and Melero, F. J., *Conference Proceedings, XXXVIII Computer Applications & Quantitative Methods in Archaeology 2010.*
————. Forthcoming(b). Modeling hypotheses in Pompeian archaeology: The House of the Faun, in Contreras, F., and Melero, F. J., *Conference Proceedings, XXXVIII Computer Applications & Quantitative Methods in Archaeology 2010.*
Dong, Y., Wang, J., Pellacini, F., Tong, X., and Guo, B. 2010. Fabricating Spatially Varying Subsurface Scattering, *ACM Transactions on Graphics* 29, 62:1–62:10.
Earl, G., Beale, G., Martinez, K., and Pagi, H. 2010. Polynomial texture mapping and related imaging technologies for the recording, analysis and presentation of archaeological materials, in *ISPRS Commission V Midterm Symposium*, Newcastle.
Earl, G., Cox, G., Cox, S., Frankland, T., Parchment, O., Takeda, K., and Thorne, S. Forthcoming. High-performance computing and the construction of archaeological computer graphic simulations, in *Conference Proceedings, XXXIX Computer Applications & Quantitative Methods in Archaeology 2011*, April 12–16, Beijing.
Earl, G., Keay, S. J., and Beale, G. 2012. Archaeological computing for recording and presentation of Roman Portus, in Keay, S. J., and Paroli, L. (Eds.), *Portus and Its Hinterland: Recent Archaeological Research, Archaeological Monographs of the British School at Rome 18.* London: The British School at Rome.

Earl, G. P. 2005. Texture viewsheds: Spatial summaries of built archaeological spaces derived from global light mapping, *Proceedings of the 11th International Conference on Virtual Systems and Multimedia* (VSMM), pp. 303–12.

Earl, G. P., Martinez, K., and Malzbender, T. 2010. Archaeological applications of polynomial texture mapping: Analysis, conservation, and representation, *Journal Archaeological Science* 37(8): 2040–50.

Engel, J., and Döllner, J. 2009. Approaches towards visual 3D analysis for digital landscapes and its applications, *Digital Landscape Architecture Proceedings 2009*, Malta, May 21–22, pp. 33–41. Anhalt: Anhalt University of Applied Sciences.

Fevrier, P. A. 1958. Ostie et Porto à la fin de l'Antiquité : Topographie religieuse et vie sociale, *MEFRA* 70: 295–330.

Garg, K., and Nayar, S. K. 2006. Photorealistic rendering of rain streaks, *ACM Transactions on Graphics*, 25(3): 996.

Giles, K. 2007. Seeing and believing: Visuality and space in premodern England, *World Archaeology* 39(1): 105–21.

Goodison, L. 2001. From Tholos tomb to throne room: Perceptions of the sun in Minoan ritual, in Laffineur, R., and Hägg, R. (Eds.), *Conference Proceedings, Aegaeum 22 POTNIA, Deities and Religion in the Aegean Bronze Age, the 8th International Aegean Conference*, Göteborg, Göteborg University, April 12–15, 2000.

Gutierrez, D., Sundstedt, V., Gomez, F., and Chalmers, A. 2008. Modeling light scattering for virtual heritage, *Journal on Computing and Cultural Heritage* 1(2): 1–15.

Haegler, S., Müller, P., and Van Gool, L. 2009. Procedural modeling for digital cultural heritage, in *EURASIP Journal on Image and Video Processing*, http://jivp.eurasipjournals.com/content/2009/1/852392.

Happa, J., Artusi, A., Czanner, S., and Chalmers, A. 2010. High dynamic range video for cultural heritage documentation and experimental archaeology, *VAST10: The 11th International Symposium on Virtual Reality, Archaeology, and Intelligent Cultural Heritage*, September 21–24, 2010, Paris, France:17–24.

Happa, J., Williams, M., Turley, G., Earl, G. P., Dubla, P., Beale, G., Gibbons, G., Debattista, K., and Chalmers, A. 2009. Virtual relighting of a roman statue head from Herculaneum: A case study, in *AFRIGRAPH '09: Proceedings of the 6th International Conference on Computer Graphics, Virtual Reality, Visualisation, and Interaction in Africa (2009)*, Franschoek, South Africa.

Happa, J., et al. 2009a. The virtual reconstruction and daylight illumination of the Panagia Angeloktisti, in Debattista, K., Perlingieri, C., Pitzalis, D., and Spina, S. (Eds.), *Conference Proceedings, 10th VAST International Symposium on Virtual Reality, Archaeology and Cultural Heritage 2009*, September 22–September 25, 2009, Valletta, Malta: 5–12.

________. 2009b. Illuminating the past: State of the art, in Debattista, K., Perlingieri, C., Pitzalis, D., and Spina, S. (Eds.), *Conference Proceedings, 10th VAST International Symposium on Virtual Reality, Archaeology and Cultural Heritage 2009, September 22–September 25, 2009*, Valletta, Malta: 155–82.

Hašan, M., et al. 2010. *Physical Reproduction of Materials with Specified Subsurface Scattering*. New York: ACM Press.

Hasic, J., Chalmers, A., and Sikudova, E. 2010. Perceptually guided high-fidelity rendering exploiting movement bias in visual attention, *ACM Transactions on Applied Perception* 8(1): 1–19.

Hašan, M., Fuchs, M., Matusi, W., and Rusinkiewicz, S. 2010. Physical reproduction of materials with specified subsurface scattering, *ACM Transactions on Graphics* 29(4), Article 61: 1–10.

Hawkins, T., Einarsson, P., and Debevec, P. 2005. *Acquisition of Time-Varying Participating Media*. New York: ACM Press.

Huelsen, C., Cecchelli, C., Giovannoni, G., Monneret de Villard, U., and Munoz, A. 1924. *Monografie sulle chiese di Roma, Sant'Agata dei Goti*. Rome: Sansaini.

Humphrey, C., and Vitebsky, P. 1997. *Sacred Architecture*. Boston: Little Brown & Co.

Iliadeis, I. 2001. The natural lighting of the mosaics in the Rotunda at Thessaloniki, *Lighting Research and Technology* 33: 13–24.

________. 2005. The natural lighting in the Dome of Rotunda at Thessaloniki, *Lighting Research and Technology* 37: 183–98.

Jay, M. 1996. Vision in context: Reflections and refractions, in Brennan, T., and Jay, M. (Eds.), *Vision in Context: Historical and Contemporary Perspectives on Sight*, London: Routledge.

Kalligas, M. 1946. *H Aisthitiki tou Horou tis Ellinikis Ekklhsias sto* Mesaiona (The Aesthetics of Greek Church in Medieval Times). Athens.

Keay, S., Millet, M., Paroli, L., and Strutt, K. 2005. *The Portus Project*. London: Archaeological Monographs of the British School at Rome.

Kider, J. T., Fletcher, R. L., Yu, N., Holod, R., Chalmers, A., and Badler, N. I. 2009. Recreating early Islamic glass lamp lighting, in VAST '09: *Proceedings of the Symposium on Virtual Reality, Archaeology and Cultural Heritage (2009)*, pp. 33–40.

Krautheimer, R. 1965. *Early Christian and Byzantine Architecture*. Baltimore: Penguin Books.

Lanciani, R. 1865. *Ricerche topografiche sulla città di Porto*. Rome.

________. 1866. Lo Xenodochio di Pammachio in Porto, *BACrist* 4: 99–103.

Ledda, P., Santos, L. P., and Chalmers, A. 2004. A local model of eye adaptation for high dynamic range images, in *AFRIGRAPH '04: Proceedings of the 3rd International Conference on Computer Graphics, Virtual Reality, Visualisation, and Interaction in Africa (2004)*, November 3–5, 2004, Stellenbosch, South Africa: 151–60.

Liritzis, I., and Vassiliou, H. 1997. Does sunrise day correlate with eastern orientation of Byzantine churches on significant solar dates and saint's days? A preliminary study, *Byzantinische Zeitschrift* 99(2): 523–34.

Llobera, M. 2003. Extending GIS-based visual analysis: The concept of visualscapes, *International Journal of Geographical Information Science* 17(1): 25–48.

Lock, G. 2003. *Using Computers in Archaeology: Towards Virtual Pasts*. London: Routledge.

MacGregor, C. 1999. Making sense of the past in the present: A sensory analysis of carved stone balls, in *World Archaeology* 31(2): 258–71.

Maggidis, C. 1994. *Burial Building 19 at Archanes: A Study of Prepalatial and Early Protopalatial Funerary Architecture and Ritual*. Ph.D. dissertation, University of Pennsylvania.

Martinez, K., and Cupitt, J. 2005. VIPS—A highly tuned image processing software architecture, in *Conference Proceedings of the 2005 IEEE International Conference on Image Processing*, September 11–14, Genoa: 574–77.

Masuda, T., and Yamada, Y. 2004. Sunlight illumination simulation for archaeological investigation: Case study of the Fugoppe Cave, in *Conference Proceedings, 10th International Conference on Virtual Systems and Multimedia.*

Matusik, W., Pfister, H., Brand, M., and McMillan, L. 2002. A data-driven reflectance model, *ACM Transactions on Graphics* 22(3): 759–70.

Merrell, P. 2007. *Example-Based Model Synthesis.* New York: Association for Computing Machinery Press.

Michelis, P. A. 1955. *An Aesthetic Approach to Byzantine Art.* London: B. T. Batsford Ltd.

Müller, G., Bendels, G. H., and Klein, R. 2005. Rapid synchronous acquisition of geometry and BTF for cultural heritage artefacts, in *Proceedings of the 6th International Symposium on Virtual Reality, Archaeology, and Cultural Heritage (VAST)*, Eurographics Association, November 2005: 13–20.

Müller, P., Vereenooghe, T., Wonka, P., Paap, I., and Van Gool, L. 2006. Procedural 3D reconstruction of Puuc buildings in Xkipché, in Ioannides, M., Arnold, D., Niccolucci, F., and Mania, K. (Eds.), *Conference Proceedings, 7th VAST International Symposium on Virtual Reality, Archaeology and Cultural Heritage 2006*, Baltimore, Maryland: Eurographics.

Müller, P., Wonka, P., Haegler, S., Ulmer, A., and Van Gool, L. 2006. Procedural modeling of buildings, *ACM Transactions on Graphics* 25: 614–23.

Müller, P., Zeng, G., Wonka, P., and Van Gool, L. 2007. Image-based procedural modeling of facades, in *Proceedings of ACM SIGGRAPH 2007/ACM Transactions on Graphics (TOG),* ACM Press 26(3), article 85.

Nicolai, V. F. 2001. *Strutture funerarie ed edifici di culto paleocristiani di Roma dal IV al VI secolo.* Rome: Pontificia Commissione di Archeologia Sacra.

Paliou, E., and Wheatley, D. 2007. Integrating spatial analysis and 3D modelling approaches to the study of visual space: Late Bronze Age Akrotiri, in *Conference Proceedings, The World Is in Your Eyes: The XXXIII CAA 2005 Computer Applications and Quantitative Methods in Archaeology, International Conference,* March 2005, Tomar Potugal: 307–12.

Paliou, E., Wheatley, D., and Earl, G. P. 2011. Three-dimensional visibility analysis of architectural spaces: Iconography and visibility of the wall paintings of Xeste 3 (Late Bronze Age Akrotiri), *Journal of Archaeological Science* 38: 375–86.

Papadatos, Y. 1999. *Mortuary Practices and Their Importance for the Reconstruction of Society and Life in Prepalatial Crete: The Evidence from Tholos Tomb Γ, in Archanes-Phourni.* Ph.D. dissertation, University of Sheffield.

Papadopoulos, C. 2010. *Death Management and Virtual Pursuits: A Virtual Reconstruction of the Minoan Cemetery at Phourni, Archanes: Examining the Use of Tholos Tomb C and Burial Building 19 and the Role of Illumination, in Relation to Mortuary Practices and the Perception of Life and Death by the Living.* Oxford: Archaeopress, British Archaeological Reports 2082.

Papadopoulos, C., and Earl, G. 2009. Structural and lighting models for the Minoan cemetery at Phourni, Crete, in Debattista, K., Perlingieri, C., Pitzalis, D., and Spina, S. (Eds.), *Conference Proceedings, 10th VAST International Symposium on Virtual Reality, Archaeology and Cultural Heritage 2009*, September 22–September 25, 2009, Valletta, Malta: 57–64.

Papadopoulos, C., and Sakellarakis, Y. Forthcoming. Virtual windows to the past: Reconstructing the ceramics workshop at Zominthos, Crete, in Contreras, F., and

Melero, F. J., *Conference Proceedings, The XXXVIII Conference on Computer Applications & Quantitative Methods in Archaeology 2010, Granada, April 6–9.*

Parisinou, E. 2007. Lighting dark rooms: Some thoughts about the use of space in early Greek domestic architecture, in Westgate, R., Fisher, N., and Whitley, J. (Eds.), *Conference Proceedings, Building Communities: House, Settlement and Society in the Aegean and Beyond*, April 17–21, 2001, Cardiff University, British School at Athens Studies 15.

Parker Pearson, M. 1999. *The Archaeology of Death and Burial*. Stroud, Gloucestershire: Sutton Publishing Limited.

Paroli, L. 2005. The Basilica Portuense, in Keay, S., Millet, M., Paroli, L., and Strutt, K. (Eds.), *Portus: An Archaeological Survey of the Port of the Imperial Rome*. London: Archaeological Monographs of the British School at Rome.

Pensabene, P., Lazzarini, L., Preite Martinez, A., and Turi, B. 2007. *Ostiensium marmorum decus et decor: Studi architettonici, decorativi e archeometrici*. Rome: L'Erma di Bretschneider.

Potamianos, J. 1996. *Light into Architecture: Evocative Aspects of Natural Light as Related to Liturgy in Byzantine Churches*. Ph.D. dissertation, University of Michigan.

________. 2000. *To Phos sti Byzantini Ekklisia* (Light in Byzantine Church). Thessaloniki: University Studio Press.

Richardson, A. 2003. Corridors of power, *Antiquity* 77: 296, 373–84.

Robin, B., Hilpert, T., Miyazawa, K., Callet, P., de Contencin, F-X., Zymla, A. 2010. Virtual replication of the Idalion Tablet, in Artusi, A., Joly-Parvex, M., Lucet, G., Ribes, A., and Pitzalis, D. (Eds.), *Proceedings of the 11th InternationalSymposium on Virtual Reality, Archaeology, and Cultural Heritage VAST (2010)*, pp.17–10.

Roussos, I. 2003a. *Image-Based Flame Lighting*. Ph.D. dissertation, University of Bristol.

________. 2003b. High-Fidelity Lighting of Knossos, in Arnold, D., Chalmers, A., and Niccolucci, F. (Eds.), *Conference Proceedings, 4th VAST International Symposium on Virtual Reality, Archaeology and Intelligent Cultural Heritage 2003*, Brighton: 47–56.

Sakellarakis, Y., and Panagiotopoulos, D. 2006. Minoan Zominthos, in Gavrilaki, E., and Tzifopoulos, Y. (Eds.), *Conference Proceedings, O Mylopotamos apo tin Archaiotita os Simera*. Perivallon, Archaiologia, Istoria, Laografia, Koinoniologia, Rethymno.

Sakellarakis, Y., and Sapouna-Sakellaraki, E. 1976. Anaskafh Archanon, *Praktika ths en Athinais Arxaiologikhs Etaireias* 131: 351–85.

________. 1997. *Archanes: Minoan Crete in a New Light*, Vols. 1–2. Athens: Ammos Publications & Eleni Nakou Foundation.

Scarre, C., and Lawson, G. (Eds.). 2006. *Archaeoacoustics*. Cambridge: McDonald Institute Monographs.

Shaw, J., and Shaw, M. 1996. *Kommos I: The Kommos Region and Houses of the Minoan Town*. Princeton, NJ: Princeton University Press.

Soles, J. 1992. The prepalatial cemeteries at Mochlos and Gournia and the House Tombs of Bronze Age Crete, in *Hesperia Supplements* 24. Princeton, NJ: American School of Classical Studies at Athens.

Sundstedt, V., Chalmers, A., and Martinez, P. 2004. High-fidelity reconstruction of the ancient Egyptian Temple of Kalabsha, in *Conference Proceedings, 3rd International*

*Conference on Computer Graphics, Virtual Reality, Visualisation, and Interaction in Africa*, November 3–5, 2004, Stellenbosch, South Africa: 107–13.

Triantafyllides, G. 1964. *Stoicheia Physikou Photismou ton Vyzatninon Ekklisiwn* (Elements of Physical Lighting in Byzantine Churches). Athens: Ypiresia Archaiotiton kai Anastiloseos. Dimosieumata tou Archaiologikou Deltiou. 3.

Trigger, G. B. 2006. *A History of Archaeological Thought*. New York: Cambridge University Press.

Unver, R., and Enarun, D. 1999. A comparative investigation of lighting of mosques and churches in Istanbul, in *Proceedings of the 24th Session of the CIE*, June 24–30, 1999, Warsaw.

Wang, L., Lin, Z., Fang, T., Yang, X., Yu, X., and Kang, S. B. 2006. *Real-Time Rendering of Realistic Rain*, Microsoft Research, Microsoft Corporation MSR-TR-2006-102. Richmond, WA.

Weidlich, A., and Wilkie, A. 2008. Modeling aventurescent gems with procedural textures, in *Proceedings of the 24th Spring Conference on Computer Graphics (SCCG)*, ACM: 51–58.

Whittemore, T. 1933. *The Mosaics of Hagia Sophia at Istanbul: Preliminary Report on the First Year's Work 1931–1932*, Vol. 1. London: Oxford University Press for the Byzantine Institute.

Wilkie, A., Weidlich, A., Magnor, M., and Chalmers, A. 2009. *SIGGRAPH Asia 2009 Course Notes: Predictive Rendering*. New York: Association for Computing Machinery Press.

Wonka, P., Wimmer, M., et al. 2003. Instant architecture, *ACM Transactions on Graphics* 22(3): 669–77.

Woodson, W. 1992. *Human Factors Design Handbook: Information and Guidelines for the Design of Systems, Facilities, Equipment, and Products for Human Use* (2nd ed.). New York: McGraw-Hill.

Yang, P. P. J., Putra, S. Y., and Li, W. 2007. Viewsphere: A GIS-based 3D visibility analysis for urban design evaluation, *Planning and Design: Environment and Planning B* 34(6): 971–92.

Zacharias, J. 2001. Pedestrian behavior and perception in urban walking environments, *Journal of Planning Literature* 16(1): 3–18.

Zányi, E. 2007. High dynamic range display of authentically illuminated Byzantine art from Cyprus, in Arnold, D., Niccolucci, F., Chalmers, A. (Eds.), *Conference Proceedings, 8th VAST International Symposium on Virtual Reality, Archaeology and Cultural Heritage 2007*, Brighton: 87–92.

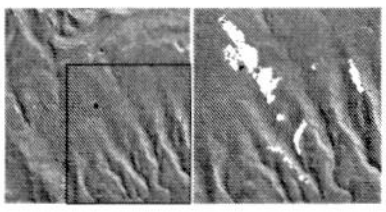

# 12 Reproducible Data Analysis and the Open Source Paradigm in Archaeology

*Benjamin Ducke*

## Archaeology as a Science

Archaeology can be practiced as a modern science. There is a wealth of material evidence to be sampled, quantified, and subjected to statistical and computational analysis. Some of the discipline's greatest innovations derive from its overlap with other scientific fields, including DNA analysis, radiocarbon dating, dendrochronology, provenance analysis, geophysical surveying, remote sensing, agent-based models, and GIS.

However, such cross-disciplinary fertilization brings with it an increased responsibility to respect and uphold the principles of good scientific practice. The foundation of modern science is a regime of self-checks, peer review, and transparent work practice, as has been established across many disciplines, in both theory and practice, throughout the 20th century. Although there is (naturally) no general agreement among academics on what exactly science is or what its epistemology should be, data and material-based research commonly adopt a distinctive working paradigm that can be labeled *scientific method* (Kuhn, 1970). This scientific method is not, as popular belief would have it, restricted to empirical studies. Rather, it represents a universal scholarly ideal that aims to lend efficiency, transparency, and credibility to any type of research.

In the 21st century, science goes hand in hand with computing, and archaeology is no exception to this rule. Software-based research and software as research output have become regular features of the discipline's academic fabric. In fact, studies such as those published by Evans, Knappet, and Rivers

*Computational Approaches to Archaeological Spaces* by Andrew Bevan and Mark Lake, Eds., 307–318 

(2008), Ducke and Kröfges (2008), and Barton, Ullah, and Bergin (2010) are so heavy on the software side that traditional printed publications no longer allow an adequate appreciation of the creativity and knowledge that went into them. This new role of software—a role that no longer coincides with the traditional "software as a tool" paradigm—necessitates a new awareness for the importance of establishing traditional means of academic rigor, such as peer review and transparency, within the new medium of software. Following is a brief discussion of how digital science and archaeological computing interact with each other at the interface of software technology. It argues against the use of commercially licensed, proprietary software in academic research and education, based on the view that the latter transgresses the principles of good scientific practice.

## Open Source Software

Modern software development uses a high-level approach based on human-readable, heavily structured programming languages such as C, C++, Java, and Python. Program code written in these languages, the so-called *source code*, must then be translated (the technical term is *compiled*) into a low-level *binary* code form before the computer can run the program. This binary code is mostly unintelligible to human interpreters.

To be able to learn from each other's work and peer review—and to modify and improve their work—programmers and researchers must have access to the original, human-readable source code. But being able to run the binary code in order to use the software is not enough for these purposes. However, since software companies often regard their source code as a trade secret and as profitable intellectual property, they are frequently unwilling to release it into the public domain.

Opposition to the restrictions and limitations of such proprietary software resulted in the Free Software movement (www.fsf.org). Free software (not to be confused with *freeware* or *shareware*) is program code that has been released into the public domain with the explicit permission for users to read, modify, and distribute it. The term *open source* was introduced later to express the same technical idea but avoid the moralistic undertones of the Free Software Foundation that have come to be perceived as "evangelical" by some of the movement's stakeholders (www.opensource.org).

A common misconception is to think of open source software as being the result of hobbyist work that is tossed into the public domain in complete ignorance of technological standards, quality assurance, and intellectual property rights. In reality, almost all large and successful open source projects are driven at least partly by professional, paid programmers, with much emphasis on open standards compliance, code quality, sustainability, and interoperability:

> A reported 29% of Europe's open-source programmers are paid for developing free software at work, and 24% are not paid for doing so but do it on company time anyway. Add to this the 17% of developers who are students, and only a few remain who are developing open-source software in their spare time. The European Commission study states that "the development of Open Source/Free Software is not at all a matter of leisure 'work' at home. Ninety-five percent of the sample claim that they use OS/FS at work, school, or university." (Klemens, 2006: 96)

Prime examples of high-quality, successful open source projects are the Linux kernel (www.kernel.org), the Apache web server (www.apache.org), Open/LibreOffice (www.documentfoundation.org), GRASS GIS (grass.osgeo.org), and R (www.r-project.org). They are legally protected by copyright licenses, with the GNU General Public License (GPL) and its derivatives being among the most popular. The decisive twist to this form of copyright, of course, is that the licensor explicitly allows the licensee to freely share and modify the source code. The Software Freedom Law Center (www.softwarefreedom.org) provides legal advice to the open source community on matters regarding the design, application, and enforcement of open source licenses.

From a technological point of view, the main ideas behind the open source model of software creation are quite straightforward, as illustrated by Klemens:

> If a company is not hoping to make big profits from a piece of software, its best bet is to go to the other extreme and open the code base entirely, allowing for free and open collaboration. There are programmers in hundreds of basements who need a good database client. One writes the core of a database client and puts all the code out for inspection. Then another programmer, ensconced in another basement, finds that the code does what her company needs but has a few bugs, which she fixes. In another basement, another wage slave finds that the code works well, except it is missing support for BLOBs (binary large objects), so she adds that. The process continues, as everybody contributes the feature that makes the code perfect in their eyes, until—for the time needed to write a few functions—everyone has a full-featured and well-tested database. (2006: 94)

How such seemingly simple approaches can result in high-quality program code has most prominently been studied and explained by Raymond (2001). The dynamics of the development and the speed at which an open source software project may progress are often nonlinear and very different from the steady growth expected from proprietary software. However, the scale effects of decentralized, global team work and increasing synergies will eventually lead to rapid increase in project size and available functionality, as can currently be witnessed in the arena of open source GIS (for example, Neteler and Mitasova, 2008; Sherman, 2008; see also www.freegis.org).

There are many ways in which an open source project may evolve. In fact, the preceding quotation does little justice to many larger open source projects, which feature highly evolved team work structures, programming technologies, rigid quality assurance, and full time, hired programmers (who do *not* work as wage slaves in basements; cf. Ghosh, 2005). But the ideas of agile and decentralized collaboration, external peer review, and building on existing components can be found at the core of every truly open software project. These characteristics, of course, bear great similarity to those of free and open science—a fact that makes it worthwhile to take a closer look at the relationship between software and science.

## Software Is Knowledge

Among archaeologists, open source software has long been known for the fact that it is generally distributed free of cost (although most open source licenses do not explicitly rule out the right to charge a fee); and these economic merits are frequently weighted against the presumably richer functionality of integrated, proprietary offers (for example, Pescarin, 2009: 127–40).

But those archaeologists who are themselves involved in the development and improvement of software have for a long time been afflicted by a different aspect of proprietary software in research—one that Hafer and Kirkpatrick (2009: 126) describe most eloquently: "Academic computer science has an odd relationship with software: publishing papers about software is considered a distinctly stronger contribution than publishing the software. The historical reasons for this paradox no longer apply, but their legacy remains."

Alternatively, one may imagine the members of a mathematical society publishing their research in a monthly journal. But with a twist: instead of publishing the way in which a mathematical problem was solved (that is, the algebraic formulae), they restrict themselves to only stating the problem and then publishing the solution at which they arrived. Although this approach would seem unacceptable to any mathematician, it is not very far removed from the practice, so common in archaeology, of using closed source, often prohibitively expensive, software to process some data and then publishing only the results in the form of screenshots, graphs, maps, or other digital depictions.

Although this form of publication may be satisfactory for computing laypeople within the discipline who simply want to know *if* a certain solution to a problem exists, it is insufficient for those who want to understand *how* such a solution works in detail and have an interest to try it out themselves, to learn from and improve on the computational aspects involved. They must be able to reproduce the entirety of the digital analysis. Three main obstacles frequently stand in the way of this: (1) the lack of access to the original data, (2) the cost of licensing the software used in the original research, and (3) the impossibility of understanding the inner workings of software that is distributed in closed source, binary form only. The first obstacle, withheld research data, belongs to a different although related problem domain. There is a general awareness for the importance of publishing data among archaeologists that is addressed by institutions such as the Archaeology Data Service (ADS; http://archaeologydataservice.ac.uk). The discipline can also take pride in being one of the first to introduce an open access, peer-reviewed journal dedicated to data publication (http://openarchaeologydata.metajnl.com). The other obstacles, related to the cost and the opacity of proprietary software, have so far not been a prominent part of the archaeological discourse—perhaps because they concern a minority of archaeologists, those who have the technical and mathematical curiosity needed to be interested in computational details.

The limits of the usefulness of software are defined by the paradigms under which the software is developed and used. In archaeology, software has long been viewed as a tool that serves a narrow and well-defined purpose. But even the simplest tool is more than just a useful item. Its shape, design, and composition reflect a deeper and clearer human understanding of the problems one hopes to solve by applying it. The same is true for software: complex programs consist of hundreds of thousands or even millions of lines of source code. They represent the cumulative result of countless hours spent on brainstorming sessions, elaborate project design, and fundamental and applied research. Indeed, software is a prime carrier of knowledge in the digital age.

Broadly speaking, knowledge is embedded in software in two principle forms: algorithms and data structures. Algorithms are well-defined sequences of mathematical operations applied to data to solve an equally well-defined problem (Cormen et al., 2009). A well-known archaeological algorithm is Ford's chronological seriation method (Ford, 1962), which was originally published as a manual procedure but is easily implemented in any programming language. Data structures are ways of arranging numerical information—in the form of tables, matrices, or lists—so as to describe the objects of interest and their relationships digitally and make them suitable for processing using algorithms. Whereas algorithms are essentially knowledge about problem solutions, data structures represent knowledge about the abstract structure of a problem and the analysis's objects of interest.

Figure 12.1 illustrates these concepts using Dijkstra's Shortest Path as an example. This famous algorithm is a way of finding, with provable correctness, the shortest path between two points in a network of locations. It is used in route finders, GPS car navigation devices, and logistical planning systems. It is also used for movement simulation, shortest path modeling, and stratigraphic analysis in archaeology (for example, Herzog and Posluschny, 2011; Pouncett and Lock, 2008).

The data structure used as input to the algorithm is a directed graph (as is the so-called "Harris Matrix" of stratigraphic relations). The shortest path from the starting point (node) A to the target node B is the one traversing the

```
Dijkstra(G, w, s)
Initialize(G, s)
S ← θ
Q ← V[G]
while Q ≠ θ
    do u ← Extract-Min(Q)
        S ← S ∪ {u}
        for each vertex v ∈ adjacent[u]
            do Relax(u, v, w)
```

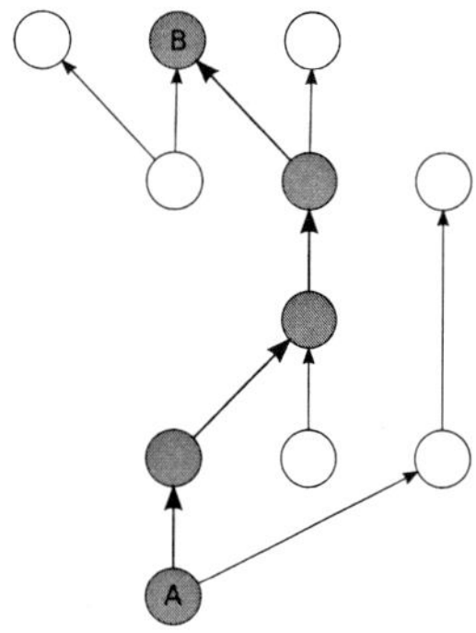


**Figure 12.1** Dijkstra's Shortest Path algorithm: (left) pseudocode (after Cormen et al., 2009); (right) directed graph with a shortest path between nodes A and B marked gray

minimal number of connections (edges) through the intermediate nodes. To find it, this algorithm uses a so-called greedy search strategy, which involves a check list of already traversed path segments and an "appetite" for always trying the next shortest segment first. A more detailed and accurate description is available in standard textbooks such as Cormen and associates (2009: 527–30).

For an actual software implementation, a programmer must translate the pseudocode in Figure 12.1 into a programming language, such as C, C++, or Java. Then, because real-world use cases are always more complex than textbook examples, a small implementation of Dijkstra's Shortest Path would grow into a full-featured application with many thousands of lines of code that provide facilities for data input and output, visualization, and a user interface. At some point, the code will inevitably reach a level of complexity that allows critical errors (bugs) to sneak in. As a result, a user who does not have access to the full source code of the software will not be able to find the reasons for unexplained results or even data loss, no matter how good her understanding of Dijkstra's Shortest Path may be (Hatton, 1997; Merali, 2010). To make matters more confusing, any part of the application, including the implementation of Dijkstra's Shortest Path, may be supplied in closed source form by a third party.

Complex software will increasingly be a product of archaeological research, because beyond the tool paradigm, it has become a research environment in archaeology just as in so many other disciplines. Statistics, data mining, spatial analysis, and numerical simulation are just a few of the digital research methods routinely used to view, describe, and analyze data, to discover structures and relations, and to test hypotheses and infer causalities. As software technology asserts a central place in modern research, its effects on fundamental aspects of this research, such as transparency, openness, and affordability of research, need to be critically examined.

## Black Boxes vs. Reproducibility

The increasing importance of software in archaeological research needs to be accompanied by increased scrutiny regarding the transparency of data processing. Despite the mathematical nature of computing and the possibility of making exact, lossless digital copies, the often enormous complexity of software introduces a significant amount of uncertainty into the analysis. Programming errors, fuzzy technical specifications, bad user manuals, and alternative implementations all contribute to the effect that the results of all but the most simple computations may vary between different software implementations. As Hatton (1997) observes: "Taken with other evidence, these two experiments suggest that the results of scientific calculations involving significant amounts of software should be treated with the same measure of disbelief as an unconfirmed physical experiment." In this respect, using proprietary software for data analysis is particularly problematic. Since it is impossible to read the actual program code that works on the data it is also impossible to account for unexpected results, because all software has errors and an unexpected outcome of a calculation may be due to one of these as much as to flawed input data.

Figure 12.2 shows the results of a viewshed analysis computed using the same dataset and two different GIS. The viewshed computation is a very common method that a researcher might employ to check which parts of the terrain are visible from a given location (for example, Conolly and Lake, 2006: 225–33). The elevation model was downloaded from the UK Ordnance Survey's free data store (www.ordnancesurvey.co.uk). It is tile "SO00" of the PANORAMA

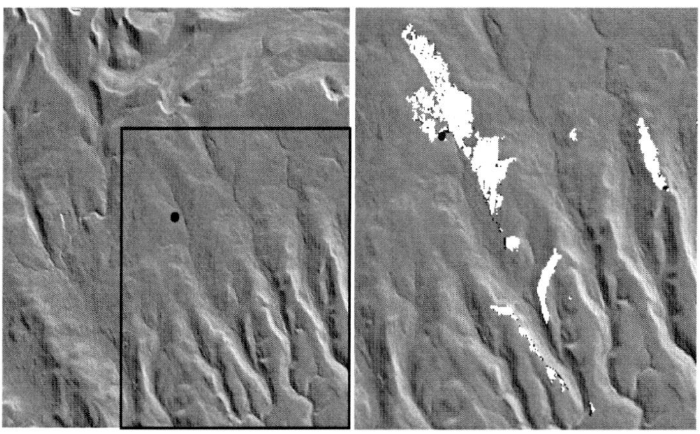

**Figure 12.2** Comparison of viewsheds calculated with two different GIS: (left) extent of elevation model and observer location; (right) enlarged area with two viewsheds. The viewshed generated by ArcGIS 10.0 (white cells) has been superimposed on the GRASS GIS 6.4.3 viewshed (black cells), which tends to be slightly larger (contains Ordnance Survey data © Crown copyright and database right 2012).

height dataset and covers an area of 20 by 20 km, at a ground resolution of 50 m. The observer's position was set in the center of the elevation model, at coordinates 310000 east and 210000 north. A viewshed was then computed with both the r.los module of GRASS GIS 6.4.3 and the Viewshed tool that can be found in the 3D Analyst extension for ArcGIS 10. In both cases, the observer height was set to ArcGIS's default of 1 m, and earth curvature correction was turned off. As the figure shows, differences between the outcomes are small (which should be expected, given that the viewshed operation is a very well-defined procedure), but they are present, especially at the edges of ridges. Since one of the software packages involved is closed source, there is no sure way to determine what causes them (possible explanations include limited precision of mathematical operations or erroneous calculations that show only in extreme cases). The only solution is to accept them and to subjectively decide which output should be preferred. Similar findings by archaeologists using other common methods, such as the computation of least-cost surfaces (Gietl, Doneus, and Fera, 2008) and spread maps (Indruszewski and Barton, 2008), suggest that most basic GIS operations suffer from this problem.

But the *black box effect* that closed source code introduces into the analysis also has a wider impact on academic practice and education (Morin et al., 2012). First and foremost, it prevents researchers and students from ever fully understanding every detail of an analysis. Second, it transgresses against what is arguably the most fundamental ideal of science, that of *reproducibility*, which is why full publication of software has been a growing demand in science for some time (cf. Barnes, 2010; Buckheit and Donoho, 1995; Donoho et al., 2009; Ince, Hatton, and Graham-Cumming, 2012; see also www.rrplanet.com).

In theory, given the same data and methods, independent researchers should be able to recreate one another's studies with the aim of verifying them and testing them for universal applicability. Reproducibility of research is especially critical in archaeology, which has no clear and generally agreed concept of proof by which to assess the credibility of one particular analytical model or hypothesis versus another one. In fact, in scientific research of any kind, the ability to prove assumptions wrong is more fundamental than that to prove them right—which will be *de facto* impossible for many archaeological research themes. To reproduce a digital analysis, however, free access to both the original data (as mentioned, this is another, equally important topic worthy of a separate treatise) and program code is a requirement. Although there are some aspects of archaeological research that seem to defy this principle, be it because the original evidence is lost as part of the research activity (archaeological excavations) or because it would be too time consuming or expensive (archaeological surveys), reproducibility must be ensured wherever possible, and technology and practice must evolve steadily to improve the level of reproducibility.

Ultimately then, the use of closed source software violates a primary principle of the scientific method. Archaeologists who decide to conduct research using such software technology should be aware of the consequences this has for independent third-party review and free dissemination of research results. In addition, although many academic archaeologists are shielded from the real cost of proprietary software via campus licensing schemes, those conducting work and research as part of public services or private enterprises are not. For them, proprietary licensing costs represent significant barriers to participation in ongoing research. Open source software technology offers a clear solution to these problems, not just in an idealistic manner but within a well-defined legal and technological framework. Individuals, research teams, and institutions should keep this in mind and evaluate whether acquiring closed source licenses should really take precedence over funding open source development.

## Conclusions

This discussion has focused on just one, albeit fundamental, aspect of open source software: its role in overcoming the obstacles of cost and opacity that prevent effective reproduction of software research under the proprietary, closed source model.

As regards spatial analysis in archaeology, one cannot help the feeling that the rapid developments of the last two decades, particularly involving GIS-based analysis, have largely been driven by a technology-first attitude that has left its propagators little time to ensure sustainable, good scientific practice. The examples of shortest path and viewshed algorithms stand for a multitude of data-processing methods implemented in modern software such as GIS. Many more, such as least-cost path computation, Kriging, and IDW interpolation, are now routinely used by archaeologists without spending a second thought on differences between implementations or software bugs. Yet the effects of problems in closed source—that is, essentially non-peer-reviewed code—can have significant consequences for the reliability of analysis results.

Thus it is for good reasons that the field of agent-based modeling (see the contribution by Michael Barton in this volume) is entirely dominated by open source software. After all, this is an approach that generates insight by experiment and observation, and complete openness is a critical ingredient in the process. Similar developments, for similar reasons, can be seen in spatial statistics, where the open source R project (www.r-project.org) has become the premier platform for development, application, and teaching (for example, Bivand, Pebesma, and Gomez-Rubio, 2008). In the case of GIS, proprietary software is much more firmly entrenched in the academic system. To a certain degree, this entrenchment has been due to the challenging learning effort imposed by open source alternatives, such as GRASS GIS (http://grass.osgeo.org). But the latest generation of user-friendly, open source desktop GIS, such as gvSIG CE (http://gvsigce.org) and QGIS (www.qgis.org) does not suffer from such problems.

There are other, closely related issues, such as that of proprietary data formats, enforced software updates through file format deprecation, and other aspects of what is commonly called *vendor lock-in*, which has devastating effects on team work and long-term accessibility of data. And then there is a further multitude of philosophical, technological, economical, educational, and social factors that cannot be discussed here for lack of space. They must be considered by archaeologists, as the discipline becomes fully aware of open source (and related concepts such as *open access*) as a key technological paradigm.

Meanwhile, as global customer demand for open source software grows, traditional software vendors are under pressure to adjust their business strategies to the new market realities. At the time of writing, all major software houses (for instance, Microsoft: www.microsoft.com/opensource) have some form of open source strategy and offering. Although this new openness is generally a welcome development, individual company's approaches have to be carefully evaluated regarding their nature and intent. For one thing, free-of-charge licenses are not the same as free software: such licenses do not open up the source code and only increase the risk of vendor lock-in via proprietary, nondisclosed file formats. For another, source code give-aways without a clear idea about how to involve the wider community in its maintenance and further development are bound to end in frustration on all sides.

To conclude, progress in science is the cumulative result of hard work, information sharing, and collaboration over generations, not the achievement of solitary geniuses with great ideas and "intellectual property." As a consequence, it is in the best interest of all those involved in academic research to identify barriers to free information flow and remove them.

## References

Barnes, N. 2010. Publish your computer code: It is good enough, *Nature* 467: 753.

Barton, C. M., Ullah, I. I., and Bergin, S. 2010. Land use, water, and Mediterranean landscapes: Modelling long-term dynamics of complex socio-ecological systems, *Philosophical Transactions of the Royal Society A: Mathematical, Physical, and Engineering Sciences* 368(1931): 5275–97.

Bivand, R., Pebesma, E. J., and Gomez-Rubio, V. 2008. *Applied Spatial Data Analysis with R* (Use R). New York: Springer.

Buckheit, J. B., and Donoho, D. L. 1995. *WaveLab and Reproducible Research*, Department of Statistics, Stanford University, Tech. Rep. 474.

Conolly, J., and Lake, M. 2006. *Geographical Information Systems in Archaeology*. Cambridge: Cambridge University Press.

Cormen, T. H., Leiserson, C. E., Rivest, R. L., and Stein, C. 2009. *Introduction to Algorithms* (3rd ed.). Cambridge, MA: MIT Press.

Donoho, L., Maleki, A., Rahman, I. U., Shahram, M., and Stodden, V. 2009. Reproducible research in computational harmonic analysis, *Computing in Science & Engineering* 11(1): 8–18.

Ducke, B., and Kröfges, P. 2008. From points to areas: Constructing territories from archaeological site patterns using an enhanced Xtent model, in Posluschny, A., Lambers, K., and Herzog, I. (Eds.), *Layers of Perception, Proceedings of the 35th International Conference on Computer Applications and Quantitative Methods in Archaeology (CAA), Berlin, April 2–6, 2007*, pp. 245–51. Bonn: Dr. Rudolf Habelt GmbH.

Evans, T., Knappet, C., and Rivers, R. 2008. Modelling maritime interaction in the Aegean Bronze Age, *Antiquity* 82: 1009–24.

Ford, J. 1962. *A Quantitative Method for Deriving Cultural Chronology*, Technical Bulletin No. 1. Washington, D.C.: Pan American Union.

Ghosh, R. A. 2005. Understanding free software developers: Findings from the FLOSS study, in Feller, J., Fitzgerald, B., Hissam, S. A., and Lakhami, K. R. (Eds.), *Perspectives on Free and Open Source Software*, pp. 23–46. Cambridge, MA: The MIT Press.

Gietl, R., Doneus, M., and Fera, M. 2008. Cost distance analysis in an Alpine environment: Comparison of different cost surface modules, in Posluschny, A., Lambers, K., and Herzog, I. (Eds.), *Layers of Perception, Proceedings of the 35th International Conference on Computer Applications and Quantitative Methods in Archaeology (CAA), Berlin, April 2–6, 2007*. Bonn: Dr. Rudolf Habelt GmbH.

Hafer, L., and Kirkpatrick, A. E. 2009. Assessing open source software as a scholarly contribution, *Communications of the ACM* 52(12): 126–29.

Hatton, L. 1997. The T experiments: Errors in scientific software, *IEEE Computational Science and Engineering* 4(2): 27–38.

Herzog, I., and Posluschny, A. 2011. Tilt: Slope-dependent least cost path calculations revisited, in Jerem, R., Redo, F., and Szeverenyi, V. (Eds.), On the Road to Reconstructing the Past, *Proceedings of the 36th Annual Conference on Computer Applications and Quantitative Methods in Archaeology, Budapest, April 2–6, 2008*. Budapest: Archaeolingua.

Ince, D. C., Hatton, L., and Graham-Cumming, J. 2012. The case for open computer programs, *Nature* 482: 485–88. doi: 10.1038/nature10836

Indruszewski, G., and Barton, C. M. 2008. Simulating sea surfaces for modeling Viking Age seafaring in the Baltic Sea, in Clark, J. T., and Hagemeister, E. M. (Eds.), *Digital Discovery: Exploring New Frontiers in Human Heritage, CAA 2006, Computer Applications and Quantitative Methods in Archaeology, Proceedings of the 34th Conference, Fargo, April 2006*, pp. 616–30. Budapest: Archaeolingua.

Klemens, K. 2006. *Math You Can't Use: Patents, Copyright, and Software*. Washington, D.C.: Brookings Institution Press.

Kuhn, T. S. 1970. *The Structure of Scientific Revolutions*. Chicago: University of Chicago Press.

Merali, Z. 2010. Computational science: . . . error, *Nature* (News) 467: 775–77.

Morin, A., Urban, J., Adams, P. D., Foster, I., Sali, A., Baker, D., and Sliz, P. 2012. Shining light into black boxes, *Science* 336(13): 159–60.

Neteler, M., and Mitasova, H. 2008. *Open Source GIS: A GRASS GIS Approach* (3rd ed.). New York: Springer.

Pescarin, S. 2009. *Reconstructing Ancient Landscape*. Budapest: Archaeolingua.

Pouncett, J., and Lock, G. 2008. Closest facility analysis: Integration of geophysical and test-pitting data from the South Cadbury Environs Project, in Posluschny, A., Lambers, K., and Herzog, I. (Eds.), *Layers of Perception, Proceedings of the 35th International Conference on Computer Applications and Quantitative Methods in Archaeology (CAA), Berlin, April 2–6, 2007*, pp. 1–6. Bonn: Dr. Rudolf Habelt GmbH.

Raymond, E. S. 2001. *The Cathedral and the Bazaar: Musings on Linux and Open Source by an Accidental Revolutionary* (rev. ed.). Sebastopol, CA: O'Reilly.

Sherman, G. E. 2008. *Desktop GIS: Mapping the Planet with Open Source Tools*. Raleigh, NC: The Pragmatic Bookshelf.

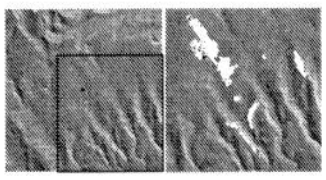

# Index

# About the Editors

**Andrew Bevan** is Senior Lecturer at the Institute of Archaeology, University College London, and a coordinator of the MSc in GIS and Spatial Analysis in Archaeology there. His primary research interests include computational modeling, landscape ecology, human demography, settlement patterns, large-scale artifact traditions and political geography. He is an author of two books on Mediterranean archaeology and has contributed to a range of archaeology and ecology journals on spatial modeling topics.

**Mark Lake** is Senior Lecturer at the Institute of Archaeology, University College London, and a coordinator of the graduate program in GIS and Spatial Analysis in Archaeology there. A specialist in GIS and computer simulation, he studies spatial patterning in prehistoric monuments and models the origins of culture. He is author of several simulation and GIS programs, coauthor of Geographical Information Systems in Archaeology and coeditor of Simulating Change, and author of numerous research articles.

## About the Contributors

*Michael Barton*, Center for Social Dynamics and Complexity, School of Human Evolution and Social Change, Arizona State University

*Gareth Beale*, Archaeological Computing Research Group, Faculty of Humanities, University of Southampton

*Terence Clark*, Canadian Museum of Civilization

*Enrico Crema*, Institute of Archaeology, University College London

*Benjamin Ducke*, Berlin, Germany, currently an independent scholar/software engineer.

*Graeme Earl*, Archaeological Computing Research Group, Faculty of Humanities, University of Southampton

*Timothy Evans*, Department of Physics, South Kensington Campus, Imperial College London
*Matthew Harrison*, Archaeological Computing Research Group, Faculty of Humanities, University of Southampton
*Irmela Herzog*, Rheinisches Amt für Bodendenkmalpflege (The Rhineland Commission for Archaeological Monuments and Sites), Bonn
*Simon Keay*, Classical and Historical Archaeology Research Group, Faculty of Humanities, University of Southampton
*Carl Knappett*, Department of Art, University of Toronto
*Kenneth Kvamme*, Department of Anthropology, University of Arkansas
*Xiuzhen Li*, Institute of Archaeology, University College London, and Museum of Emperor Qin Shihuang's Mausoleum, Xi'an
*Damon Ortega*, London, currently an independent scholar/software engineer.
*Hembo Pagi*, Archaeological Computing Research Group, Faculty of Humanities, University of Southampton
*Eleftheria Paliou*, Institute of Classical Archaeology, University of Heidelberg
*Alessio Palmisano*, Institute of Archaeology, University College London
*Constantinos Papadopoulos*, Archaeological Computing Research Group, Faculty of Humanities, University of Southampton
*Vito Porcelli*, Archaeological Computing Research Group, Faculty of Humanities, University of Southampton
*L. S. Premo*, Department of Anthropology, Washington State University
*Ray Rivers*, Department of Physics, South Kensington Campus, Imperial College London